STATISTICAL MECHANICS OF NONEQUILIBRIUM LIQUIDS

SECOND EDITION

In recent years the interaction between dynamical systems theory and nonequilibrium statistical mechanics has been enormous. The discovery of fluctuation theorems as a fundamental structure common to almost all nonequilibrium systems, and the connections with the free-energy calculation methods of Jarzynski and Crooks, have excited both theorists and experimentalists. This book charts the development and theoretical analysis of molecular dynamics as applied to equilibrium and nonequilibrium systems.

Substantially updated and revised, this book is designed both for experts in the field and beginning graduate students of physics. It connects molecular-dynamics simulation with mathematical theory to understand nonequilibrium steady states. It also provides a link between the atomic, nano, and macro worlds, showing how these length scales relate. The book ends with an introduction to the use of nonequilibrium statistical mechanics to justify a thermodynamic treatment of nonequilibrium steady states, and gives a direction to further avenues of exploration.

DENIS J. EVANS is Professor of Theoretical Chemistry at the Australian National University (ANU), Dean of the Research School of Chemistry and Convenor of the ANU College of Science. He has won several prizes, including the Moyal Medal from Macquarie University for distinguished contributions to mathematics, physics or statistics, the Centenary Medal from the Australian Government, and the H. G. Smith Memorial Medal from the Royal Australian Chemical Institute.

GARY MORRISS is Associate Professor and Undergraduate Director in the School of Physics at the University of New South Wales, Australia. He is a Fellow of the Institute of Physics and a member of the American Physical Society. His research areas include nonequilibrium statistical mechanics and dynamical systems.

STATISTICAL MECHANICS OF NONEQUILIBRIUM LIQUIDS

SECOND EDITION

DENIS J. EVANS

Research School of Chemistry, Australian National University, Canberra

GARY MORRISS

School of Physics, University of New South Wales, Sydney

CAMBRIDGE
UNIVERSITY PRESS

CAMBRIDGE
UNIVERSITY PRESS

University Printing House, Cambridge CB2 8BS, United Kingdom

Cambridge University Press is part of the University of Cambridge.

It furthers the University's mission by disseminating knowledge in the pursuit of education, learning and research at the highest international levels of excellence.

www.cambridge.org
Information on this title: www.cambridge.org/9781107424531

© D. Evans and G. Morris 2008

First published 1990
Second Edition 2008
First paperback edition 2014

A catalogue record for this publication is available from the British Library

ISBN 978-0-521-85791-8 Hardback
ISBN 978-1-107-42453-1 Paperback

Contents

Preface to the second edition

Since 1990, when the first edition appeared, there has been a significant advance in the development of nonequilibrium systems. The centerpiece of the first edition was the nonequilibrium molecular-dynamics methods and their theoretical analysis, the connections between linear and nonlinear response theory, and the design of the simulation methods. This is now a mature field with only one significant addition, which is the new method for elongational flows.

Chapter 10 in the first edition was called "Towards a thermodynamics of steady states." This contained an introduction to deterministic chaotic systems. The second edition has the same title for Chapter 10, but the contents are now completely different. The application of the ideas of modern dynamical-systems theory to nonequilibrium systems has grown enormously with all of Chapter 8 devoted to this. However, this still constitutes the barest of introductions with whole books (Gaspard, 1998; Dorfman, 1999; Ott, 2002; and Sprott, 2003) devoted to this theme. The theoretical advances in this area are some of the biggest. The development of methods to study the time evolution using periodic orbits, and the use of periodic orbits to develop SRB measures for nonequilibrium systems are exciting steps forward.

Based on the dynamical properties, Lyapunov exponents in particular, there have been great strides made in the development of the study of fluctuations in nonequilibrium systems. The fluctuation theorems, and methods for calculating free-energy differences using nonequilibrium paths, have dominated conferences for the last 6–7 years. The additional fact that these can be tested in real (rather than computer) experiments and used to measure free-energy differences in the unfolding of biological molecules will have a large impact.

Thanks are due to many people. Customarily we thank our wives, and remark that they are still the same! To those that have led the development of statistical mechanics and inspired and mentored those that followed we owe a great debt: Eddie Cohen, Bob Dorfman, Siegfried Hess, Christian Gruber, and many others.

A number of people were especially generous with their time for which we are very grateful, in particular, Peter Daivis, Carl Dettmann, Tooru Taniguchi, and Debra Bernhardt (Searles). Special thanks also to Billy Todd, Tom Hunt, Lamberto Rondoni, Chris Angstmann, David Kruss, Anthony Whelan, Dean Robinson, David Monaghan and Tony Chung. I am very pleased to acknowledge the contribution of Ian Watson, who as an undergraduate student read and learnt from this book, in the process discovering (and helping to correct) many of its faults. As ever, Eddie Cohen was a frustrating inspiration to us all!

I feel the pull of the white ship at the Grey Havens in the long firth of Lune.

Preface to the first edition

During the 1980s there have been many new developments regarding the nonequilibrium statistical mechanics of dense classical systems. These developments have had a major impact on the computer simulation methods used to model nonequilibrium fluids. Some of these new algorithms are discussed in the recent book by Allen and Tildesley (1987), *Computer Simulation of Liquids*. However, that book was never intended to provide a detailed statistical mechanical backdrop to the new computer algorithms. As the authors commented in their preface, their main purpose was to provide a working knowledge of computer simulation techniques. The present volume is, in part, an attempt to provide a pedagogical discussion of the statistical mechanical environment of these algorithms.

There is a symbiotic relationship between nonequilibrium statistical mechanics on the one hand and the theory and practice of computer simulation on the other. Sometimes, the initiative for progress has been with the pragmatic requirements of computer simulation and at other times, the initiative has been with the fundamental theory of nonequilibrium processes. Although progress has been rapid, the number of participants who have been involved in the exposition and development, rather than with application, has been relatively small.

The formal theory is often illustrated with examples involving shear flow in liquids. Since a central theme of this volume is the nonlinear response of systems, this book could be described as a text on theoretical rheology. However our choice of rheology as a test-bed for theory is merely a reflection of personal interest. The statistical mechanical theory that is outlined in this book is capable of far wider application.

All but two pages of this book are concerned with atomic rather than molecular fluids. This restriction is one of economy. The main purpose of this text is best served by choosing simple applications.

Many people deserve thanks for their help in developing and writing this book. Firstly we must thank our wives, Val and Jan, for putting up with our absences, our irritability, and our exhaustion. We would also like to thank Dr. David MacGowan for reading sections of the manuscript. Thanks must also go to Mrs. Marie Lawrence for help with indexing. Finally special thanks must go to Professors Cohen, Hanley, and Hoover for incessant argument and interest.

1

Introduction

Mechanics provides a complete microscopic description of the state of a system. When the equations of motion are combined with initial conditions and boundary conditions, the subsequent time evolution of a classical system can be predicted. In systems with more than just a few degrees of freedom such an exercise is impossible. There is simply no practical way of measuring the initial microscopic state of, for example, a glass of water, at some instant in time. In any case, even if this was possible we could not then solve the equations of motion for a coupled system of 10^{23} molecules.

In spite of our inability to fully describe the microstate of a glass of water, we are all aware of useful macroscopic descriptions for such systems. Thermodynamics provides a theoretical framework for correlating the equilibrium properties of such systems. If the system is not at equilibrium, fluid mechanics is capable of predicting the macroscopic nonequilibrium behaviour of the system. In order for these macroscopic approaches to be useful, their laws must be supplemented, not only with a specification of the appropriate boundary conditions, but with the values of thermophysical constants such as equation-of-state data and transport coefficients. These values cannot be predicted by macroscopic theory. Historically this data has been supplied by experiments. One of the tasks of statistical mechanics is to predict these parameters from knowledge of the interactions of the system's constituent molecules. This then is a major purpose for statistical mechanics. How well have we progressed?

Equilibrium classical statistical mechanics is relatively well developed. The basic ground rules – Gibbsian ensemble theory – have been known for the best part of a century (Gibbs, 1902). The development of electronic computers in the 1950s provided unambiguous tests of the theory of simple liquids leading to a consequently rapid development of integral equation and perturbation treatments of liquids (Barker and Henderson, 1976). With the possible exceptions of phase equilibria and interfacial phenomena (Rowlinson and Widom, 1982) one could say that the equilibrium statistical mechanics of atomic fluids is a solved problem. Much of the emphasis has moved to molecular, even macromolecular, liquids.

The nonequilibrium statistical mechanics of dilute atomic gases – kinetic theory – is, likewise, essentially complete (Ferziger and Kaper, 1972). However, attempts to extend kinetic theory to higher densities have been fraught with severe difficulties. One might have imagined being able to develop a power-series expansion of the transport coefficients in much the same way that one expands the equilibrium equation of state in the virial series. Dorfman and Cohen (1965; 1972) proved that such an expansion does not exist. The Navier–Stokes transport coefficients are nonanalytic functions of density.

It was at about this time that computer simulations began to have an impact on the field. In a celebrated paper, Kubo (1957) showed that linear transport coefficients could be calculated from a knowledge of the equilibrium fluctuations in the flux associated with the particular transport coefficient. For example the shear viscosity η, is defined as the ratio of the shear stress, $-P_{xy}$, to the shear rate, $\partial u_x / \partial y \equiv \gamma$:

$$P_{xy} \equiv -\eta\gamma. \tag{1.1}$$

The Kubo relation predicts that the limiting, small shear rate, viscosity, is given by:

$$\eta = \beta V \int_0^\infty ds \, \langle P_{xy}(0)P_{xy}(s)\rangle, \tag{1.2}$$

where β is the reciprocal of the absolute temperature T, multiplied by Boltzmann's constant k_B, V is the system volume and the angle brackets denote an *equilibrium* ensemble average. The viscosity is then the infinite time integral of the equilibrium, autocorrelation function of the shear stress. Similar relations are valid for the other Navier–Stokes transport coefficients such as the self diffusion coefficient, the thermal conductivity, and the bulk viscosity (see Chapter 4).

Alder and Wainwright (1956) were the first to use computer simulations to compute the transport coefficients of atomic fluids. What they found was unexpected. It was believed that at sufficiently long time, equilibrium autocorrelation functions should decay exponentially. Alder and Wainwright discovered that in two-dimensional systems, the velocity autocorrelation function which determines the self-diffusion coefficient, only decays as t^{-1}. Since the diffusion coefficient is thought to be the integral of this function, we were forced to the reluctant conclusion that the self diffusion coefficient does not exist for two-dimensional systems. It is presently believed that each of the Navier–Stokes transport coefficients diverge in two dimensions (Pomeau and Resibois, 1975).

This does *not* mean that two-dimensional fluids are infinitely resistant to shear flow. Rather, it means that the Newtonian constitutive relation Equation (1.1) is an inappropriate definition of viscosity in two dimensions. There is no linear regime close to equilibrium where Newton's law (Equation 1.1), is valid. It is thought that at small strain rates, $P_{xy} \cong \gamma \log \gamma$. If this were the case then the limiting value of the shear viscosity $\lim_{\gamma \to 0} -\partial P_{xy}/\partial \gamma$ would be infinite. All this presupposes that steady laminar shear flow is stable in two dimensions. This would be an entirely natural presumption on the basis of our three-dimensional experience. However there is some evidence that even this assumption may be wrong (Evans and Morriss, 1983b). Recent computer simulation data suggests that in two dimensions, laminar flow may be unstable at *small* strain rates.

In three dimensions the situation is better. The Navier–Stokes transport coefficients appear to exist. However the nonlinear Burnett coefficients, higher-order terms in the Taylor series expansion of the shear stress in powers of the strain rate (Section 2.3, Section 9.5), are thought to diverge (Kawasaki and Gunton, 1973). These divergences are sometimes summarized in Dorfman's Lemma (Zwanzig, 1982): *all relevant fluxes are nonanalytic functions of all relevant variables!* The transport coefficients are thought to be nonanalytic functions of density, frequency, and the magnitude of the driving thermodynamic force, the strain rate, or the temperature gradient etc.

In this book we will discuss the framework of nonequilibrium statistical mechanics. We will not discuss in detail, the practical results that have been obtained. Rather we seek to derive a nonequilibrium analog of the Gibbsian basis for equilibrium statistical mechanics. At equilibrium we have a number of idealizations which serve as standard models for experimental systems. Among these are the well-known microcanonical, canonical, and grand canonical ensembles. The real system of interest will not correspond exactly to any one particular ensemble, but such models furnish useful and reliable information about the experimental system. We have become so accustomed to mapping each real experiment onto its nearest Gibbsian ensemble that we sometimes forget that the canonical ensemble, for example, does not exist in Nature. It is an idealization.

A nonequilibrium system can be modeled as a perturbed equilibrium ensemble; we will therefore need to add the perturbing field to the statistical mechanical description. The perturbing field does work on the system – this prevents the system from relaxing to equilibrium. This work is converted to heat, and the heat must be removed in order to obtain a well-defined steady state. Therefore thermostats will also need to be included in our statistical mechanical models. A major theme of this book is the development of a set of idealized nonequilibrium systems which can play the same role in nonequilibrium statistical mechanics as the Gibbsian ensembles play at equilibrium.

After a brief discussion of linear irreversible thermodynamics in Chapter 2, we address the Liouville equation in Chapter 3. The Liouville equation is the fundamental vehicle of nonequilibrium statistical mechanics. We introduce its formal solution using mathematical operators called propagators (Section 3.3). In Chapter 3, we also outline the procedures by which we identify statistical mechanical expressions for the basic field variables of hydrodynamics.

After this background in both macroscopic and microscopic theory we go on to derive the Green–Kubo relations for linear transport coefficients in Chapter 4 and the basic results of linear response theory in Chapter 5. The Green–Kubo relations derived in Chapter 4 relate *thermal* transport coefficients, such as the Navier–Stokes transport coefficients, to equilibrium fluctuations. Thermal transport processes are driven by boundary conditions. The expressions derived in Chapter 5 relate *mechanical* transport coefficients to equilibrium fluctuations. A mechanical transport process is one that is driven by a perturbing external field which actually changes the mechanical equations of motion for the system. In Chapter 5 we show how the thermostatted linear mechanical response of many body systems is related to equilibrium fluctuations.

In Chapter 6 we exploit similarities in the fluctuation formulae for the mechanical and the thermal response, by deriving computer simulation algorithms for calculating the linear Navier–Stokes transport coefficients. Although the algorithms are designed to calculate linear thermal-transport coefficients, they employ mechanical methods. The validity of these algorithms is proved using thermostatted linear-response theory (Chapter 5) and the knowledge of the Green–Kubo relations provided in Chapter 4.

A diagrammatic summary of some of the common algorithms used to compute shear viscosity is given in Figure 1.1. The Green–Kubo method simply consists of simulating an equilibrium fluid under periodic boundary conditions and making the appropriate analysis of the time-dependent stress fluctuations using Equation (1.2). Gosling *et al.* (1973) proposed performing a nonequilibrium simulation of a system subject to a sinusoidal transverse force. Monitoring the field-induced velocity profile and extrapolating the results to infinite wavelength, the viscosity can be calculated. Hoover and Ashurst (1975), used external reservoirs of particles to induce a nearly planar shear in a model fluid. In the reservoir technique, the viscosity is calculated by measuring the average ratio of the shear stress to the strain rate, in the bulk of the fluid, away from the reservoir regions. The presence of the reservoir regions gives rise to significant inhomogeneities in the thermodynamic properties of the fluid and in the strain rate in particular. This leads to obvious difficulties in the calculation of the shear viscosity. Lees and Edwards (1972), showed that if one used "sliding brick" periodic boundary conditions, one could induce *planar* Couette flow in a simulation. The so-called Lees–Edwards periodic

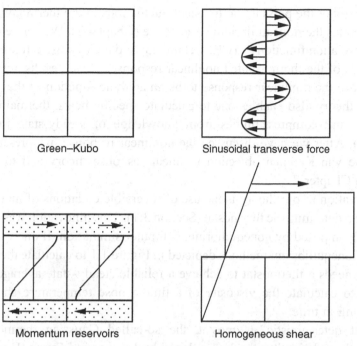

Figure 1.1 Methods of determining the shear viscosity

boundary conditions enable one to perform homogeneous simulations of shear flow in which the low Reynolds-number velocity profile is linear.

With the exception of the Green–Kubo method, these simulation methods all involve nonequilibrium simulations. The Green–Kubo technique is useful in that all linear transport coefficients can, in principle, be calculated from a single simulation. It is restricted though, to *only* calculating linear transport coefficients. The nonequilibrium methods, on the other hand, provide information about the non-linear as well as the linear response of systems. They therefore provide a direct link with rheology.

The use of nonequilibrium computer simulation algorithms, so-called nonequilibrium molecular dynamics (NEMD), leads inevitably to the question of the large field, nonlinear response. Indeed the calculation of linear transport coefficients using NEMD proceeds by calculating the nonlinear response and extrapolating the results to zero field. One of our main aims will be to derive a number of nonlinear generalizations of the Kubo relations which give an exact framework within which one can calculate and characterize transport processes far from equilibrium (Chapter 7). Because of the divergences alluded to above, the nonlinear theory cannot rely on power-series expansions about the equilibrium state. A major system of interest is the nonequilibrium steady state. Theory enables us to relate the nonlinear transport coefficients and mechanical quantities, like the internal energy or the pressure, to transient fluctuations in the thermodynamic flux which generates the nonequilibrium steady state (Chapter 7). We derive the transient time correlation function (TTCF, Section 7.3) and the Kawasaki representations (Section 7.2) of the thermostatted nonlinear response. These results are exact and do not require the nonlinear response to be an analytic function of the perturbing fields. The theory also enables one to calculate specific heats, thermal-expansion coefficients and compressibilities from knowledge of steady-state fluctuations (Chapter 9). After we have discussed the nonlinear response, we present a resolution of the van Kampen objection to linear response theory and to the Kubo relations in Chapter 7.

An innovation in our theory is the use of reversible equations of motion which incorporate a deterministic thermostat (Section 3.1). This innovation was motivated by the needs imposed by nonequilibrium computer simulation. If one wants to use any of the nonequilibrium methods depicted in Figure 1.1 to calculate the shear viscosity, one needs a thermostat to achieve a reliable steady-state average. It is not clear how to calculate the viscosity of a fluid whose temperature and pressure are increasing in time.

The first deterministic thermostat, the so-called Gaussian thermostat, was independently and simultaneously developed by Hoover and Evans (Hoover *et al.*, 1982) and Evans (1983a). It permitted homogeneous simulations of nonequilibrium steady states using molecular-dynamics techniques. Hitherto molecular dynamics had involved solving Newton's equations for systems of interacting particles. As work was performed on such a system in order to drive it away from equilibrium, the system inevitably heated with the irreversible conversion of work into heat.

Hoover and Evans showed that if such a system evolved under their thermostatted equations of motion, the so-called Gaussian isokinetic equations of motion, the dissipative heat could be removed by a thermostatting force which was part of the equations of motion themselves. Now, computer simulators had been simulating nonequilibrium steady states for some years, but in the past the dissipative heat was removed by simple ad hoc rescaling of the second moment of the appropriate velocity. The significance of the Gaussian isokinetic equations of motion was that since the thermostatting was part of the equations of motion it could be analyzed theoretically using response theory. Earlier ad hoc rescaling or Andersen's stochastic thermostat (Andersen, 1980), could not be so easily analyzed. In Chapter 5 we prove that while the adiabatic (i.e. unthermostatted) linear response of a system can be calculated as the integral of an unthermostatted (i.e. Newtonian) equilibrium time-correlation function, the thermostatted linear response is related to the corresponding thermostatted equilibrium time-correlation function. These results are quite new and can be proved only because the thermostatting mechanism is reversible and deterministic.

It is natural to ask whether the "thermostatted" response depends upon the details of the thermostatting mechanism. Provided the amount of heat Q, removed by a thermostat within the characteristic microscopic relaxation time τ, of the system is small compared to the enthalpy I, of the fluid (i.e. ($\tau \, dQ/dt)/$ $I < 1$), we expect that the microscopic details of the thermostat will be unimportant. In the linear regime, close to equilibrium, this will always be the case. Even for systems far (but not too far), from equilibrium this condition is often satisfied. In Section 5.4 we give a mathematical proof of the independence of the linear response to the thermostatting mechanism.

Although originally motivated by the needs of nonequilibrium simulations, we have now reached the point where we can simulate equilibrium systems at constant internal energy E, at constant enthalpy I, or at constant temperature T, and pressure p. If we employ the so-called Nosé–Hoover (Nosé, 1984b; Hoover, 1985) thermostat, we can allow fluctuations in the state-defining variables while controlling their mean values. These methods have had a major impact on computer simulation methodology and practice.

To illustrate the point: in an ergodic system at equilibrium, Newton's equations of motion *generate* the molecular dynamics ensemble in which the number of particles, the total energy, the volume, and the total linear momentum are all precisely fixed $(N, E, V, \Sigma \mathbf{p}_i)$. Previously this was the only equilibrium ensemble accessible to molecular dynamics simulation. Now however we can use Gaussian methods to generate equilibrium ensembles in which the precise value of say, the enthalpy and pressure are fixed $(N, E, p, \Sigma \mathbf{p}_i)$. Alternatively, Nosé–Hoover equations of motion could be used which generate the canonical ensemble ($\exp[-\beta H]$). Gibbs proposed the various ensembles as idealized statistical *distributions* in phase space. Now we have the *dynamics* that is capable of generating each of those distributions.

A new element in the theory of nonequilibrium steady states is the abandonment of Hamiltonian dynamics. The Hamiltonian, of course, plays a central role in Gibbs' equilibrium statistical mechanics. It leads to a compact and elegant description. However the existence of a Hamiltonian which generates dynamical trajectories is, as we will see, not essential.

In the space of relevant variables, neither the Gaussian thermostatted equations of motion nor the Nosé–Hoover equations of motion can be derived from a Hamiltonian. This is true even in the absence of external perturbing fields. This implies, in turn, that the usual form of the Liouville equation, $df/dt = 0$, for the N-particle distribution function f, is invalid. Thermostatted equations of motion necessarily imply a compressible phase space.

The abandonment of a Hamiltonian approach to particle dynamics had, in fact, been forced on us somewhat earlier. The Evans–Gillan equations of motion for heat flow (Section 6.5), which predate both the Gaussian and Nosé–Hoover thermostatted dynamics, cannot be derived from a Hamiltonian. The Evans–Gillan equations provide the most efficient presently known dynamics for describing heat flow in systems close to equilibrium. A synthetic external field was invented so that its interaction with a N-particle system precisely mimics the impact a real temperature gradient would have on the system. Linear response theory is then used to prove that the response of a system to a real temperature gradient is identical to the response to the synthetic Evans–Gillan external field.

We use the term *synthetic* to note the fact that the Evans–Gillan field does not exist in Nature. It is a mathematical device used to transform a difficult boundary condition problem, the flow of heat in a system bounded by walls maintained at differing temperatures, into a much simpler mechanical problem. The Evans–Gillan field acts upon the system in a homogeneous way permitting the use of periodic rather than inhomogeneous boundary conditions. This synthetic field exerts a force on each particle which is proportional to the difference of the particle's enthalpy from the mean enthalpy per particle. The field thereby induces a flow of heat in the absence of either a temperature gradient or of any mass flow. No Hamiltonian is known which can generate the resulting equations of motion.

In a similar way Kawasaki showed that the boundary condition that corresponds to planar Couette shear flow can be incorporated exactly into the equations of motion. These equations are known as the SLLOD equations (Section 6.3). They give an exact description of the shearing motion of systems arbitrarily far from equilibrium. Again, no Hamiltonian can be found which is capable of generating these equations.

When external fields or boundary conditions perform work on a system we have at our disposal a very natural set of mechanisms for constructing nonequilibrium ensembles in which different sets of thermodynamic state variables are used to constrain, or define, the system. Thus we can generate on the computer, or analyze theoretically, nonequilibrium analogs of the canonical, microcanonical, or isobaric–isoenthalpic ensembles.

At equilibrium one is used to the idea of pairs of conjugate thermodynamic variables generating conjugate equilibrium ensembles. In the canonical ensemble particle number N, volume V, and temperature T, are the state variables whereas in the isothermal–isobaric ensemble the role played by the volume is replaced by the pressure, its thermodynamic conjugate. In the same sense one can generate conjugate pairs of nonequilibrium ensembles. If the driving thermodynamic force is X, it could be a temperature gradient or a strain rate, and then one could consider the (N, V, T, X) ensemble or alternatively the conjugate (N, p, T, X) ensemble.

However in nonequilibrium steady states one can go much further than this. The dissipation, the heat removed by the thermostat per unit time dQ/dt, can always be written as a product of a thermodynamic force, X, and a thermodynamic flux, $J(\Gamma)$.

If, for example, the force is the shear rate, γ, then the conjugate flux is the shear stress, $-P_{xy}$. One can then consider nonequilibrium ensembles in which the thermodynamic flux rather than the thermodynamic force is the independent state variable. For example we could define the nonequilibrium steady state as an (N, V, T, J) ensemble. Such an ensemble is, by analogy with electrical circuit theory, called a Norton ensemble, while the case where the force is the state variable, (N, V, T, X), is called a Thévenin ensemble. A major postulate in this work is the macroscopic equivalence of corresponding Norton and Thévenin ensembles.

The equations of motion for a system of particles which undergo collisions are usually chaotic (although there are examples, like the wind-tree model, that are not). The application of the ideas of modern dynamical systems theory has had a large impact on nonequilibrium statistical mechanics in the last 15 to 20 years. The books by Gaspard (1998), Dorfman (1999), Ott (2002) and Sprott (2003) are more comprehensive than the development that we present in Chapter 8. However, our approach is to begin with the characterization of chaos in a dynamical system, and then to use the ideas of Ruelle (1978), Cvitanovic, and others (2005) to show how to develop an understanding of the time evolution of both the probability distribution and an arbitrary phase variable. Surprisingly, just the structure of the theory gives a simple argument to show that the transport coefficients must be non-negative.

The discovery of relations satisfied by the fluctuations in nonequilibrium steady states has become a major area of activity in the last decade. The discovery of the fluctuation theorem by Evans *et al.* (1993a), the derivations by Evans and Searles (1994) and Gallavotti and Cohen (1995a) have sparked a great deal of interest and controversy. The subsequent discovery of methods of calculating free-energy differences using arbitrary nonequilibrium paths by Jarzynski (1997) and Crooks (1998) has stimulated many experiments designed both to confirm the theoretical predictions and to use the techniques in physical and biological systems.

In the last chapter we introduce material which is quite recent and perhaps controversial. We attempt to develop a thermodynamics of nonequilibrium steady states which may be considered a nonlinear generalization of the conventional linear irreversible thermodynamics treated in Chapter 2. The difficulty is extending our notions of temperature and entropy to nonequilibrium systems. We take as an axiom, the observation from computer simulation studies, that the internal energy of the system is also a function of the field that perturbs the system from equilibrium. Thus the internal energy U is a function of temperature, volume, and the shear rate for a system undergoing Couette flow. We consider two approaches; the first using the assumption of linear viscoelasticity, and the second using a purely statistical mechanical treatment.

What is surprising is that the steady-state nonequilibrium distribution function is a singular and fractal object. This implies that the fine grained Gibbs entropy:

$$S = -k_B \int_{\text{all } \Gamma \text{ space}} d\Gamma\, f(\Gamma, t) \ln f(\Gamma, t), \qquad (1.3)$$

diverges to negative infinity. (If no thermostat is employed, the nonequilibrium entropy is a constant of the motion Gibbs (1902)). The question of the expression for the nonequilibrium entropy, and how to calculate it, remain unresolved.

2

Linear irreversible thermodynamics

2.1 The conservation equations

At the hydrodynamic level we are interested in the macroscopic evolution of densities of conserved extensive variables such as mass, energy, and momentum. Because these quantities are conserved, their respective densities can only change by a process of *redistribution*. As we shall see, this means that the relaxation of these densities is slow, and therefore the relaxation plays a macroscopic role. If this relaxation were fast (i.e. if it occurred on a molecular timescale for instance) it would be unobservable at a macroscopic level. The macroscopic equations of motion for the densities of conserved quantities are called the Navier–Stokes equations. We will now give a brief description of how these equations are derived. It is important to understand this derivation because one of the objects of statistical mechanics is to provide a microscopic or molecular justification for the Navier–Stokes equations. In the process, statistical mechanics sheds light on the limits of applicability of these equations. Similar treatments can be found in de Groot and Mazur (1962) and Kreuzer (1981).

Let $M(t)$ be the total mass contained in an arbitrary volume V, then

$$M = \int_V d\mathbf{r}\rho(\mathbf{r}, t), \tag{2.1}$$

where $\rho(\mathbf{r}, t)$ is the mass density at position $\mathbf{r}$ and time t, and $d\mathbf{r}$ is the volume element at $\mathbf{r}$. Since mass is conserved, the only way that the mass in the volume V can change is by flowing through the enclosing surface, S (see Figure 2.1).

$$\frac{dM}{dt} = -\int_S d\mathbf{S} \cdot \rho(\mathbf{r}, t)\mathbf{u}(\mathbf{r}, t) = -\int_V d\mathbf{r}\boldsymbol{\nabla} \cdot [\rho(\mathbf{r}, t)\mathbf{u}(\mathbf{r}, t)]. \tag{2.2}$$

Here $\mathbf{u}(\mathbf{r}, t)$ is the fluid streaming velocity at position $\mathbf{r}$ and time t. The vector $d\mathbf{S}$ denotes the outward area element of the enclosing surface S, and $\boldsymbol{\nabla}$ is the spatial gradient vector operator, $\left(\frac{\partial}{\partial x}, \frac{\partial}{\partial y}, \frac{\partial}{\partial z}\right)$. It is clear that the rate of change

11

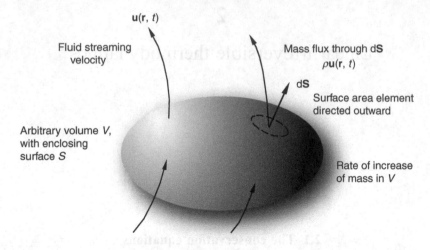

Figure 2.1 The change in the mass contained in an arbitrary closed volume V can be calculated by integrating the mass flux through the enclosing surface S $dM/dt = -\int d\mathbf{S} \cdot \rho\mathbf{u}(\mathbf{r}, t)$

of the enclosed mass can also be written in terms of the change in mass density $\rho(\mathbf{r}, t)$, as:

$$\frac{dM}{dt} = \int_V d\mathbf{r} \frac{\partial \rho(\mathbf{r}, t)}{\partial t}. \tag{2.3}$$

If we equate these two expressions for the rate of change of the total mass we find that since the volume V was arbitrary:

$$\frac{\partial \rho(\mathbf{r}, t)}{\partial t} = -\boldsymbol{\nabla} \cdot [\rho(\mathbf{r}, t)\mathbf{u}(\mathbf{r}, t)]. \tag{2.4}$$

This is called the mass continuity equation and is essentially a statement that mass is conserved. We can write the mass continuity equation in an alternative form if we use the relation between the total or streaming derivative, and the various partial derivatives. For an arbitrary function of position $\mathbf{r}$ and time t, for example $a(\mathbf{r}, t)$, we have:

$$\frac{d}{dt} a(\mathbf{r}, t) = \frac{\partial}{\partial t} a(\mathbf{r}, t) + \mathbf{u} \cdot \boldsymbol{\nabla} a(\mathbf{r}, t). \tag{2.5}$$

If we let $a(\mathbf{r}, t) \equiv \rho(\mathbf{r}, t)$ in Equation (2.5), and combine this with Equation (2.4) then the mass continuity equation can be written as:

$$\frac{d\rho(\mathbf{r}, t)}{dt} = -\rho(\mathbf{r}, t)\boldsymbol{\nabla} \cdot \mathbf{u}(\mathbf{r}, t). \tag{2.6}$$

In an entirely analogous fashion we can derive an equation of continuity for momentum. Let $\mathbf{G}(t)$ be the total momentum of the arbitrary volume V, then the rate of change of momentum is given by:

$$\frac{d\mathbf{G}}{dt} = \int_V d\mathbf{r}\, \frac{\partial[\rho(\mathbf{r}, t)\mathbf{u}(\mathbf{r}, t)]}{\partial t}. \tag{2.7}$$

The total momentum of volume V can change in two ways. Firstly it can change by convection. Momentum can flow through the enclosing surface. This convective term can be written as:

$$\frac{d\mathbf{G}_c}{dt} = -\int_S d\mathbf{S}\cdot\rho(\mathbf{r}, t)\mathbf{u}(\mathbf{r}, t)\mathbf{u}(\mathbf{r}, t). \tag{2.8}$$

Here we use the dyadic product of two first-rank tensors (or ordinary vectors) $\mathbf{u}$ and $\mathbf{u}$ to obtain a second-rank tensor $\mathbf{uu}$. The second way that the momentum could change is by the action of a force. There are two possible forces: the pressure exerted on V by the surrounding fluid, and some external force that acts directly on the fluid. We refer to the pressure exerted on V as the stress contribution. The force $d\mathbf{F}$, exerted by the fluid across an element of area $d\mathbf{F}$, which is moving at the streaming velocity of the fluid, must be proportional to the magnitude of the area $d\mathbf{S}$. The most general such linear relation is:

$$d\mathbf{F} \equiv -d\mathbf{S} \cdot \mathsf{P}. \tag{2.9}$$

This is in fact the definition of the pressure tensor P. It is also the negative of the stress tensor. That the pressure tensor is a second-rank tensor rather than a simple

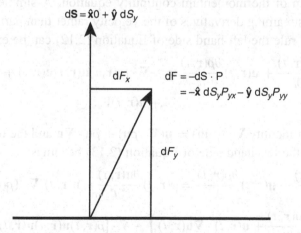

Figure 2.2 The pressure tensor P is the infinitesimal force $d\mathbf{F}$ across an infinitesimal area element $d\mathbf{S} = \hat{x}0 + \hat{y}dS_y$

scalar is a reflection of the fact that the force $d\mathbf{F}$ and the area vector $d\mathbf{S}$ need not be parallel. In fact for molecular fluids the pressure tensor is not symmetric in general.

As P is second-rank tensor it is appropriate to define the notational conventions that we will use. P requires two subscripts to specify an element. In Einstein notation Equation (2.9) reads $dF_\alpha = -dS_\beta P_{\beta\alpha}$, where the repeated index β implies a summation. Notice that the contraction (or dot product) involves the first index of P and that the vector character of the force $d\mathbf{F}$ is determined by the second index of P. We will use bold sans serif characters to denote tensors of rank two or more. Figure 2.2 gives a diagrammatic representation of the tensorial relations in the definition of the pressure tensor. Using this definition the total force contribution to the momentum change can be seen to be:

$$\frac{d\mathbf{G}_c}{dt} = -\int_S d\mathbf{S}\cdot\mathsf{P} + \int_V d\mathbf{r}\, n(\mathbf{r},t)\mathbf{F}_e, \tag{2.10}$$

where $n(\mathbf{r},t)\mathbf{F}_e$ is the external force density (force times the number density $n(\mathbf{r},t)$). Combining Equations (2.8, 2.10) and using the divergence theorem to convert surface integrals to volume integrals gives:

$$\frac{d\mathbf{G}}{dt} = \int_V d\mathbf{r}\, \frac{\partial[\rho(\mathbf{r},t)\mathbf{u}(\mathbf{r},t)]}{\partial t}$$

$$= -\int_V d\mathbf{r}(\mathbf{\nabla}\cdot[\rho(\mathbf{r},t)\mathbf{u}(\mathbf{r},t)\mathbf{u}(\mathbf{r},t) + \mathsf{P}] - n(\mathbf{r},t)\mathbf{F}_e). \tag{2.11}$$

Since this equation is true for arbitrary V we conclude that,

$$\frac{\partial[\rho(\mathbf{r},t)\mathbf{u}(\mathbf{r},t)]}{\partial t} = -\mathbf{\nabla}\cdot[\rho(\mathbf{r},t)\mathbf{u}(\mathbf{r},t)\mathbf{u}(\mathbf{r},t) + \mathsf{P}] + n(\mathbf{r},t)\mathbf{F}_e. \tag{2.12}$$

This is one form of the momentum continuity equation. A simpler form can be obtained using streaming derivatives of the velocity rather than partial derivatives. Using the chain rule the left-hand side of Equation (2.12) can be expanded as:

$$\rho(\mathbf{r},t)\frac{\partial\mathbf{u}(\mathbf{r},t)}{\partial t} + \mathbf{u}(\mathbf{r},t)\frac{\partial\rho(\mathbf{r},t)}{\partial t} = -\mathbf{\nabla}\cdot[\rho(\mathbf{r},t)\mathbf{u}(\mathbf{r},t)\mathbf{u}(\mathbf{r},t) + \mathsf{P}]$$

$$+ n(\mathbf{r},t)\mathbf{F}_e. \tag{2.13}$$

Using the vector identity $\mathbf{\nabla}\cdot(\rho\mathbf{u}\mathbf{u}) = \mathbf{u}(\mathbf{\nabla}\cdot\rho\mathbf{u}) + \rho\mathbf{u}\cdot\mathbf{\nabla}\mathbf{u}$ and the mass continuity Equation (2.4), the left-hand side of Equation (2.13) becomes:

$$\rho(\mathbf{r},t)\frac{\partial\mathbf{u}(\mathbf{r},t)}{\partial t} + \mathbf{u}(\mathbf{r},t)\frac{\partial\rho(\mathbf{r},t)}{\partial t} = \rho(\mathbf{r},t)\frac{\partial\mathbf{u}(\mathbf{r},t)}{\partial t} - \mathbf{u}(\mathbf{r},t)(\mathbf{\nabla}\cdot(\rho(\mathbf{r},t)\mathbf{u}(\mathbf{r},t)))$$

$$= \rho(\mathbf{r},t)\left(\frac{\partial\mathbf{u}(\mathbf{r},t)}{\partial t} + \mathbf{u}(\mathbf{r},t)\cdot\mathbf{\nabla}\mathbf{u}(\mathbf{r},t)\right) - \mathbf{\nabla}\cdot[\rho(\mathbf{r},t)\mathbf{u}(\mathbf{r},t)\mathbf{u}(\mathbf{r},t)].$$

$$\tag{2.14}$$

Now, from Equation (2.5):

$$\rho(\mathbf{r}, t)\frac{d\mathbf{u}(\mathbf{r}, t)}{dt} = \rho(\mathbf{r}, t)\left(\frac{\partial\mathbf{u}(\mathbf{r}, t)}{\partial t} + \mathbf{u}(\mathbf{r}, t)\cdot\nabla\mathbf{u}(\mathbf{r}, t)\right). \tag{2.15}$$

so that combining Equations (2.13), (2.14), and (2.15) gives:

$$\rho(\mathbf{r}, t)\frac{d\mathbf{u}(\mathbf{r}, t)}{dt} = -\nabla\cdot\mathsf{P} + n(\mathbf{r}, t)\mathbf{F}_e. \tag{2.16}$$

The final conservation equation we will derive is the energy equation. If we denote the total energy per unit mass or the specific total energy as $e(\mathbf{r}, t)$, then the total energy density is $\rho(\mathbf{r}, t)e(\mathbf{r}, t)$. If the fluid is convecting there is obviously a simple convective kinetic energy component in $e(\mathbf{r}, t)$. If this is removed from the energy density then what remains should be a thermodynamic internal energy density, $\rho(\mathbf{r}, t)U(\mathbf{r}, t)$:

$$\rho(\mathbf{r}, t)e(\mathbf{r}, t) = \rho(\mathbf{r}, t)\frac{1}{2}\mathbf{u}(\mathbf{r}, t)^2 + \rho(\mathbf{r}, t)U(\mathbf{r}, t). \tag{2.17}$$

Here we have identified the first term on the right-hand side as the convective kinetic energy. Using Equation (2.16) we can show that:

$$\rho(\mathbf{r}, t)\frac{d}{dt}\frac{\mathbf{u}(\mathbf{r}, t)^2}{2} = \rho(\mathbf{r}, t)\mathbf{u}(\mathbf{r}, t)\cdot\frac{d\mathbf{u}(\mathbf{r}, t)}{dt}$$

$$= -\mathbf{u}(\mathbf{r}, t)\cdot[\nabla\cdot\mathsf{P}] = -\mathbf{u}\nabla:\mathsf{P}. \tag{2.18}$$

The second equality is a consequence of the momentum conservation Equation (2.16). We use the dyadic product of two first-rank tensors $\mathbf{u}$ and ∇ to obtain a second-rank tensor $\mathbf{u}\nabla$. In Einstein notation $(\mathbf{u}\nabla)_{\alpha\beta} \equiv u_\alpha\nabla_\beta$. In the first form given in Equation (2.18), ∇ is contracted into the first index of P and then $\mathbf{u}$ is contracted into the second remaining index. This defines the meaning of the double contraction notation after the second equals sign – inner indices are contracted first, then outer indices – that is $\mathbf{u}\nabla:\mathsf{P} = (\mathbf{u}\nabla)_{\alpha\beta}\mathsf{P}_{\beta\alpha} \equiv u_\alpha\nabla_\beta\mathsf{P}_{\beta\alpha}$.

For any variable a, using Equation (2.5) we have:

$$\rho(\mathbf{r}, t)\frac{da(\mathbf{r}, t)}{dt} = \rho(\mathbf{r}, t)\frac{\partial a(\mathbf{r}, t)}{\partial t} + \rho(\mathbf{r}, t)\mathbf{u}(\mathbf{r}, t)\cdot\nabla a(\mathbf{r}, t) = \frac{\partial[\rho(\mathbf{r}, t)a(\mathbf{r}, t)]}{\partial t}$$

$$+ \rho(\mathbf{r}, t)\mathbf{u}(\mathbf{r}, t)\cdot\nabla a(\mathbf{r}, t) - a(\mathbf{r}, t)\frac{\partial\rho(\mathbf{r}, t)}{\partial t}. \tag{2.19}$$

Using the mass continuity Equation (2.4):

$$\rho(\mathbf{r}, t)\frac{\mathrm{d}a(\mathbf{r}, t)}{\mathrm{d}t} = \frac{\partial[\rho(\mathbf{r}, t)a(\mathbf{r}, t)]}{\partial t} + \rho(\mathbf{r}, t)\mathbf{u}(\mathbf{r}, t) \cdot \nabla a(\mathbf{r}, t)$$

$$+ a(\mathbf{r}, t)\nabla \cdot [\rho(\mathbf{r}, t)\mathbf{u}(\mathbf{r}, t)]$$

$$= \frac{\partial[\rho(\mathbf{r}, t)a(\mathbf{r}, t)]}{\partial t} + \nabla \cdot [\rho(\mathbf{r}, t)\mathbf{u}(\mathbf{r}, t)a(\mathbf{r}, t)]. \qquad (2.20)$$

If we let the total energy inside a volume V be E, then clearly:

$$\frac{\mathrm{d}E}{\mathrm{d}t} = \int_V \mathrm{d}\mathbf{r} \, \frac{\partial[\rho(\mathbf{r}, t)e(\mathbf{r}, t)]}{\partial t}. \qquad (2.21)$$

Because the energy is conserved we can make a detailed account of the energy balance in the volume V. The energy can simply convect through the containing surface, it could diffuse through the surface and the surface stresses or external forces could do work on the volume V. In order, these terms can be written:

$$\frac{\mathrm{d}E}{\mathrm{d}t} = -\int_S \mathrm{d}\mathbf{S} \cdot [\rho(\mathbf{r}, t)e(\mathbf{r}, t)\mathbf{u}(\mathbf{r}, t) + \mathbf{J}_Q(\mathbf{r}, t)]$$

$$-\int_S \mathrm{d}\mathbf{S} \cdot \mathsf{P}(\mathbf{r}, t) \cdot \mathbf{u}(\mathbf{r}, t) - \int_V \mathrm{d}\mathbf{r} n(\mathbf{r}, t)\mathbf{F}_e \cdot \mathbf{u}(\mathbf{r}, t). \qquad (2.22)$$

In Equation (2.22) $\mathbf{J}_Q$ is called the heat flux vector. It gives the energy flux across a surface which is moving with the local fluid streaming velocity. Using the divergence theorem, Equation (2.22) can be written as:

$$\frac{\mathrm{d}E}{\mathrm{d}t} = -\int_V \mathrm{d}\mathbf{r}\nabla \cdot [\rho(\mathbf{r}, t)e(\mathbf{r}, t)\mathbf{u}(\mathbf{r}, t) + \mathbf{J}_Q(\mathbf{r}, t) + \mathsf{P}(\mathbf{r}, t) \cdot \mathbf{u}(\mathbf{r}, t)]$$

$$-\int_V \mathrm{d}\mathbf{r} \, n(\mathbf{r}, t)\mathbf{u}(\mathbf{r}, t) \cdot \mathbf{F}_e. \qquad (2.23)$$

Comparing Equations (2.21) and (2.23), we derive the continuity equation for total energy:

$$\frac{\partial[\rho(\mathbf{r}, t)e(\mathbf{r}, t)]}{\partial t} = -\nabla \cdot [\rho(\mathbf{r}, t)e(\mathbf{r}, t)\mathbf{u}(\mathbf{r}, t) + \mathbf{J}_Q(\mathbf{r}, t) + \mathsf{P}(\mathbf{r}, t) \cdot \mathbf{u}(\mathbf{r}, t)]$$

$$+ n(\mathbf{r}, t)\mathbf{u}(\mathbf{r}, t) \cdot \mathbf{F}_e. \qquad (2.24)$$

We can use Equation (2.20) to express this equation in terms of streaming derivatives of the total specific energy:

$$\rho(\mathbf{r}, t)\frac{de(\mathbf{r}, t)}{dt} = -\nabla \cdot [\mathbf{J}_Q(\mathbf{r}, t) + \mathsf{P}(\mathbf{r}, t) \cdot \mathbf{u}(\mathbf{r}, t)] + n(\mathbf{r}, t)\mathbf{u}(\mathbf{r}, t) \cdot \mathbf{F}_e. \quad (2.25)$$

Finally, Equations (2.17) and (2.18) can be used to derive a continuity Equation for the specific internal energy:

$$\rho(\mathbf{r}, t)\frac{dU(\mathbf{r}, t)}{dt} = -\nabla \cdot \mathbf{J}_Q(\mathbf{r}, t) - \mathsf{P}(\mathbf{r}, t)^{\mathrm{T}} : \nabla\mathbf{u}(\mathbf{r}, t) + n(\mathbf{r}, t)\mathbf{u}(\mathbf{r}, t) \cdot \mathbf{F}_e, \quad (2.26)$$

where the superscript T denotes transpose. The transpose of the pressure tensor appears as a result of our double contraction notation because in Equation (2.25), ∇ is contracted into the first index of P.

The three continuity equations (2.6), (2.16), and (2.26) are continuum expressions of the fact that mass, momentum, and energy are conserved. These equations are exact.

2.2 Entropy production

Thus far, our description of the equations of hydrodynamics has been exact. We will now derive an equation for the rate at which entropy is produced spontaneously in a nonequilibrium system. The second law of thermodynamics states that entropy is not a conserved quantity. In order to complete this derivation we must assume that we can apply the laws of equilibrium thermodynamics, at least on a local scale, in nonequilibrium systems. This assumption is called the *local thermodynamic equilibrium postulate*. We expect that this postulate should be valid for systems that are sufficiently close to equilibrium (de Groot and Mazur, 1962). This macroscopic theory provides no information on how *small* these deviations from equilibrium should be in order for local thermodynamic equilibrium to hold. It turns out, however, that the local thermodynamic equilibrium postulate is satisfied for a wide variety of systems over a wide range of conditions. One obvious condition that must be met is that the characteristic distances over which inhomogeneities occur in the nonequilibrium system must be large in terms of molecular dimensions. If this is not the case then the thermodynamic state variables will change so rapidly in space that a local thermodynamic state cannot be defined. Similarly the timescale for nonequilibrium change in the system must be large compared to the timescales required for the attainment of local equilibrium.

We let the entropy per unit mass be denoted as $s(\mathbf{r}, t)$ and the entropy of an arbitrary volume V, be denoted by S. Clearly:

$$\frac{\mathrm{d}S}{\mathrm{d}t} = \int_V \mathrm{d}\mathbf{r} \, \frac{\partial(\rho(\mathbf{r}, t)s(\mathbf{r}, t))}{\partial t}. \tag{2.27}$$

In contrast to the derivations of the conservation laws, we do not expect that, by taking account of convection and diffusion, we can totally account for the entropy of the system. The *excess* change of entropy is what we are seeking to calculate. We shall call the entropy produced per unit time per unit volume, the entropy source strength, $\sigma(\mathbf{r}, t)$:

$$\frac{\mathrm{d}S}{\mathrm{d}t} = \int_V \mathrm{d}\mathbf{r}\sigma(\mathbf{r}, t) - \int_S \mathrm{d}S \cdot \mathbf{J}_{ST}(\mathbf{r}, t). \tag{2.28}$$

In this equation $\mathbf{J}_{ST}(\mathbf{r}, t)$ is the total entropy flux. As before we use the divergence theorem and the arbitrariness of V to calculate,

$$\frac{\partial[\rho(\mathbf{r}, t)s(\mathbf{r}, t)]}{\partial t} = \sigma(\mathbf{r}, t) - \boldsymbol{\nabla} \cdot \mathbf{J}_{ST}(\mathbf{r}, t). \tag{2.29}$$

We can decompose $\mathbf{J}_{ST}(\mathbf{r}, t)$ into a streaming or convective term $\rho(\mathbf{r}, t)s(\mathbf{r}, t)\mathbf{u}(\mathbf{r}, t)$, in analogy with Equation (2.8), and a diffusive term $\mathbf{J}_S(\mathbf{r}, t)$. Using these terms Equation (2.29) can be written as:

$$\frac{\partial[\rho(\mathbf{r}, t)s(\mathbf{r}, t)]}{\partial t} = \sigma(\mathbf{r}, t) - \boldsymbol{\nabla} \cdot [\mathbf{J}_S(\mathbf{r}, t) + \rho(\mathbf{r}, t)s(\mathbf{r}, t)\mathbf{u}(\mathbf{r}, t)]. \tag{2.30}$$

Using Equation (2.5) to convert to total time derivatives, we have:

$$\rho(\mathbf{r}, t)\frac{\mathrm{d}s(\mathbf{r}, t)}{\mathrm{d}t} = \sigma(\mathbf{r}, t) - \boldsymbol{\nabla} \cdot \mathbf{J}_S(\mathbf{r}, t). \tag{2.31}$$

At this stage we introduce the assumption of local thermodynamic equilibrium. We *postulate* a local version of the Gibbs relation, $T\mathrm{d}S = \mathrm{d}U + p\mathrm{d}V$. Converting this relation to a local version, with extensive quantities replaced by the specific entropy, energy, and volume respectively, and noting that the specific volume V/M is simply $\rho(\mathbf{r}, t)^{-1}$, we find that:

$$\begin{aligned}T(\mathbf{r}, t)\frac{\mathrm{d}s(\mathbf{r}, t)}{\mathrm{d}t} &= \frac{\mathrm{d}U(\mathbf{r}, t)}{\mathrm{d}t} + p(\mathbf{r}, t)\frac{\mathrm{d}}{\mathrm{d}t}\rho(\mathbf{r}, t)^{-1} \\ &= \frac{\mathrm{d}U(\mathbf{r}, t)}{\mathrm{d}t} - \frac{p(\mathbf{r}, t)}{\rho(\mathbf{r}, t)^2}\frac{\mathrm{d}\rho(\mathbf{r}, t)}{\mathrm{d}t}.\end{aligned} \tag{2.32}$$

We can now use the mass continuity equation to eliminate the density derivative:

$$T(\mathbf{r}, t)\frac{ds(\mathbf{r}, t)}{dt} = \frac{dU(\mathbf{r}, t)}{dt} + \frac{p(\mathbf{r}, t)}{\rho(\mathbf{r}, t)}\nabla \cdot \mathbf{u}(\mathbf{r}, t). \tag{2.33}$$

Multiplying Equation (2.33) by $\rho(\mathbf{r}, t)$ and dividing by $T(\mathbf{r}, t)$ gives:

$$\rho(\mathbf{r}, t)\frac{ds(\mathbf{r}, t)}{dt} = \frac{\rho(\mathbf{r}, t)}{T(\mathbf{r}, t)}\frac{dU(\mathbf{r}, t)}{dt} + \frac{p(\mathbf{r}, t)}{T(\mathbf{r}, t)}\nabla \cdot \mathbf{u}(\mathbf{r}, t). \tag{2.34}$$

Assuming that the external field is zero, we can substitute the energy continuity expression (2.26) for dU/dt into Equation (2.34) giving:

$$\rho(\mathbf{r}, t)\frac{ds(\mathbf{r}, t)}{dt} = -\frac{1}{T(\mathbf{r}, t)}[\nabla \cdot \mathbf{J}_Q(\mathbf{r}, t) + \mathsf{P}(\mathbf{r}, t)^{\mathrm{T}} : \nabla\mathbf{u}(\mathbf{r}, t) - p(\mathbf{r}, t)\nabla \cdot \mathbf{u}(\mathbf{r}, t)]. \tag{2.35}$$

We now have two expressions for the streaming derivative of the specific entropy, $\rho(\mathbf{r}, t)ds(\mathbf{r}, t)/dt$, Equation (2.31), and Equation (2.35). The diffusive entropy flux $\mathbf{J}_S(\mathbf{r}, t)$, using the time derivative of the local equilibrium postulate $dQ = Tds$, is equal to the heat flux divided by the absolute temperature and therefore:

$$\nabla \cdot \mathbf{J}_S(\mathbf{r}, t) = \nabla \cdot \left[\frac{\mathbf{J}_Q(\mathbf{r}, t)}{T(\mathbf{r}, t)}\right] = \frac{\nabla \cdot \mathbf{J}_Q(\mathbf{r}, t)}{T(\mathbf{r}, t)} - \frac{\mathbf{J}_Q(\mathbf{r}, t) \cdot \nabla T(\mathbf{r}, t)}{T(\mathbf{r}, t)^2}. \tag{2.36}$$

Equating (2.31) and (2.35) using (2.36) gives:

$$\sigma(\mathbf{r}, t) = -\frac{1}{T(\mathbf{r}, t)}\left[\mathsf{P}(\mathbf{r}, t)^{\mathrm{T}} : \nabla\mathbf{u}(\mathbf{r}, t) - p(\mathbf{r}, t)\nabla \cdot \mathbf{u}(\mathbf{r}, t) + \frac{\mathbf{J}_Q(\mathbf{r}, t) \cdot \nabla T(\mathbf{r}, t)}{T(\mathbf{r}, t)}\right]$$

$$= -\frac{\mathbf{J}_Q(\mathbf{r}, t) \cdot \nabla T(\mathbf{r}, t)}{T(\mathbf{r}, t)^2} - \frac{\mathsf{P}(\mathbf{r}, t)^{\mathrm{T}} : \nabla\mathbf{u}(\mathbf{r}, t) - p(\mathbf{r}, t)\nabla \cdot \mathbf{u}(\mathbf{r}, t)}{T(\mathbf{r}, t)}. \tag{2.37}$$

We define the viscous pressure tensor $\mathbf{\Pi}$ as the nonequilibrium part of the pressure tensor.

$$\mathbf{\Pi}(\mathbf{r}, t) = \mathsf{P}(\mathbf{r}, t) - p_{eq}(\mathbf{r}, t)\mathbf{I}, \tag{2.38}$$

where $p_{eq}(\mathbf{r}, t)$ comes from the equilibrium equation of state at the local temperature and density, that is $p_{eq}(T(\mathbf{r}, t), \rho(\mathbf{r}, t))$. Using this definition the entropy source strength can be written as:

$$\sigma(\mathbf{r}, t) = -\frac{\mathbf{J}_Q(\mathbf{r}, t) \cdot \nabla T(\mathbf{r}, t)}{T(\mathbf{r}, t)^2} - \frac{\mathbf{\Pi}(\mathbf{r}, t)^{\mathrm{T}} : \nabla\mathbf{u}(\mathbf{r}, t)}{T(\mathbf{r}, t)}. \tag{2.39}$$

A second postulate of nonlinear irreversible thermodynamics is that the entropy source strength always takes the canonical form (de Groot and Mazur, 1962):

$$\sigma = \sum_i \mathbf{J}_i \cdot \mathbf{X}_i. \tag{2.40}$$

This canonical form defines what are known as thermodynamic fluxes, $\mathbf{J}_i$, and their conjugate thermodynamic forces, $\mathbf{X}_i$. We can see immediately that our Equation (2.39) takes this canonical form, provided we make the identifications that the thermodynamic fluxes are the various Cartesian elements of the heat flux vector, $\mathbf{J}_Q(\mathbf{r}, t)$, and the viscous pressure tensor, $\mathbf{\Pi}(\mathbf{r}, t)$. The thermodynamic forces conjugate to these fluxes are the *corresponding* Cartesian components of the temperature gradient divided by the square of the absolute temperature, $T(\mathbf{r}, t)^{-2}\nabla T(\mathbf{r}, t)$, and the strain rate tensor divided by the absolute temperature, $T(\mathbf{r}, t)^{-1}\nabla\mathbf{u}(\mathbf{r}, t)$, respectively. We use the term *corresponding* quite deliberately; the αth element of the heat flux is conjugate to the αth element of the temperature gradient. There are no cross couplings. Similarly the α, β element of the pressure viscous pressure tensor is conjugate to the α, β element of the strain rate tensor.

There is clearly some ambiguity in defining the thermodynamic fluxes and forces. There is no fundamental thermodynamic reason why we included the temperature factors, $T(\mathbf{r}, t)^{-2}$ and $T(\mathbf{r}, t)^{-1}$, into the forces rather than into the fluxes. Either choice is possible. Ours is simply one of convention. More importantly there is no thermodynamic way of distinguishing between the fluxes and the forces. At a macroscopic level it is simply a convention to identify the temperature gradient as a thermodynamic force rather than a flux. The canonical form for the entropy source strength and the associated postulates of irreversible thermodynamics do not permit a distinction to be made between what we should identify as fluxes and what should be identified as a force. Microscopically it is clear that the heat flux is a *flux*. It is the diffusive energy flow across a comoving surface. At a macroscopic level, however, no such distinction can be made.

Perhaps the simplest example of this macroscopic duality is the Norton constant-current electrical circuit, and the Thevénin constant-voltage equivalent circuit. We can talk of the resistance of a circuit element or of a conductance. At a macroscopic level, the choice is simply one of practical convenience or convention.

2.3 Curie's theorem

Consistent with our use of the local thermodynamic equilibrium postulate, which is assumed to be valid sufficiently close to equilibrium, a linear relation should hold between the conjugate thermodynamic fluxes and forces. We therefore postulate the existence of a set of linear phenomenological transport coefficients $\{L_{ij}\}$ which relate the set forces $\{X_j\}$ to the set of fluxes $\{J_i\}$. We use the term *phenomenological* to indicate that these transport coefficients are to be defined within the framework of linear irreversible thermodynamics and as we shall see there may be slight differences between the phenomenological transport coefficients L_{ij} and practical transport coefficients such as the viscosity coefficients or the usual thermal conductivity.

We postulate that *all* the thermodynamic forces appearing in the equation for the entropy source strength, Equation (2.40), are related to the various fluxes by a linear equation of the form:

$$\mathbf{J}_i = \sum_j \mathbf{L}_{ij}\mathbf{X}_j. \qquad (2.41)$$

This equation could be thought of as arising from a Taylor series expansion of the fluxes in terms of the forces. Such a Taylor series will only exist if the flux is an analytic function of the force at $\mathbf{X} = 0$.

$$\mathbf{J}_i(\mathbf{X}) = \mathbf{J}_i(0) + \sum_j \frac{\partial \mathbf{J}_i}{\partial \mathbf{X}_j}\bigg|_{\mathbf{X}=0} \cdot \mathbf{X}_j + \sum_{j,k} \frac{1}{2!}\frac{\partial^2 \mathbf{J}_i}{\partial \mathbf{X}_j \partial \mathbf{X}_k}\bigg|_{\mathbf{X}=0} : \mathbf{X}_j\mathbf{X}_k + O(\mathbf{X}^3). \qquad (2.42)$$

Clearly the first term is zero as the fluxes vanish when the thermodynamic forces are zero. The term which is linear in the forces is evidently derivable, at least formally, from the equilibrium properties of the system as the functional derivative of the fluxes with respect to the forces computed at equilibrium, $\mathbf{X} = 0$. The quadratic term is related to what are known as the nonlinear Burnett coefficients. They represent nonlinear contributions to the linear theory of irreversible thermodynamics.

If we substitute the linear phenomenological relations into the equation for the entropy source strength, Equation (2.40), we find that:

$$\sigma = \sum_{i,j} \mathbf{X}_i \mathbf{L}_{ij} \mathbf{X}_j. \qquad (2.43)$$

A postulate of linear irreversible thermodynamics is that the entropy source strength is always non-negative. There is always an increase in the entropy of a system, so the transport coefficients are positive. Since this is also true for the mirror image of any system, we conclude that the entropy source strength is a positive polar scalar quantity. (A polar scalar is invariant under a mirror inversion of the coordinate axes. A pseudo scalar, on the other hand, changes its sign under a mirror inversion. The same distinction between polar and pseudo quantities also applies to vectors and tensors.)

Suppose that we are studying the transport processes taking place in a fluid. In the absence of any external non-dissipative fields (such as gravitational or magnetic fields), the fluid is at equilibrium and assumed to be isotropic. Clearly, since the linear transport coefficients can be formally calculated as a zero-field functional derivative, they should have the symmetry characteristic of an isotropic system. Furthermore they should be invariant under a mirror reflection of the coordinate axes.

Suppose that all the fluxes and forces are scalars. The most general linear relation between the forces and fluxes is given by Equation (2.41). Since the transport coefficients must be polar scalars, there cannot be any coupling between a pseudo scalar flux and a polar force or between a polar flux and a pseudo scalar force. This is a simple application of the quotient rule in tensor analysis. Scalars of like parity only, can be coupled by the transport matrix L_{ij}.

If the forces and fluxes are vectors, the most general linear relation between the forces and fluxes which is consistent with isotropy is:

$$\mathbf{J}_i = \sum_j \mathbf{L}_{ij} \cdot \mathbf{X}_j = \sum_j L_{ij} \mathbf{I} \cdot \mathbf{X}_j = \sum_j L_{ij} \mathbf{X}_j. \qquad (2.44)$$

In this equation $\mathbf{L}_{ij}$ is a second-rank polar tensor, because the transport coefficients must be invariant under mirror inversion, just like the equilibrium system itself. If the equilibrium system is isotropic, then $\mathbf{L}_{ij}$ must be expressible as a scalar L_{ij} times the only isotropic second-rank tensor $\mathbf{I}$, (the Kronecker delta tensor $\mathbf{I} = \delta_{\alpha\beta}$). The thermodynamic forces and fluxes which couple together must either all be pseudo vectors or polar vectors. Otherwise, since the transport coefficients are polar quantities, the entropy source strength could be pseudo scalar. By comparing the trace of $\mathbf{L}_{ij}$ with the trace of $L_{ij}\mathbf{I}$, we see that the polar scalar transport coefficients are given as:

$$L_{ij} = \tfrac{1}{3}\mathrm{Tr}(\mathbf{L}_{ij}) = \tfrac{1}{3}\mathbf{L}_{ij} : \mathbf{I}. \qquad (2.45)$$

If the thermodynamic forces and fluxes are all symmetric traceless second-rank tensors $\overset{0}{\mathbf{J}}_i$, $\overset{0}{\mathbf{X}}_i$, where $\overset{0}{\mathbf{J}}_i = \tfrac{1}{2}(\mathbf{J}_i + \mathbf{J}_i^{\mathrm{T}}) - \tfrac{1}{3}\mathrm{Tr}(\mathbf{J}_i)\mathbf{I}$, (we denote symmetric traceless tensors with a centred superscript zero), then:

$$\overset{0}{\mathbf{J}}_i^{(2)} = \sum_j \mathbf{L}_{ij}^{(4)} : \overset{0}{\mathbf{X}}_j^{(2)}, \qquad (2.46)$$

is the most linear general linear relation between the forces and fluxes. $\mathbf{L}_{ij}^{(4)}$ is a symmetric fourth-rank transport tensor. Unlike second-rank tensors there are three linearly independent isotropic fourth-rank polar tensors. (There are no isotropic pseudo tensors of the fourth rank.) These tensors can be related to the Kronecker delta tensor, and we depict these tensors by the forms:

$$\mathbf{UU}_{\alpha\beta\gamma\eta} = \delta_{\alpha\beta}\delta_{\gamma\eta}$$

$$\mathbf{W}_{\alpha\beta\gamma\eta} = \delta_{\alpha\gamma}\delta_{\beta\eta}$$

$$\overset{\mathsf{U}}{\mathbf{U}}_{\alpha\beta\gamma\eta} = \delta_{\alpha\eta}\delta_{\beta\gamma}. \qquad (2.47)$$

Since $L_{ij}^{(4)}$ is an isotropic tensor, it must have a representation as a linear combination of isotropic fourth-rank tensors. It is convenient to write:

$$L_{ij}^{(4)} = L_{ij}^s \left[\frac{1}{2} \left(\overset{\scriptscriptstyle U}{U} + W \right) - \frac{1}{3} UU \right] + L_{ij}^a \frac{1}{2} \left(\overset{\scriptscriptstyle U}{U} - W \right) + L_{ij}^{Tr} \frac{1}{3} UU. \qquad (2.48)$$

It is easy to show that for any second-rank tensor A:

$$L_{ij} : A^{(2)} = L_{ij}^s \overset{0}{A} + L_{ij}^a \overset{a}{A} + L_{ij}^{Tr} A\, l, \qquad (2.49)$$

where $\overset{0}{A} = \frac{1}{2}(A + A^T) - \frac{1}{3}\mathrm{Tr}(A)l$ is the symmetric traceless part of $A^{(2)}$, $\overset{a}{A} = \frac{1}{2}(A - A^T)$ is the antisymmetric part of $A^{(2)}$ (we denote antisymmetric tensors with a centred superscript a), and $A = \frac{1}{3}\mathrm{Tr}(A)$. This means that the three isotropic fourth-rank tensors decouple the linear force flux relations into three separate sets of equations which relate, respectively, the symmetric second-rank forces and fluxes, the antisymmetric second-rank forces and fluxes, and the traces of the forces and fluxes. These equations can be written as:

$$\overset{0}{J}_i = \sum_j L_{ij}^s \overset{0}{X}_j, \qquad (2.50a)$$

$$\overset{a}{J}_i = \sum_j L_{ij}^a \overset{a}{X}_j, \qquad (2.50b)$$

$$J_i = \sum_j L_{ij}^{Tr} X_j, \qquad (2.50c)$$

where $\overset{a}{J}_i$ is the antisymmetric part of J, and $J = \frac{1}{3}\mathrm{Tr}(J)$. As $\overset{a}{J}_i$ has only three independent elements, it turns out that $\overset{a}{J}_i$ can be related to a pseudo vector. This relationship is conveniently expressed in terms of the Levi–Civita isotropic third-rank tensor $\varepsilon^{(3)}$. (Note: $\varepsilon^{(3)}_{\alpha\beta\gamma} = +1$ if $\alpha\beta\gamma$ is an even or cyclic permutation, -1 if $\alpha\beta\gamma$ is an odd permutation, and is zero otherwise.) If we denote the pseudo vector dual of $\overset{a}{J}_i$ as J_i^{ps} then:

$$J_i^{ps} = -\tfrac{1}{2}\varepsilon^{(3)} : J_i^{(2)} \quad \text{and} \quad \overset{a}{J}_i^{\,(2)} = J_i^{ps} \cdot \varepsilon^{(3)}. \qquad (2.51)$$

This means that the second equation, Equation (2.50b), can be rewritten as:

$$J_i^{ps} = \sum_j L_{ij}^a X_j^{ps}. \qquad (2.52)$$

Looking at Equations (2.50) and (2.52), we see that we have decomposed the 81 elements of the (three-dimensional) fourth-rank transport tensor $L_{ij}^{(4)}$, into three scalar quantities, L_{ij}^s, L_{ij}^a, and L_{ij}^{Tr}. Furthermore we have found that there are three *irreducible* sets of forces and fluxes. Couplings only exist within the sets.

There are no couplings of forces of one set with fluxes of another set. The sets naturally represent the symmetric traceless parts, the antisymmetric part, and the trace of the second-rank tensors. The three irreducible components can be identified with a symmetric irreducible second-rank polar tensor component, an irreducible pseudo vector, and an irreducible polar scalar. Curie's principle states that linear transport couples can only occur between irreducible tensors of the same rank and parity.

If we return to our basic equation for the entropy source strength, Equation (2.40), we see that our irreducible decomposition of Cartesian tensors allows us to make the following decomposition for second-rank fields and fluxes:

$$
\sigma = \sum_i \left((J_i \mathbf{l}) : (X_i \mathbf{l}) + \overset{a}{\mathsf{J}}_i : \overset{a}{\mathsf{X}}_i + \overset{0}{\mathsf{J}}_i : \overset{0}{\mathsf{X}}_i \right)
$$

$$
= \sum_i \left(3 J_i X_i - 2 J_i^{ps} \cdot X_i^{ps} + \overset{0}{\mathsf{J}}_i : \overset{0}{\mathsf{X}}_i \right). \tag{2.53}
$$

The conjugate forces and fluxes appearing in the entropy source equation separate into irreducible sets. This is easily seen when we realise that all cross couplings between irreducible tensors of different rank vanish, $\mathbf{l} : \overset{a}{\mathsf{J}}_i = \mathbf{l} : \overset{a}{\mathsf{X}} = \overset{a}{\mathsf{J}}_i : \overset{0}{\mathsf{X}} = 0$, etc. Conjugate thermodynamic forces and fluxes must have the same irreducible rank and parity.

We can now apply Curie's principle to the entropy source Equation (2.39):

$$
\sigma(\mathbf{r}, t) = \frac{-1}{T(\mathbf{r}, t)} \left[\frac{\mathbf{J}_Q(\mathbf{r}, t) \cdot \nabla T(\mathbf{r}, t)}{T(\mathbf{r}, t)} - \overset{0}{\mathsf{P}}(\mathbf{r}, t) : \overset{0}{\nabla \mathbf{u}}(\mathbf{r}, t) - \mathbf{\Pi}^{ps}(\mathbf{r}, t) \cdot \nabla \right.
$$

$$
\times \mathbf{u}(\mathbf{r}, t) - \mathbf{\Pi}(\mathbf{r}, t) \nabla \cdot \mathbf{u}(\mathbf{r}, t)]. \tag{2.54}
$$

In writing this equation we have used the fact that the transpose of $\overset{0}{\mathsf{P}}$ is equal to $\overset{0}{\mathsf{P}}$, and we have used Equation (2.51) and the definition of the cross product $\nabla \times \mathbf{u} = -\boldsymbol{\varepsilon}^{(3)} : \overset{0}{\nabla \mathbf{u}}$ to transform the antisymmetric part of P^T. Note that the transpose of P^a is equal to $-\mathsf{P}^a$. There is no conjugacy between the vector $\mathbf{J}_Q(\mathbf{r}, t)$ and the pseudo vector $\nabla \times \mathbf{u}(\mathbf{r}, t)$ because they differ in parity. It can be easily shown that for atomic fluids the antisymmetric part of the pressure tensor is zero so that the terms in Equation (2.54) involving the vorticity $\nabla \times \mathbf{u}(\mathbf{r}, t)$ are identically zero. For molecular fluids, terms involving the vorticity do appear, but we also have to consider another conservation equation – the conservation of angular momentum. In our description of the conservation equations we have ignored angular momentum conservation. The complete description of the hydrodynamics of molecular fluids must include this additional conservation law.

For single component atomic fluids we can now use Curie's principle to define the phenomenological transport coefficients:

$$\mathbf{J}_Q = L_Q \mathbf{X}_Q = -L_Q \frac{\mathbf{\nabla} T}{T^2}, \tag{2.55a}$$

$$\overset{0}{\mathbf{\Pi}} = \overset{0}{L_\Pi} \overset{0}{X_\Pi} = - \overset{0}{L_\Pi} \frac{\overset{0}{\mathbf{\nabla} \mathbf{u}}}{T}, \tag{2.55b}$$

$$\Pi = L_\Pi X_\Pi = -L_\Pi \frac{3\mathbf{\nabla} \cdot \mathbf{u}}{T}. \tag{2.55c}$$

The positive sign of the entropy production implies that each of the phenomenological transport coefficients must be positive. As mentioned before these phenomenological definitions differ slightly from the usual definitions of the Navier–Stokes transport coefficients.

$$\mathbf{J}_Q = -\lambda \mathbf{\nabla} T, \tag{2.56a}$$

$$\overset{0}{\mathbf{\Pi}} = -2\eta \overset{0}{\mathbf{\nabla} \mathbf{u}}, \tag{2.56b}$$

$$\Pi = -\eta_V \mathbf{\nabla} \cdot \mathbf{u}. \tag{2.56c}$$

These equations were postulated long before the development of linear irreversible thermodynamics. The first equation is known as Fourier's law of heat conduction. It gives the definition of the thermal conductivity λ. The second equation is known as Newton's law of viscosity (illustrated in Figure 2.3). It gives a definition of the shear viscosity coefficient η. The third equation is a more recent development. It defines the bulk viscosity coefficient η_V. These equations are known

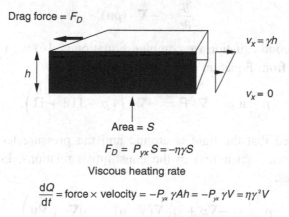

Drag force = F_D

$v_x = \gamma h$

h

$v_x = 0$

Area = S

$F_D = P_{yx} S = -\eta\gamma S$

Viscous heating rate

$$\frac{dQ}{dt} = \text{force} \times \text{velocity} = -P_{yx}\gamma Ah = -P_{yx}\gamma V = \eta\gamma^2 V$$

Figure 2.3 Newton's constitutive relation for shear viscosity

collectively as linear constitutive equations. When they are substituted into the con-servation equations they yield the Navier–Stokes equations of hydrodynamics. The conservation equations relate thermodynamic fluxes and forces. They form a system of equations in two unknown fields – the force fields and the flux fields. The constitutive equations relate the forces and the fluxes. By combining the two systems of equations we can derive the Navier–Stokes equations, which in their usual form give us a closed system of equations for the thermodynamic forces. Once the boundary conditions are supplied, the Navier–Stokes equations can be solved to give a complete macroscopic description of the nonequilibrium flows expected in a fluid close to equilibrium in the sense required by linear irreversible thermodynamics. It is worth restating the expected conditions for the linearity to be observed:

(1) The thermodynamic forces should be sufficiently small so that linear constitutive relations are accurate.
(2) The system should likewise be sufficiently close to equilibrium for the local thermo-dynamic equilibrium condition to hold. For example, the nonequilibrium equation of state must be the same function of the local position and time-dependent thermo-dynamic state variables (such as the temperature and density) that it is at equilibrium.
(3) The characteristic distances over which the thermodynamic forces vary should be suffi-ciently large so that these forces can be viewed as being constant over the microscopic length scale required to properly define a local thermodynamic state.
(4) The characteristic times over which the thermodynamic forces vary should be suffi-ciently long that these forces can be viewed as being constant over the microscopic times required to properly define a local thermodynamic state.

After some tedious but quite straightforward algebra (de Groot and Mazur, 1962), the Navier–Stokes equations for a single component atomic fluid are obtained. The first of these is simply the mass conservation Equation (2.4):

$$\frac{\partial \rho}{\partial t} = -\mathbf{\nabla} \cdot (\rho \mathbf{u}).$$

(2.57)

To obtain the second equation we combine Equation (2.16) with the definition of the stress tensor from Equation (2.12) which gives:

$$\rho \frac{\mathrm{d}}{\mathrm{d}t} \mathbf{u} = -\mathbf{\nabla} \cdot \mathsf{P} = -\mathbf{\nabla} \cdot \left((p + \mathsf{\Pi})\mathsf{I} + \overset{0}{\mathsf{\Pi}} \right).$$

(2.58)

We have assumed that the fluid is atomic and the pressure tensor contains no antisymmetric part. Substituting in the constitutive relations, Equations (2.56b) and (2.56c), gives:

$$\rho \frac{\mathrm{d}\mathbf{u}}{\mathrm{d}t} = -\mathbf{\nabla}p + \eta_{\mathrm{V}}\mathbf{\nabla}(\mathbf{\nabla} \cdot \mathbf{u}) + 2\eta\mathbf{\nabla} \cdot \left(\overset{0}{\mathbf{\nabla}\mathbf{u}} \right).$$

(2.59)

Here we explicitly assume that the transport coefficients η_V and η are simple constants, independent of position $\mathbf{r}$, time and flow rate $\mathbf{u}$. The $\alpha\beta$ component of the symmetric traceless tensor $\overset{0}{\nabla}\mathbf{u}$ is given by:

$$\left(\overset{0}{\nabla}\mathbf{u}\right)_{\alpha\beta} = \tfrac{1}{2}\left(\frac{\partial u_\beta}{\partial x_\alpha} + \frac{\partial u_\alpha}{\partial x_\beta}\right) - \delta_{\alpha\beta}\tfrac{1}{3}\frac{\partial u_\gamma}{\partial x_\gamma}, \tag{2.60}$$

where, as usual, the repeated index γ implies a summation with respect to γ. It is then straightforward to see that:

$$\nabla \cdot \left(\overset{0}{\nabla}\mathbf{u}\right) = \tfrac{1}{2}\nabla^2\mathbf{u} + \tfrac{1}{6}\nabla(\nabla \cdot \mathbf{u}), \tag{2.61}$$

and it follows that the momentum flow Navier–Stokes equation is:

$$\rho\frac{d\mathbf{u}}{dt} = -\nabla p + \eta\nabla^2\mathbf{u} + \left(\tfrac{1}{3}\eta + \eta_V\right)\nabla(\nabla \cdot \mathbf{u}). \tag{2.62}$$

The Navier–Stokes equation for energy flow can be obtained from Equation (2.26) and the constitutive relations, Equation (2.56). Again we assume that the pressure tensor is symmetric, and the second term on the right-hand side of Equation (2.26) becomes:

$$\begin{aligned}
\mathbf{P}^T : \nabla\mathbf{u} &= \left((p + \Pi)\mathbf{I} + \overset{0}{\mathbf{\Pi}}\right) : \left(\tfrac{1}{3}(\nabla \cdot \mathbf{u})\mathbf{I} + \overset{0}{\nabla}\mathbf{u}\right) \\
&= \left((p - \eta_V(\nabla \cdot \mathbf{u}))\mathbf{I} - 2\eta\overset{0}{\nabla}\mathbf{u}\right) : \left(\tfrac{1}{3}(\nabla \cdot \mathbf{u})\mathbf{I} + \overset{0}{\nabla}\mathbf{u}\right) \\
&= p(\nabla \cdot \mathbf{u}) - \eta_V(\nabla \cdot \mathbf{u})^2 - 2\eta\overset{0}{\nabla}\mathbf{u} : \overset{0}{\nabla}\mathbf{u}.
\end{aligned} \tag{2.63}$$

It is then straightforward to see that:

$$\rho\frac{dU}{dt} = \lambda\nabla^2 T - p(\nabla \cdot \mathbf{u}) + \eta_V(\nabla \cdot \mathbf{u})^2 + 2\eta\overset{0}{\nabla}\mathbf{u} : \overset{0}{\nabla}\mathbf{u}. \tag{2.64}$$

2.4 Non-Markovian constitutive relations: viscoelasticity

Consider a fluid undergoing planar Couette flow. This flow is defined by the streaming velocity:

$$\mathbf{u}(\mathbf{r}, t) = (u_x, u_y, u_z) = (\gamma y, 0, 0). \tag{2.65}$$

According to Curie's principle the only nonequilibrium flux that will be excited by such a flow is the pressure tensor. According to the constitutive relation

Equation (2.56) the pressure tensor is:

$$P(\mathbf{r}, t) = \begin{pmatrix} p & -\eta\gamma & 0 \\ -\eta\gamma & p & 0 \\ 0 & 0 & p \end{pmatrix}, \tag{2.66}$$

where η is the shear viscosity and γ is the shear rate. If the strain rate is time dependent, then the shear stress, $-P_{xy} = -P_{yx} = \eta\gamma(t)$. It is known that many fluids do not satisfy this relation regardless of how small the strain rate is. There must therefore be a *linear*, but time dependent, constitutive relation for shear flow, which is more general than the Navier–Stokes constitutive relation.

Poisson (1829) pointed out that there is a deep correspondence between the shear stress induced by a strain rate in a fluid, and the shear stress induced by a strain in an elastic solid. The strain tensor is $\nabla\varepsilon$ where $\varepsilon(\mathbf{r}, t)$ gives the displacement of atoms at $\mathbf{r}$ from their equilibrium lattice sites. It is clear that:

$$\frac{d\nabla\varepsilon}{dt} = \nabla\mathbf{u}. \tag{2.67}$$

Maxwell (1873) realized that if a displacement were applied to a liquid, then for a short time the liquid must behave as if it were an elastic solid. After a *Maxwell relaxation time* the liquid would relax to equilibrium since *by definition* a liquid cannot support a strain Frenkel (1955).

It is easier to analyze this matter by transforming to the frequency domain. Maxwell said that at low frequencies the shear stress of a liquid is generated by the Navier–Stokes constitutive relation for a Newtonian fluid (Equation 2.66). In the frequency domain this states that:

$$\tilde{P}_{xy}(\omega) = -\eta\tilde{\gamma}(\omega), \tag{2.68}$$

where:

$$\tilde{A}(\omega) = \int_0^\infty dt \exp[-i\omega t] A(t), \tag{2.69}$$

denotes the Fourier–Laplace transform of $A(t)$.

At very high frequencies we should have:

$$\tilde{P}_{xy}(\omega) = -G\frac{\partial\tilde{\varepsilon}_x(\omega)}{\partial y}, \tag{2.70}$$

where G is the infinite frequency shear modulus. From Equation (2.67) we can transform the terms involving the strain into terms involving the strain rate (we assume that at $t = 0$, the strain $\varepsilon(0) = 0$). At high frequencies therefore,

$$\tilde{P}_{xy}(\omega) = -\frac{G}{i\omega}\frac{\partial \tilde{u}_x}{\partial y} = -\frac{G}{i\omega}\tilde{\gamma}(\omega). \tag{2.71}$$

The Maxwell model of viscoelasticity is obtained by simply summing the high and low frequency expressions for the compliances $i\omega/G$ and η^{-1}, thus Equations (2.68) and (2.71), to give:

$$\tilde{\gamma}(\omega) = -\left(\frac{i\omega}{G} + \frac{1}{\eta}\right)\tilde{P}_{xy}(\omega) = -\frac{\tilde{P}_{xy}(\omega)}{\tilde{\eta}_M(\omega)}. \tag{2.72}$$

The expression for the frequency dependent Maxwell viscosity is (Figure 2.4):

$$\tilde{\eta}_M(\omega) = \frac{\eta}{1 + i\omega\tau_M}. \tag{2.73}$$

It is easily seen that this expression smoothly interpolates between the high and low frequency limits. The Maxwell relaxation time $\tau_M = \eta/G$ controls the transition frequency between low frequency viscous behaviour and high frequency elastic behavior.

The Maxwell model provides a rough approximation to the viscoelastic behavior of so-called viscoelastic fluids such as polymer melts or colloidal suspensions. It is important to remember that viscoelasticity is a linear phenomenon. The resulting shear stress is a linear function of the strain rate. It is also important to point out

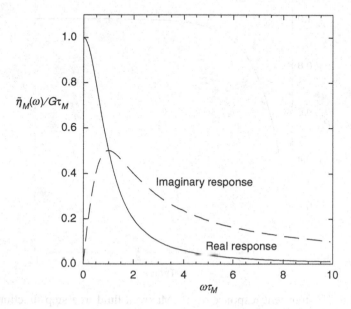

Figure 2.4 Frequency dependent viscosity of the Maxwell model

that Maxwell believed that all fluids are viscoelastic. The reason why polymer melts are observed to exhibit viscoelasticity is that their Maxwell relaxation times are macroscopic, of the order of seconds. On the other hand, the Maxwell relaxation time for argon at its triple point is approximately 10^{-12} seconds! Using standard viscometric techniques elastic effects are completely unobservable in argon.

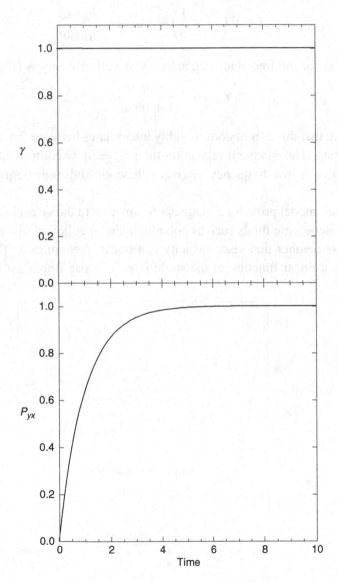

Figure 2.5 The transient response of the Maxwell fluid to a step-function shear rate is the integral of the memory function for the model, $\eta_M(t)$

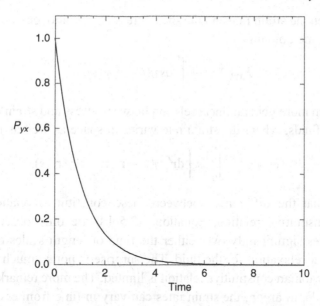

Figure 2.6 The transient response of the Maxwell model to a delta function strain rate $\gamma(t) = \gamma_0 \delta(0)$ is the memory function itself, $\eta_M(t)$

If we rewrite the Maxwell constitutive relation in the time domain using an inverse Fourier–Laplace transform we see that:

$$P_{xy}(t) = -\int_0^t \mathrm{d}s \ \eta_M(t-s)\gamma(s). \qquad (2.74)$$

In this equation $\eta_M(t)$ is called the Maxwell memory function (Figures 2.5 and 2.6). It is called a memory function because the shear stress at time t is not simply linearly proportional to the strain rate at the current time t, but to the entire strain rate *history*, over times s where $0 \le s \le t$. History dependent constitutive relations are called nonMarkovian. A Markovian process is one in which the present state of the system is all that is required to determine its future. The Maxwell model of viscoelasticity describes nonMarkovian behaviour. The Maxwell memory function is easily identified as an exponential:

$$\eta_M(t) = G \exp[-t/\tau_M] \qquad (2.75)$$

Although the Maxwell model of viscoelasticity is approximate, the basic idea that liquids take a finite time to respond to changes in strain rate, or equivalently that liquids remember their strain rate histories, is correct. The most general linear relation between the strain rate and the shear stress for a homogeneous fluid can be written in the time domain as:

$$P_{xy}(t) = - \int_0^t ds\, \eta(t-s)\gamma(s). \tag{2.76}$$

There is an even more general linear relation between stress and strain rate which is appropriate in fluids, where the strain rate varies in space as well as in time:

$$P_{xy}(\mathbf{r}, t) = - \int_0^t ds \int d\mathbf{r}'\ \eta(\mathbf{r}-\mathbf{r}', t-s)\gamma(\mathbf{r}', s). \tag{2.77}$$

We reiterate that the differences between these constitutive relations and the Newtonian constitutive relation, Equations (2.56b), are only observable if the strain rate varies significantly over either the time or length scales characteristic of the molecular relaxation for the fluid. The surprise is not so much that the validity of the Newtonian constitutive relation is limited. The more remarkable thing is that, for example, in argon, the strain rates can vary in time from essentially zero frequency to 10^{12} Hz, or in space from zero wavevector to 10^{-9} m^{-1}, before non-Newtonian effects are observable. It is clear from this discussion that analogous corrections will be needed for all the other Navier–Stokes transport coefficients, if their corresponding thermodynamic fluxes vary on molecular time or distance scales.

3

The microscopic connection

3.1 Classical mechanics

In nonequilibrium statistical mechanics we seek to model transport processes beginning with an understanding of the motion and interactions of individual atoms or molecules. The laws of classical mechanics govern the motion of atoms and molecules, so in this chapter we begin with a brief description of the mechanics of Newton, Lagrange, and Hamilton. It is often useful to be able to treat *constrained* mechanical systems. We will use a principle due to Gauss to treat many different types of constraint – from simple bond-length constraints, to constraints on kinetic energy. As we shall see, kinetic energy constraints are useful for constructing various constant temperature ensembles. We will then discuss the Liouville equation and its formal solution. This equation is the central vehicle of nonequilibrium statistical mechanics. We will then need to establish the link between the microscopic dynamics of individual atoms and molecules and the macroscopic hydrodynamical description discussed in the last chapter. We will discuss two procedures for making this connection. The Irving and Kirkwood procedure relates hydrodynamic variables to nonequilibrium *ensemble averages* of microscopic quantities. A more direct procedure, which we will describe, succeeds in deriving *instantaneous* expressions for the hydrodynamic field variables.

Newtonian mechanics

Classical mechanics (Goldstein, 1980) is based on Newton's three laws of motion. This theory introduced the concepts of a force and an acceleration. Prior to Newton's work, the connection had been made between forces and velocities. Newton's laws of motion were supplemented by the notion of *force* acting at a distance. With the identification of the force of gravity and an appropriate initial condition – initial coordinates and velocities – trajectories could be computed.

Philosophers of science have debated the content of Newton's laws, but when augmented with a force which is expressible as a function of time, position, or possibly of velocity, those laws lead to the equation:

$$m\ddot{\mathbf{r}} = \mathbf{F}(\mathbf{r}, \dot{\mathbf{r}}, t), \tag{3.1}$$

which is well posed and possesses a unique solution.

Lagrangian mechanics

After Newton, scientists discovered different sets of equivalent laws or axioms upon which classical mechanics could be based. More *elegant* formulations based on D'Alembert's principle are due to Lagrange and Hamilton. Newton's laws are less general than they might seem. For instance the position $\mathbf{r}$, that appears in Newton's equation must be a Cartesian vector in a Euclidean space. One does not have the freedom of, say, using angles as measures of position. Lagrange solved the problem of formulating the laws of mechanics in a form which is valid for *generalized* coordinates.

Let us consider a system with generalized coordinates q. These coordinates may be Cartesian positions, angles, or any other convenient parameters that can be found to uniquely specify the configuration of the system. The kinetic energy T will, in general, be a function of the coordinates and their time derivatives $\dot{q}$. If $V(q)$ is the potential energy, we define the Lagrangian to be $L(q, \dot{q}, t) \equiv T(q, \dot{q}, t) - V(q)$. The fundamental dynamical postulate states that the motion of a system is such that the *action*, S, is an extremum:

$$\delta S = \delta \int_{t_0}^{t_1} L(q, \dot{q}, t)\mathrm{d}t = 0. \tag{3.2}$$

Let $q(t)$ be the coordinate trajectory that satisfies this condition then consider $q(t) + \delta q(t)$ where $\delta q(t)$ is an arbitrary variation of the coordinate trajectory. The varied motion must be consistent with the fixed initial and final positions so that, $\delta q(t_1) = \delta q(t_0) = 0$. We consider the change in the action due to this variation:

$$\delta S = \int_{t_0}^{t_1} L(q + \delta q, \dot{q} + \delta \dot{q}, t)\mathrm{d}t - \int_{t_0}^{t_1} L(q, \dot{q}, t)\mathrm{d}t$$

$$= \int_{t_0}^{t_1} \left(\frac{\partial L}{\partial q} \delta q + \frac{\partial L}{\partial \dot{q}} \delta \dot{q} \right)\mathrm{d}t. \tag{3.3}$$

Integrating the second term by parts gives:

$$\delta S = \left[\frac{\partial L}{\partial \dot{q}} \delta q \right]_{t_0}^{t_1} + \int_{t_0}^{t_1} \left(\frac{\partial L}{\partial q} - \frac{\mathrm{d}}{\mathrm{d}t} \left(\frac{\partial L}{\partial \dot{q}} \right) \right) \delta q \, \mathrm{d}t \tag{3.4}$$

The first term vanishes because δq is zero at both endpoints. Since $\delta q(t)$ is arbitrary for $t_0 < t < t_1$, the only way that the variation in the action δS can vanish is if the integrand is zero, thus:

$$\frac{\partial L}{\partial q} - \frac{\mathrm{d}}{\mathrm{d}t}\left(\frac{\partial L}{\partial \dot{q}}\right) = 0. \tag{3.5}$$

This is Lagrange's equation of motion. If the coordinates are Cartesian, it is easy to see that Lagrange's equation reduces to Newton's.

Hamiltonian mechanics

Although Lagrange's equation has removed the special status attached to Cartesian coordinates, it has introduced a new difficulty. The Lagrangian is a function of generalized coordinates, their time derivatives and possibly time. There is still a strong physical significance attached to coordinates and velocities. Hamilton derived an equivalent set of equations in which the roles played by coordinates and *velocities* can be interchanged. Hamilton defined the *canonical momentum p*:

$$p \equiv \frac{\partial L(q, \dot{q}, t)}{\partial \dot{q}}, \tag{3.6}$$

and introduced the function:

$$H(q, p, t) \equiv \dot{q}\frac{\partial L}{\partial \dot{q}} - L = \dot{q}p - L. \tag{3.7}$$

This function is, of course, now known as the Hamiltonian. Consider a change in the Hamiltonian, which can be written as:

$$\mathrm{d}H = \dot{q}\mathrm{d}p + p\mathrm{d}\dot{q} - \mathrm{d}L. \tag{3.8}$$

The Lagrangian is a function of q, $\dot{q}$, and t so that the change $\mathrm{d}L$, can be written as:

$$\mathrm{d}L = \frac{\partial L}{\partial q}\mathrm{d}q + \frac{\partial L}{\partial \dot{q}}\mathrm{d}\dot{q} + \frac{\partial L}{\partial t}\mathrm{d}t. \tag{3.9}$$

Using the definition of the canonical momentum p, and substituting for $\mathrm{d}L$, the expression for $\mathrm{d}H$ becomes:

$$\mathrm{d}H = \dot{q}\mathrm{d}p - \frac{\partial L}{\partial q}\mathrm{d}q - \frac{\partial L}{\partial t}\mathrm{d}t. \tag{3.10}$$

Lagrange's equation of motion (3.5), rewritten in terms of the canonical momenta is:

$$\dot{p} = \frac{\partial L}{\partial q}, \tag{3.11}$$

so that the change in H is:

$$\mathrm{d}H = \dot{q}\mathrm{d}p - \dot{p}\mathrm{d}q - \frac{\partial L}{\partial t}\mathrm{d}t. \tag{3.12}$$

Since the Hamiltonian is a function of q, p, and t, it is easy to see that Hamilton's equations of motion are:

$$\dot{q} = \frac{\partial H}{\partial p} \quad \text{and} \quad \dot{p} = -\frac{\partial H}{\partial q}. \tag{3.13}$$

If H has no explicit time dependence, its value is a constant of the motion.

A particularly nice formulation of Hamiltonian mechanics is obtained using symplectic notation (José and Saletan, 1998), that is defining the $2n$-dimensional vector $\boldsymbol{\Gamma} \equiv (q, p)$, where q and p are n-dimensional vectors of coordinates and momenta. We will use this notation extensively throughout the book. The equations of motion can then be written as:

$$\dot{\boldsymbol{\Gamma}} \equiv \begin{pmatrix} \dot{q} \\ \dot{p} \end{pmatrix} = \begin{pmatrix} 0 & 1 \\ -1 & 0 \end{pmatrix} \begin{pmatrix} \partial H/\partial q \\ \partial H/\partial q \end{pmatrix} = J\frac{\partial H}{\partial \boldsymbol{\Gamma}}, \tag{3.14}$$

where in the $2n \times 2n$ matrix J, 0 is the $n \times n$ null vector and 1 is the $n \times n$ identity matrix.

We consider the transformation to a new set of canonical variables $(q, p) \Rightarrow (Q, P)$, or equivalently $\boldsymbol{\Gamma} = \boldsymbol{\Gamma}^{(new)}$, and take η_i to be an arbitrary element of the vector $\boldsymbol{\Gamma}$ and ζ_i to be an arbitrary element of $\boldsymbol{\Gamma}^{(new)}$, then the time derivatives are related by:

$$\dot{\zeta}_i = \sum_{j=1}^{2n} \frac{\partial \zeta_i}{\partial \eta_j}\dot{\eta}_j \quad \text{so} \quad \dot{\boldsymbol{\Gamma}}^{(new)} = M\dot{\boldsymbol{\Gamma}} = MJ\frac{\partial H}{\partial \boldsymbol{\Gamma}}, \tag{3.15}$$

using Equation (3.14). The i, j element of $M_{ij} \equiv \partial \zeta_i/\partial \eta_j$. The existence of the inverse transformation $H(\boldsymbol{\Gamma}) \to H(\boldsymbol{\Gamma}^{(new)})$ implies that:

$$\frac{\partial H}{\partial \eta_i} = \sum_j \frac{\partial \zeta_j}{\partial \eta_i}\frac{\partial H}{\partial \zeta_j} \quad \text{so} \quad \frac{\partial H}{\partial \boldsymbol{\Gamma}} = M^{\mathrm{T}}\frac{\partial H}{\partial \boldsymbol{\Gamma}^{(new)}}, \tag{3.16}$$

where M^{T} is the transpose of M. Combining Equations (3.15) and (3.16) gives:

$$\dot{\boldsymbol{\Gamma}}^{(new)} = MJM^{\mathrm{T}}\frac{\partial H}{\partial \boldsymbol{\Gamma}^{(new)}} \quad \text{or} \quad \dot{\zeta} = MJM^{\mathrm{T}}\frac{\partial H}{\partial \zeta}. \tag{3.17}$$

For the transformation to be canonical we require that $\dot{\zeta} = J(\partial H/\partial \zeta)$ thus:

$$MJM^{\mathrm{T}} = J. \tag{3.18}$$

This is the symplectic condition for the transformation to be canonical and if *M* satisfies this condition, the dynamics is said to be symplectic. A particularly important canonical transformation is that which corresponds to the time evolution of the system and then the time evolution matrix derived from *M* is symplectic.

Thermodynamic constraints

Apart from relativistic or quantum corrections, classical mechanics is thought to give an exact description of motion. In this section our point of view will change somewhat. Newtonian or Hamiltonian mechanics imply a certain set of constants of the motion: energy, and linear and angular momentum. In thermodynamically interesting systems, the natural fixed quantities are the thermodynamic state variables, the number of molecules *N*, the volume *V*, and the temperature *T*. At other times it may be convenient to keep the pressure fixed rather than the volume. Thermodynamically interesting systems usually exchange energy, momentum, and mass with their surroundings. This means that within thermodynamic systems none of the classical constants of the motion are actually constant.

Typical *thermodynamic* systems are characterized by fixed values of thermo-dynamic variables: temperature, pressure, chemical potential, density, enthalpy, or internal energy. The system is maintained at a fixed thermodynamic state (say temperature) by placing it in contact with a reservoir, with which it exchanges energy (heat) in such a manner as to keep the temperature of the system of interest fixed. The heat capacity of the reservoir must be much larger than that of the system, so that the heat exchanged with the reservoir does not affect the reservoir temperature.

Classical mechanics is an awkward vehicle for describing this type of system. The only way that thermodynamic systems can be treated in Newtonian or Hamiltonian mechanics is by explicitly modeling the system, the reservoir, and the exchange processes. This is complex, tedious, and, as we will see below, it is also unnecessary. We will now describe a little-known principle of classical mechanics which is extremely useful for *designing* equations of motion which are more useful thermodynamic systems. This principle does indeed allow us to modify classical mechanics so that thermodynamic variables may be constants of the motion.

The fundamental principles of mechanics

Just over 150 years ago, Gauss formulated a mechanics more general than Newton's. This mechanics has as its foundation Gauss' principle of least constraint.

Gauss (1829) referred to this as *the most fundamental dynamical principle* (Whittacker, 1961; Pars, 1968). Pars constructs mechanics using one of three different forms of what he calls the *fundamental equation*. Consider a general dynamical system with N equations of motion, subject to m constraints. We need to introduce the forces of constraint f_i, so the equations of motion become $m_i \ddot{x}_i = F_i + f_i$. If we consider a virtual displacement δx_i, at fixed time, which is consistent with the constraints, and that the forces of constraint f_i do no work on the system, we obtain the *first* fundamental equation:

$$\sum_{i=1}^{N} (m_i \ddot{x}_i - F_i) \delta x_i = 0, \qquad (3.19)$$

which is valid for an arbitrary virtual displacement. This is d'Alembert's principle which leads to Lagrange's equations of motion.

To obtain the *second* form of the fundamental equation, consider the same configuration, and the same instant of time, two different velocities for the system $\dot{x}_1, \dot{x}_2, \ldots, \dot{x}_N$ and $\dot{x}_1 + \Delta \dot{x}_1, \dot{x}_2 + \Delta \dot{x}_2, \ldots, \dot{x}_N + \Delta \dot{x}_N$. The (finite) velocity variations $\Delta \dot{x}_1, \ldots, \Delta \dot{x}_N$ then satisfy the *second* form of the fundamental equation (Jourdain, 1908):

$$\sum_{i=1}^{N} (m_i \ddot{x}_i - F_i) \Delta \dot{x}_i = 0. \qquad (3.20)$$

In this form both the configuration and time are given, and we consider the difference (either finite or infinitesimal) between any two possible velocities for the system.

The *third* form of the fundamental equation is obtained by considering two possible motions from the same configuration and velocity at time t, with different accelerations, $\ddot{x}$ and $\ddot{x} + \Delta \ddot{x}$. The (finite) acceleration-variations $\Delta \ddot{x}_1, \ldots, \Delta \ddot{x}_N$ satisfy the equations for the virtual displacement, and we may write:

$$\sum_{i=1}^{N} (m_i \ddot{x}_i - F_i) \Delta \ddot{x}_i = 0, \qquad (3.21)$$

(used by Gauss and Gibbs). To summarise, in the first form we consider an infinitesimal virtual displacement from a given configuration. In the second form the configuration is not varied, and we use the difference between any two possible velocities. In the third form both coordinates and velocities are unvaried, and we use the difference between any two possible accelerations.

Gauss' principle of least constraint

Suppose the position and velocity of the system are given, and consider the square of the curvature C, regarded as a function of the accelerations:

$$C(\ddot{\mathbf{x}}) = \frac{1}{2} \sum_{i=1}^{N} m_i \left(\ddot{x}_i - \frac{F_i}{m_i} \right)^2 \qquad (3.22)$$

The values of $\ddot{x}$ considered are those that are possible for the system. Gauss' principle of least constraint states that the actual acceleration is that for which C is a minimum. If we consider a variation of $\ddot{x}_i$ to $\ddot{x}_i + \Delta \ddot{x}_i$, then the proof is straightforward as:

$$\Delta C = \frac{1}{2} \sum_{i=1}^{N} m_i (\Delta \ddot{x}_i)^2 + \sum_{i=1}^{N} (m_i \ddot{x}_i - F_i) \Delta \ddot{x}_i. \qquad (3.23)$$

If $\ddot{x}_i$ is the actual physical acceleration then the last term is zero using the third fundamental form, and any variation about $\ddot{x}_i$ increases C. To find the equations of motion for a given system we need only the less powerful result that C is stationary for the actual motion $\delta C = 0$. Notice that in the application of Gauss' principle we are concerned with the simple algebraic problem of minimizing a quadratic form, the curvature C.

C is a function of the set of accelerations $\{\ddot{\mathbf{r}}\}$. Gauss' principle states that the actual physical acceleration corresponds to the minimum value of C. Clearly if the system is not subject to a constraint then $C = 0$ and the system evolves under Newton's equations of motion.

The types of constraints which might be applied to a system fall naturally into two types, *holonomic* and *nonholonomic*. A holonomic constraint is usually a constraint on the coordinates which can be integrated out or removed from the equations of motion. For instance, if a certain generalised coordinate is fixed, its conjugate momentum is zero for all time, so we can simply consider the problem in the reduced set of unconstrained variables. We need not be conscious of the fact that a force of constraint is acting upon the system to fix the coordinate and the momentum. An analysis of the two-dimensional motion of an ice skater need not refer to the fact that the gravitational force is exactly resisted by the stress on the ice surface fixing the vertical coordinate and velocity of the ice skater. We can ignore these degrees of freedom.

Nonholonomic constraints usually involve velocities. These constraints are not integrable and generally will do work on the system. Thermodynamic constraints are invariably nonholonomic (Morriss and Dettmann, 1998). We can write a general constraint as:

$$g(\mathbf{r}, \dot{\mathbf{r}}, t) = 0 \qquad (3.24)$$

where g is a function of positions, velocities, and possibly time. Either type of constraint function, holonomic or nonholonomic, can be written in this form with the equality replaced by an inequality. If this equation is differentiated with respect to time, once for nonholonomic constraints and twice for holonomic constraints we see that:

$$\mathbf{n}(\mathbf{r}, \dot{\mathbf{r}}, t) \cdot \ddot{\mathbf{r}} = s(\mathbf{r}, \dot{\mathbf{r}}, t). \qquad (3.25)$$

We refer to this equation as the *differential constraint equation* and it plays a fundamental role in Gauss' principle of least constraint. It is the equation for a plane, which we refer to as the *constraint plane*. $\mathbf{n}$ is the vector normal to the constraint plane (Figure 3.1).

Our problem is to solve Newton's equation subject to the constraint. Newton's equation gives us the acceleration in terms of the unconstrained forces. The differential constraint equation places a condition on the acceleration vector for the system. The differential constraint equation says that the constrained acceleration vector must terminate on a hyper-plane in the $3N$-dimensional acceleration space (Equation 3.25).

Imagine for the moment that at some initial time the system satisfies the constraint equation $g = 0$. In the absence of the constraint the system would evolve according to Newton's equations of motion, where the acceleration is given by:

$$m\ddot{\mathbf{r}}_i^u = \zeta_i. \qquad (3.26)$$

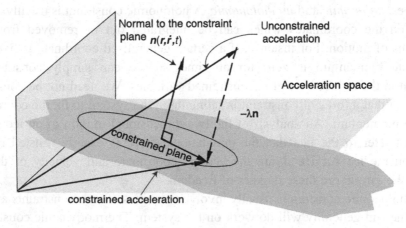

Figure 3.1 Gauss' principle of least constraint

This trajectory would, in general, not satisfy the constraint. Further, the constraint function *g* tells us that the only accelerations which do continuously satisfy the constraint are those which terminate on the constraint plane. To obtain the constrained acceleration we must project the unconstrained acceleration back into the constraint plane.

Gauss' principle of least constraint gives us a unique prescription for constructing this projection. *Gauss' principle states that the trajectories actually followed are those which deviate as little as possible, in a least squares sense, from the unconstrained Newtonian trajectories.* The projection which the system actually follows is the one which minimizes the magnitude of the constraint force. This means that the force of constraint must be parallel to the normal of the constraint surface. The Gaussian equations of motion are then:

$$\ddot{\mathbf{r}}_i = \zeta_i - \lambda \mathbf{n}, \tag{3.27}$$

where λ is a Gaussian multiplier which is a function of position, velocity, and time.

To calculate the multiplier we use the differential form of the constraint function. Substituting for the acceleration we obtain:

$$\lambda = \frac{\mathbf{n} \cdot \zeta - s}{\mathbf{n} \cdot \mathbf{n}}. \tag{3.28}$$

It is worthwhile at this stage to make a few comments about the procedure outlined above. First, notice that the original constraint equation is never used explicitly. Gauss' principle only refers to the *differential* form of the constraint equation. This means that the precise *value* of the constrained quantity is undetermined. The constraint acts only to stop its value changing. In the holonomic case Gauss' principle and the principle of least action are, of course, completely equivalent. In the nonholonomic case the equations resulting from the application of Gauss' principle cannot be derived from a Hamiltonian and the principle of least action *cannot* be used to derive constraint-satisfying equations. In the nonholonomic case, Gauss' principle does *not* yield equations of motion for which the work done by the constraint forces is a minimum.

The derivation of constrained equations of motion given above is geometric. From an operational point of view, a much simpler derivation of constrained equations of motion is possible using Lagrange multipliers. The square of the curvature *C* is a function of accelerations only (the Cartesian coordinates and velocities are considered to be given parameters). Gauss' principle reduces to finding the minimum of *C*, subject to the constraint. The constraint function must also be written as a function of accelerations, but this is easily achieved by differentiating with respect to time. If *G* is the acceleration dependent form of the constraint, then the constrained equations of motion are obtained from:

$$\frac{\partial}{\partial \ddot{\mathbf{r}}}(C - \lambda G) = 0. \tag{3.29}$$

It is easy to see that the Lagrange multiplier λ is (apart from the sign) equal to the Gaussian multiplier. We will illustrate Gauss' principle by considering some useful examples.

Gauss' principle for holonomic constraints

The most common type of holonomic constraint in statistical mechanics is probably that of fixing bond lengths and bond angles in molecular systems. The vibrational degrees of freedom typically have a period which is orders of magnitude faster than the translational degrees of freedom, and are therefore often irrelevant to the processes under study. As an example of the application of Gauss' principle of least constraint for holonomic constraints we consider a diatomic molecule with a fixed bond length. The generalisation of this method to more than one bond length is straightforward (see Edberg *et al.*, 1986) and the application to bond angles is trivial since they can be formulated as second-nearest-neighbour distance constraints. The constraint function for a diatomic molecule is that the distance between sites 1 and 2 be equal to d_{12}, that is:

$$g(\mathbf{r}, \dot{\mathbf{r}}, t) = \mathbf{r}_{12}^2 - d_{12}^2 = 0, \tag{3.30}$$

where we define $\mathbf{r}_{12}$ to be the vector from $\mathbf{r}_1$ to $\mathbf{r}_2$, $(\mathbf{r}_{12} \equiv \mathbf{r}_2 - \mathbf{r}_1)$. Differentiating twice with respect to time gives the acceleration-dependent constraint equation:

$$\mathbf{r}_{12} \cdot \ddot{\mathbf{r}}_{12} + (\dot{\mathbf{r}}_{12})^2 = 0. \tag{3.31}$$

To obtain the constrained equations of motion we minimise the function C subject to the constraint Equation (3.3). That is:

$$\frac{\partial}{\partial \ddot{\mathbf{r}}_i}\left\{ \frac{m_1}{2}\left(\ddot{\mathbf{r}}_1 - \frac{\mathbf{F}_1}{m_1}\right)^2 + \frac{m_2}{2}\left(\ddot{\mathbf{r}}_2 - \frac{\mathbf{F}_2}{m_2}\right)^2 - \lambda(\mathbf{r}_{12} \cdot \ddot{\mathbf{r}}_{12} + (\dot{\mathbf{r}}_{12})^2) \right\} = 0. \tag{3.32}$$

For i equal to 1 and 2 this gives:

$$\begin{aligned} m_1\ddot{\mathbf{r}}_1 &= \mathbf{F}_1 - \lambda\mathbf{r}_{12}, \\ m_2\ddot{\mathbf{r}}_2 &= \mathbf{F}_2 + \lambda\mathbf{r}_{12}. \end{aligned} \tag{3.33}$$

Notice that the extra constraint terms in these equations have opposite signs, thus total constraint force on each molecule is zero and the momentum of the molecule is conserved. To obtain an expression for the multiplier λ we combine these two equations to give an equation of motion for the bond vector $\mathbf{r}_{12}$:

$$\ddot{\mathbf{r}}_{12} = \left(\frac{\mathbf{F}_2}{m_2} - \frac{\mathbf{F}_1}{m_1}\right) + \lambda\left(\frac{1}{m_2} + \frac{1}{m_1}\right)\mathbf{r}_{12}. \tag{3.34}$$

Substituting this into the differential form of the constraint function (3.24) gives:

$$\lambda = -\frac{\mathbf{r}_{12} \cdot (m_1\mathbf{F}_2 - m_2\mathbf{F}_1) + m_1 m_2 \dot{\mathbf{r}}_{12}^2}{(m_1 + m_2)\mathbf{r}_{12}^2}. \tag{3.35}$$

It is very easy to implement these constrained equations of motion as the multiplier is a simple explicit function of the positions, velocities, and Newtonian forces. For more complicated systems with multiple bond-length and bond-angle constraints (all written as distance constraints) we obtain a set of coupled linear equations to solve for the multipliers (Edberg *et al.*, 1986; Morriss and Evans, 1991).

Gauss' principle for nonholonomic constraints

One of the simplest and most useful applications of Gauss' principle is to derive equations of motion for which the kinetic temperature is a constant of the motion (Evans *et al.*, 1983). Here the constraint function is:

$$g(\mathbf{r}, \dot{\mathbf{r}}, t) = \sum_{i=1}^{N} \frac{1}{2}m_i\dot{\mathbf{r}}_i^2 - \frac{3}{2}Nk_BT = 0. \tag{3.36}$$

Differentiating once with respect to time gives the equation for the constraint plane:

$$\sum_{i=1}^{N} m_i\dot{\mathbf{r}}_i \cdot \ddot{\mathbf{r}}_i = 0. \tag{3.37}$$

Therefore to obtain the constrained Gaussian equations we minimise C subject to the constraint Equation (3.37). That is:

$$\frac{\partial}{\partial\ddot{\mathbf{r}}_i}\left(\frac{1}{2}\sum_{j=1}^{N} m_j\left(\ddot{\mathbf{r}}_j - \frac{\mathbf{F}_j}{m_j}\right)^2 + \lambda\sum_{j=1}^{N} m_j\dot{\mathbf{r}}_j \cdot \ddot{\mathbf{r}}_j\right) = 0. \tag{3.38}$$

This gives:

$$m_i\ddot{\mathbf{r}}_i = \mathbf{F}_i - \lambda m_i\dot{\mathbf{r}}_i. \tag{3.39}$$

Substituting the equations of motion into the differential form of the constraint equation, we find that the multiplier is given by:

$$\lambda = \frac{\sum_{i=1}^{N} \mathbf{F}_i \cdot \dot{\mathbf{r}}_i}{\sum_{i=1}^{N} m_i\dot{\mathbf{r}}_i^2}. \tag{3.40}$$

As before, λ is a simple function of the forces and velocities so that the implementation of the constant kinetic energy constraint in a molecular dynamics computer program only requires a trivial modification of the equations of motion. Equations (3.39 and 3.40) constitute what have become known as the *Gaussian iso-kinetic equations of motion*. These equations were first proposed simultaneously and independently by Hoover *et al.* (1982) and Evans (1983a). In these original papers Gauss' principle was, however, not referred to. It was a year before the connection with Gauss' principle was made.

With regard to the general application of Gauss' principle of least constraint one should always examine the statistical mechanical properties of the resulting dynamics. If one applies Gauss' principle to the problem of maintaining a constant heat flow, then a comparison with linear response theory shows that the Gaussian equations of motion *cannot* be used to calculate thermal conductivity (Hoover, 1986). The correct application of Gauss' principle is limited to arbitrary holonomic constraints and apparently, to nonholonomic constraint functions which are *homogeneous* functions of the momenta.

3.2 Phase space

To give a complete description of the state of a three-dimensional N-particle system at any given time, it is necessary to specify the $3N$ coordinates and $3N$ momenta. The $6N$ dimensional space of coordinates and momenta is called *phase space* (or Γ-space). As time progresses the phase point $\Gamma \equiv (q, p)$ traces out a path which we call the phase-space trajectory (Figure 3.2) of the system. As the equations of motion for Γ are $6N$ first-order differential equations, there are $6N$ constants of

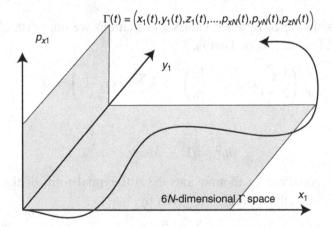

$$\Gamma(t) = \left(x_1(t), y_1(t), z_1(t), ..., p_{xN}(t), p_{yN}(t), p_{zN}(t) \right)$$

p_{x1}

y_1

6N-dimensional Γ space x_1

Figure 3.2 Phase space trajectory in $6N$-dimensional phase space is a path parameterized by the time

integration (they may be, for example, the $6N$ initial conditions $\Gamma(0)$). Rewriting the equations of motion in terms of these constants shows that the trajectory of Γ is completely determined by specifying these $6N$ constants. An alternate description of the time evolution of the system is given by the trajectory in the extended Γ'-space, where $\Gamma' = (\Gamma, t)$. As the $6N$ initial conditions uniquely determine the trajectory, two points in phase space with different initial conditions form distinct non-intersecting trajectories in Γ'-space.

To illustrate the ideas of Γ-space and Γ'-space, it is useful to consider one of the simplest mechanical systems, the harmonic oscillator. The Hamiltonian for the harmonic oscillator is $H = \frac{1}{2}(kx^2 + p^2/m)$ where m is the mass of the oscillator and k is the spring constant. The equations of motion are:

$$\dot{x} = \frac{\partial H}{\partial p} = \frac{p}{m},$$
$$\dot{p} = -\frac{\partial H}{\partial x} = -kx, \tag{3.41}$$

and the energy (or the Hamiltonian) is a constant of the motion. The Γ-space for this system is two-dimensional (x, p) and the Γ-space trajectory is given by

$$(x(t), p(t)) = \left(x_0 \cos \omega t + \frac{p_0}{m\omega} \sin \omega t, p_0 \cos \omega t - m\omega x_0 \sin \omega t\right). \tag{3.42}$$

The constants x_0 and p_0 are the two integration constants written, in this case, as an initial condition. The frequency ω is related to the spring constant and mass by $\omega^2 = k/m$. The Γ-space trajectory is an ellipse:

$$m^2\omega^2 x(t)^2 + p(t)^2 = m^2\omega^2 x_0^2 + p_0^2 = 2mE, \tag{3.43}$$

which has intercepts on the x-axis at $\pm(x_0^2 + p_0^2/m^2\omega^2)^{1/2}$ and intercepts on the p-axis at $\pm(p_0^2 + m^2\omega^2 x_0^2)^{1/2}$. The period of the motion is $T = 2\pi/\omega = 2\pi(m/k)^{1/2}$. This is the surface of constant energy for the harmonic oscillator. Any oscillator with the same energy must traverse the same Γ space trajectory, that is another oscillator with the same energy, but different initial starting points (x_0, p_0) will follow the same ellipse, but with a different initial phase angle.

The trajectory in Γ'-space is an elliptical coil, and the constant energy surface in Γ'-space is an elliptical cylinder, and oscillators with the same energy start from different points on the ellipse at time zero (corresponding to different initial phase angles) and wind around the elliptical cylinder. The trajectories in Γ'-space are nonintersecting. If two trajectories in Γ'-space meet at time t, then the two trajectories must have had the same initial condition. As the choice of time origin is arbitrary, the trajectories must be the same for all time.

In Γ-space the situation is somewhat different. The trajectory for the harmonic oscillator winds around the ellipse, returning to its initial phase point (x_0, p_0) after a time T. The period of time taken for a system to return to (or to within an ε-neighbourhood of) its initial starting phase is called the Poincaré recurrence time. For a simple system, such as the harmonic oscillator, the recurrence time is trivial to calculate, but for higher-dimensional systems, the recurrence time quickly exceeds the estimated age of the universe.

3.3 Distribution functions and the Liouville equation

In the first few sections of this chapter we have given a description of the mechanics of individual N-particle systems. The development which follows describes an *ensemble* of such systems; that is an essentially infinite number of systems characterized by identical dynamics and identical state variables (N, V, E, or T etc.) but different initial conditions ($\Gamma(0)$). We wish to consider the average behavior of a collection of macroscopically identical systems distributed over a range of initial states (microstates). In generating the ensemble we make the usual assumptions of classical mechanics (Tolman, 1979). We assume that it is possible to know all the positions and momenta of an N-particle system to arbitrary precision at some initial time, and that the motion can be calculated exactly from the equations of motion.

The ensemble contains an infinite number of individual systems so that the number of systems in a particular state may be considered to change continuously as we pass to neighboring states. This assumption allows us to define a density function $f(\Gamma, t)$, which assigns a probability to points in phase space. Implicit in this assumption is the requirement that $f(\Gamma, t)$ has continuous partial derivatives with respect to all its variables; otherwise the phase density will not change *continuously* as we move to neighboring states. If the system is Hamiltonian and all trajectories are confined to the energy surface, then $f(\Gamma, t)$ will not have continuous partial derivatives with respect to energy. Problems associated with this particular source of discontinuity can obviously be avoided by eliminating the energy as a variable, and considering $f(\Gamma, t)$ to be a density function defined on a surface of constant energy (effectively reducing the dimensionality of the system). However, it is worth pointing out that other sources of discontinuity in the phase space density may not be so easily removed.

To define a distribution function for a particular system we consider an ensemble of identical systems whose initial conditions *span* the phase space specified by the macroscopic constraints. We consider an infinitesimal element of phase space located at $\Gamma \equiv (\mathbf{q}, \mathbf{p})$. The fraction of systems δN, which at time t have coordinates and momenta within $\delta\mathbf{q}$, $\delta\mathbf{p}$ of $\mathbf{q}$, $\mathbf{p}$, is used to define the phase space distribution function $f(\mathbf{q}, \mathbf{p}, t)$, by:

$$\delta N = f(\mathbf{q}, \mathbf{p}, t)\delta\mathbf{q}\,\delta\mathbf{p}. \qquad (3.44)$$

The total number of systems in the ensemble is fixed, so integrating over the whole phase space we can normalize the distribution function:

$$1 = \int f(\mathbf{q}, \mathbf{p}, t)\delta\mathbf{q}\,\delta\mathbf{p}. \qquad (3.45)$$

If we consider a small volume element of phase space, the number of trajectories entering the rectangular volume element $\delta\mathbf{q}\,\delta\mathbf{p}$ through some face will, in general, be different from the number which leave through an opposite face. For the faces normal to the q_1-axis, located at q_1, and $q_1 + \delta q_1$, the fraction of ensemble members entering the first face is:

$$f(q_1, \ldots, t)\dot{q}_1(q_1, \ldots, t)\delta q_2, \ldots, \delta q_{3N}\,\delta\mathbf{p}. \qquad (3.46)$$

Similarly the fraction of points leaving through the second face is:

$$f(q_1 + \delta q_1, \ldots, t)\dot{q}_1(q_1 + \delta q_1, \ldots, t)\delta q_2, \ldots, \delta q_{3N}\,\delta\mathbf{p}$$

$$\approx \left(f(q_1, \ldots, t) + \frac{\partial f}{\partial q_1}\delta q_1\right)\left(\dot{q}_1(q_1 + \delta q_1, \ldots, t) + \frac{\partial \dot{q}_1}{\partial q_1}\delta q_1\right)\delta q_2, \ldots, \delta q_{3N}\,\delta\mathbf{p}.$$

$$(3.47)$$

Combining these expressions gives the change in δN due to fluxes in the q_1 direction:

$$\frac{\mathrm{d}}{\mathrm{d}t}\delta N_{q_1} = -\left(\dot{q}_1\frac{\partial f}{\partial q_1} + f\frac{\partial \dot{q}_1}{\partial q_1}\right)\delta\mathbf{q}\,\delta\mathbf{p}. \qquad (3.48)$$

Summing over all coordinate (and momentum) directions gives the total fractional change δN as:

$$\frac{\mathrm{d}}{\mathrm{d}t}\delta N = -\sum_{i=1}^{N}\left[f\left(\frac{\partial}{\partial\mathbf{q}_i}\cdot\dot{\mathbf{q}}_i + \frac{\partial}{\partial\mathbf{p}_i}\cdot\dot{\mathbf{p}}_i\right) + \dot{\mathbf{q}}_i\cdot\frac{\partial f}{\partial\mathbf{q}_i} + \dot{\mathbf{p}}_i\frac{\partial f}{\partial\mathbf{p}_i}\right]\delta\mathbf{q}\,\delta\mathbf{p}. \qquad (3.49)$$

Dividing through by the phase space volume element $\delta\mathbf{q}\,\delta\mathbf{p}$ we obtain the rate of change in density $f(\mathbf{q}, \mathbf{p})$, at the point $(\mathbf{q}, \mathbf{p})$:

$$\frac{1}{\delta\mathbf{q}\,\delta\mathbf{p}}\frac{\mathrm{d}}{\mathrm{d}t}\delta N = \frac{\partial}{\partial t}\left(\frac{\delta N}{\delta\mathbf{q}\,\delta\mathbf{p}}\right) = \frac{\partial f}{\partial t}\bigg|_{\mathbf{q},\mathbf{p}}. \qquad (3.50)$$

Using the notation, $\Gamma = (\mathbf{q}, \mathbf{p}) = (q_1, q_2, \ldots, q_{3N}, p_1, p_2, \ldots, p_{3N})$ for the $6N$-dimensional phase point, this may be written as:

$$\left.\frac{\partial f}{\partial t}\right|_\Gamma = -f \frac{\partial}{\partial \Gamma} \cdot \dot{\Gamma} - \dot{\Gamma} \cdot \frac{\partial f}{\partial \Gamma} = -\frac{\partial}{\partial \Gamma} \cdot \left(f \dot{\Gamma}\right). \tag{3.51}$$

This is the Liouville equation for the phase-space distribution function. Using the streaming or total time derivative of the distribution function, we can rewrite the Liouville equation in an equivalent form as:

$$\frac{df}{dt} = \frac{\partial f}{\partial t} + \dot{\Gamma} \cdot \frac{\partial f}{\partial \Gamma} = -f \frac{\partial}{\partial \Gamma} \cdot \dot{\Gamma} = -f \Lambda(\Gamma). \tag{3.52}$$

This equation has been obtained without reference to the equations of motion. Its correctness does not require the existence of a Hamiltonian to generate the equations of motion. The equation rests on two conditions: that ensemble members cannot be created or destroyed, and that the distribution function is sufficiently smooth that the appropriate derivatives exist. $\Lambda(\Gamma)$ is called the *phase-space compression factor* since it is equal to the negative time derivative of the logarithm of the phase-space distribution function:

$$\frac{d}{dt} \ln(f(\Gamma, t)) = -\Lambda(\Gamma). \tag{3.53}$$

The Liouville equation is usually written in a slightly simpler form. If the equations of motion can be generated from a Hamiltonian, then it is a simple matter to show that $\Lambda(\Gamma) = 0$. This is so even in the presence of external fields, which may be driving the system away from equilibrium by performing work on the system:

$$\Lambda(\Gamma) = \sum_{i=1}^{N} \left(\frac{\partial}{\partial \mathbf{q}_i} \cdot \dot{\mathbf{q}}_i + \frac{\partial}{\partial \mathbf{p}_i} \cdot \dot{\mathbf{p}}_i \right) = \sum_{i=1}^{N} \left(\frac{\partial}{\partial \mathbf{q}_i} \cdot \frac{\partial H}{\partial \mathbf{p}_i} - \frac{\partial}{\partial \mathbf{p}_i} \cdot \frac{\partial H}{\partial \mathbf{q}_i} \right) = 0. \tag{3.54}$$

The existence of a Hamiltonian is a sufficient, but not necessary, condition for the phase-space compression factor to vanish. If phase space is incompressible then the Liouville equation takes on its simplest form:

$$\frac{df}{dt} = 0. \tag{3.55}$$

Time evolution of the distribution function

The following sections will be devoted to developing a formal operator algebra for manipulating the distribution function and averages of mechanical phase variables. This development is an extension of the treatment given by (Berne, 1977), which is applicable to Hamiltonian systems only. We will use the compact operator notation:

$$\frac{\partial f}{\partial t} = -i\mathscr{L}f = -\left(\frac{\partial}{\partial \mathbf{\Gamma}} \cdot \dot{\mathbf{\Gamma}} + \dot{\mathbf{\Gamma}} \cdot \frac{\partial}{\partial \mathbf{\Gamma}}\right)f, \tag{3.56}$$

for the Liouville equation, Equation (3.51). The operator $i\mathscr{L}$ is called the distribution function (or f-) Liouvillean. Both the distribution function f, and the f-Liouvillean are functions of the initial phase $\mathbf{\Gamma}$. We assume that there is no explicit time dependence in the equations of motion. Using this notation we can write the formal solution of the Liouville equation for the time dependent N-particle distribution function $f(\mathbf{\Gamma}, t)$ as:

$$f(\mathbf{\Gamma}, t) = \exp[-i\mathscr{L}t]f(\mathbf{\Gamma}, 0), \tag{3.57}$$

where $f(\mathbf{\Gamma}, 0)$, is the initial distribution function. This representation for the distribution function contains the exponential of an operator, which is a symbolic representation for the infinite series of operators. The f-*propagator* is defined as:

$$\exp[-i\mathscr{L}t] = \sum_{n=0}^{\infty} \frac{(-t)^n}{n!} (i\mathscr{L})^n. \tag{3.58}$$

The formal solution given above can therefore be written as:

$$f(t) = \sum_{n=0}^{\infty} \frac{(-t)^n}{n!} (i\mathscr{L})^n f(0) = \sum_{n=0}^{\infty} \frac{(-t)^n}{n!} \frac{\partial^n}{\partial t^n} f(0). \tag{3.59}$$

This form makes it clear that the formal solution derived above is the Taylor series expansion of the *explicit* time dependence of $f(\mathbf{\Gamma}, t)$, about $f(\mathbf{\Gamma}, 0)$.

Time evolution of phase variables

We will need to consider the time evolution of functions of the phase of the system. Such functions are called phase variables. An example would be the phase variable for the internal energy of a system, $H_0(\mathbf{\Gamma}) = \sum_i p_i^2/2m_i + \Phi(q_1, \ldots, q_n)$. Phase variables, by definition, do not depend on time explicitly, their time dependence comes solely from the time dependence of the phase $\mathbf{\Gamma}$. Using the chain rule, the equation of motion for an arbitrary phase variable $B(\mathbf{\Gamma})$ can be written as:

$$\dot{B}(\mathbf{\Gamma}) = \dot{\mathbf{\Gamma}} \cdot \frac{\partial}{\partial \mathbf{\Gamma}} B = \sum_{i=1}^{N} \left(\dot{q}_i \cdot \frac{\partial}{\partial q_i} + \dot{p}_i \cdot \frac{\partial}{\partial p_i}\right) B(\mathbf{\Gamma}) = iL(\mathbf{\Gamma})B(\mathbf{\Gamma}). \tag{3.60}$$

The operator associated with the time derivative of a phase variable $iL(\mathbf{\Gamma})$ is referred to as the phase variable (or p-) Liouvillean. The formal solution of this equation can be written in terms of the *p-propagator*, $\exp[iLt]$. This gives the value of the phase variable as a function of time:

$$B(t) = \exp[iLt]B(0). \qquad (3.61)$$

This expression is very similar in form to that for the distribution function. It is the Taylor series expansion of the total time dependence of $B(t)$, expanded about $B(0)$. If the phase-space compression factor $\Lambda(\Gamma)$ is identically zero then the *p*-Liouvillean is equal to the *f*-Liouvillean, and the *p*-propagator is simply the adjoint or Hermitian conjugate of the *f*-propagator. In general, this is not the case.

Properties of Liouville operators

In this section we will derive some of the more important properties of the Liouville operators. These will lead us naturally to a discussion of various *representations* of the properties of classical systems. The first property we shall discuss relates the *p*-Liouvillean to the *f*-Liouvillean as follows,

$$\int d\Gamma f(0)iLB(\Gamma) = -\int d\Gamma B(\Gamma)i\mathscr{L}f(0). \qquad (3.62)$$

This is true for an arbitrary distribution function $f(0)$. To prove this identity the LHS can be written as:

$$\int d\Gamma f(\Gamma)\dot{\Gamma} \cdot \frac{\partial}{\partial\Gamma}B(\Gamma) = [f(\Gamma)\dot{\Gamma}B(\Gamma)]_S - \int d\Gamma B(\Gamma)\frac{\partial}{\partial\Gamma}\cdot(f(\Gamma)\dot{\Gamma})$$

$$= -\int d\Gamma B(\Gamma)\left(\dot{\Gamma}\cdot\frac{\partial}{\partial\Gamma} + \frac{\partial}{\partial\Gamma}\cdot\dot{\Gamma}\right)f(\Gamma)$$

$$= -\int d\Gamma B(\Gamma)i\mathscr{L}f(\Gamma). \qquad (3.63)$$

The boundary term (or surface integral) is zero because $f(S) \to 0$ as any component of the momentum goes to infinity, and f can be taken to be periodic in all coordinates. If the coordinate space for the system is bounded, then the surface S is the system boundary, and the surface integral is again zero as there can be no flow through the boundary.

Equations (3.62 and 3.63) show that L, $\mathscr{L}$ are adjoint operators. If the equations of motion are such that the phase space compression factor, (3.53) is identically zero, then obviously $L = \mathscr{L}$ and the Liouville operator is self-adjoint, or *Hermitian*.

Schrödinger and Heisenberg representations

We can calculate the value of a phase variable $B(t)$ at time t by following B as it changes along a single trajectory in phase space. The average $\langle B(\Gamma(t)) \rangle$ can then be calculated by summing the values of $B(t)$ with a weighting factor determined by the probability of starting from each initial phase Γ. These probabilities are chosen from an initial distribution function $f(\Gamma, 0)$. This is the Heisenberg picture of phase space averages:

$$\langle B(t) \rangle = \int d\Gamma B(t) f(\Gamma) = \int d\Gamma f(\Gamma) \exp[iLt] B(\Gamma). \qquad (3.64)$$

The Heisenberg picture (Figure 3.3) is exactly analogous to the Lagrangian formulation of fluid mechanics; we can imagine that the phase space *mass point* has a differential box $d\Gamma$ surrounding it which changes shape (and volume for a compressible fluid) with time as the phase point follows its trajectory. The

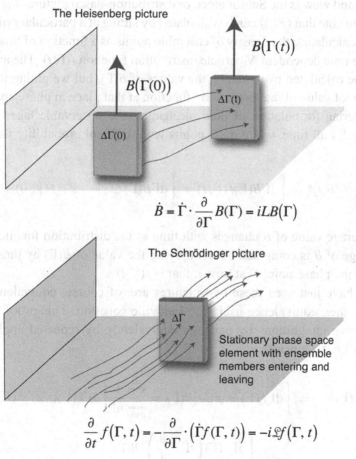

$$\dot{B} = \dot{\Gamma} \cdot \frac{\partial}{\partial \Gamma} B(\Gamma) = iLB(\Gamma)$$

$$\frac{\partial}{\partial t} f(\Gamma, t) = -\frac{\partial}{\partial \Gamma} \cdot \left(\dot{\Gamma} f(\Gamma, t) \right) = -i\mathscr{L} f(\Gamma, t)$$

Figure 3.3 The Schrödinger–Heisenberg equivalence

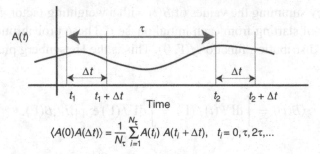

$$\langle A(0)A(\Delta t)\rangle = \frac{1}{N_\tau}\sum_{i=1}^{N_\tau} A(t_i)\,A(t_i+\Delta t), \quad t_i = 0,\tau,2\tau,\ldots$$

Figure 3.4 Equilibrium autocorrelation function of a real variable A. For samples to be independent $\langle A(0)A(\tau)\rangle \ll \langle A(0)^2\rangle$

probability of the differential element, or mass $f(\Gamma)d\Gamma$ remains constant, but the value of the observable changes implicitly in time.

The second view is the Schrödinger, or distribution-based picture (Figure 3.4). In this case we note that $\langle B(t)\rangle$ can be calculated by sitting at a particular point in phase space and calculating the density of ensemble points as a function of time. This will give us the time dependent N-particle distribution function $f(\Gamma, t)$. The average of B can now be calculated by summing the values of $B(\Gamma)$, but weighting these values by the current value of the distribution function at that place in phase space. Just as in the Eulerian formulation of fluid mechanics, the observable takes on a fixed value $B(\Gamma)$ for all time, while mass points with different probability flow through the box:

$$\langle B(t)\rangle = \int d\Gamma B(\Gamma) f(\Gamma, t) = \int d\Gamma B(\Gamma)\exp[-i\mathscr{L}t]f(\Gamma, 0). \qquad (3.65)$$

The average value of B changes with time as the distribution function changes. The average of B is computed by multiplying the value of $B(\Gamma)$ by the probability of finding the phase point Γ at time t, that is $f(\Gamma, t)$.

As we have just seen these two pictures are, of course, equivalent. One can also prove their equivalence using the Liouville equation. This proof is obtained by successive integrations by parts, or equivalently by repeated applications of Equation (3.62). Consider:

$$\int d\Gamma f(\Gamma) B(t) = \int d\Gamma f(\Gamma)\exp[iLt]B(\Gamma) = \sum_{n=0}^{\infty}\frac{1}{n!}\int d\Gamma f(\Gamma)(iLt)^n B(\Gamma)$$

$$= \sum_{n=0}^{\infty}\frac{1}{n!}\int d\Gamma f(\Gamma)\left(t\dot{\Gamma}\cdot\frac{\partial}{\partial\Gamma}\right)^n B(\Gamma). \qquad (3.66)$$

One can *unroll* each p-Liouvillean in turn from the phase variable to the distribution function (for the first transfer we consider $(iL)^{n-1} B$ to be a composite phase variable) so that Equation (3.66) becomes:

$$= \sum_{n=0}^{\infty} \frac{1}{n!} \int d\Gamma \left(-t \frac{\partial}{\partial \Gamma} \cdot (\dot{\Gamma} f(\Gamma)) \right) \left(t\dot{\Gamma} \cdot \frac{\partial}{\partial \Gamma} \right)^{n-1} B(\Gamma). \quad (3.67)$$

This is essentially the property of phase and distribution-function Liouvilleans which we have already proved, applied to the nth Liouvillean. Repeated application of this result leads to:

$$= \sum_{n=0}^{\infty} \frac{(-t)^n}{n!} \int d\Gamma \left(\left(\frac{\partial}{\partial \Gamma} \cdot \dot{\Gamma} \right)^n f(\Gamma) \right) B(\Gamma) = \int d\Gamma B(\Gamma) \exp[-i\mathscr{L} t] f(\Gamma).$$

So finally we have the result,

$$\int d\Gamma f(\Gamma) B(t) = \int d\Gamma B(\Gamma) f(\Gamma, t). \quad (3.68)$$

The derivation we have used assumes that the Liouvillean for the system has no explicit time dependence. Our present derivation makes no other references to the details of either the initial distribution function, or the equations of motion for the system. This means that these results are valid for systems subject to time-independent external fields, whether or not those equations are derivable from a Hamiltonian. These results are also independent of whether or not the phase-space compression factor vanishes identically.

A final point that can be made concerning the Schrödinger and Heisenberg pictures is that these two ways of computing phase averages by no means exhaust the range of possibilities. The Schrödinger and Heisenberg pictures differ in terms of the time chosen to calculate the distribution function, $f(\Gamma, t)$. In the Heisenberg picture that time is zero while in the Schrödinger picture the time is t. One can, of course, develop intermediate representations corresponding to any time between zero and t (e.g. the interaction representation).

3.4 Ergodicity, mixing, and Lyapunov exponents

For many systems it is apparent that after possible initial transients lasting a time t_0, the N-particle distribution function $f(\Gamma, t)$, becomes essentially time independent. This is evidenced by the fact that the macroscopic properties of the system relax to fixed average values. This obviously happens for equilibrium systems. It also occurs in some nonequilibrium systems, so-called nonequilibrium steady states. We will call all such systems stationary.

For a stationary system, we may define the ensemble average of a phase variable $B(\Gamma)$, using the stationary distribution function $f(\Gamma)$, so that:

$$\langle B \rangle = \int d\Gamma f(\Gamma) B(\Gamma). \tag{3.69}$$

On the other hand we may define a time average of the same phase variable as:

$$\langle B \rangle_t = \lim_{T \to \infty} \frac{1}{T} \int_{t_0}^{t_0 + T} dt B(t), \tag{3.70}$$

where t_0 is the relaxation time required for the establishment of the stationary state. An ergodic system is a stationary system for which the ensemble and time averages of *usual* phase variables, exist and are equal. By *usual* we mean phase-variable representations of the common macroscopic thermodynamic variables (see Section 3.7).

Example: We can give a simple example of ergodic flow if we take the energy surface to be the two-dimensional unit square $0 < p < 1$ and $0 < q < 1$. We shall assume that the equations of motion are given by:

$$\dot{p} = \alpha, \quad \dot{q} = 1, \tag{3.71}$$

and we impose periodic boundary conditions on the system. These equations of motion can be solved to give:

$$p(t) = p_0 + \alpha t, \quad q(t) = q_0 + t. \tag{3.72}$$

The phase-space trajectory on the energy surface is given by eliminating t from these two equations:

$$p = p_0 + \alpha(q - q_0). \tag{3.73}$$

If α is a rational number, $\alpha = m/n$, then the trajectory will be periodic and will repeat after a period $T = n$. If α is irrational, then the trajectory will be dense on the unit square, but will not fill it. When α is irrational the system is ergodic. To show this explicitly, consider the Fourier series expansion of an arbitrary phase function $A(q, p)$:

$$A(q,p) = \sum_{j,k=-\infty}^{\infty} A_{jk} \exp[2\pi i(jq + kp)]. \tag{3.74}$$

We wish to show that the time average and phase average of $A(q, p)$ are equal for α irrational. The time average is given by:

$$\langle A \rangle_t = \lim_{T\to\infty} \frac{1}{T} \int_{t_0}^{t_0+T} dt \sum_{j,k=-\infty}^{\infty} A_{jk} \exp[2\pi i(j(q_0 + t) + k(p_0 + \alpha t))]$$

$$= A_{00} + \lim_{T\to\infty} \frac{1}{T} \sum_{j,k\neq 0}^{\infty} A_{jk} \exp[2\pi i(j(q_0 + t_0) + k(p_0 + \alpha t_0))]$$

$$\times \frac{e^{2\pi i(j+\alpha k)T} - 1}{2\pi i(j + \alpha k)}. \tag{3.75}$$

For irrational α, the denominator can never be equal to zero, therefore:

$$\langle A \rangle_t = A_{00}. \tag{3.76}$$

Similarly we can show that the phase-space average of A is:

$$\langle A \rangle_{qp} = \int_0^1 dq \int_0^1 dp \, A(q, p) = A_{00} \tag{3.77}$$

and hence the system is ergodic. For rational α, the denominator in Equation (3.75) does become singular for a particular jk-mode. The system is in the pure state labelled by jk. There is no mixing.

Ergodicity does *not* guarantee the relaxation of a system toward a stationary state. Consider a probability density which is not constant over the unit square, for example let $f(q, p, t = 0)$ be given by:

$$f(q, p, 0) = \sin(\pi p_0) \sin(\pi q_0), \tag{3.78}$$

then at time t, under the above dynamics (with irrational α), it will be:

$$f(q, p, t) = \sin(\pi(p_0 - \alpha t)) \sin(\pi(q_0 - t)). \tag{3.79}$$

The probability distribution is not changed in shape, it is only displaced. It has also *not* relaxed to a time-independent equilibrium distribution function. However, after an infinite length of time it will have wandered uniformly over the entire energy surface. It is therefore ergodic, but it is termed *nonmixing*.

It is often easier to show that a system is not ergodic, rather than to show that it is ergodic. For example, the phase space of a system must be metrically transitive for it to be ergodic. That is, all of phase space, except possibly a set of measure zero, must be accessible to *almost all* the trajectories of the system. The reference to *almost all* is because of the possibility that a set of initial starting states of measure zero may remain forever within a subspace of phase space, which is itself of measure zero. Ignoring the more pathological cases, if it is possible to divide phase space into two (or more) finite regions of nonzero measure, so that trajectories initially in a particular region remain there forever, then the system is not ergodic. A typical example would be a system in which a particle was trapped in a certain region of configuration space. Later we shall see examples of this specific type.

Lyapunov exponents

If we consider two harmonic oscillators (see Section 3.2) which have the same frequency ω, but different initial conditions (x_1, p_1) and (x_0, p_0), we can define the *distance* between the two phase points at any time by:

$$\delta(t) = \|\Gamma\| = (\Gamma \cdot \Gamma)^{\frac{1}{2}} = \sqrt{(x_1(t) - x_0(t))^2 + \frac{(p_1(t) - p_0(t))^2}{m^2 \omega^2}}, \qquad (3.80)$$

where $x_i(t)$ and $p_i(t)$ are the position and momenta of oscillator i, at time t. Using the equation of motion for the trajectory of the harmonic oscillator (Equation 3.42), we see that this distance is given by:

$$\delta(t) = \sqrt{(x_1(0) - x_0(0))^2 + \frac{(p_1(0) - p_0(0))^2}{m^2 \omega^2}} = \delta(0). \qquad (3.81)$$

This means that the trajectories of two independent harmonic oscillators always remain the same distance apart in Γ-space. This is not the typical behavior of a *non-linear* system. The neighboring trajectories of most N-body nonlinear systems tend to move apart with time. Indeed, it is clear that if a system is to be mixing, then the separation of neighboring trajectories is a precondition. Chains of weakly coupled harmonic oscillators are an exception to the generally observed trajectory separation. This was a cause of some concern in the earliest dynamical simulations (Fermi *et al.*, 1955).

As the separation between neighboring trajectories can be easily calculated in a classical mechanical simulation, this has been used to obtain quantitative measures of the mixing properties of nonlinear many-body systems. If we consider two

N-body systems composed of particles which interact via identical sets of inter-particle forces, but whose initial conditions differ by a small amount, then the phase-space separation is observed to change exponentially as:

$$\delta(t) \equiv \sqrt{(\Gamma_1(t) - \Gamma_2(t))^2} \cong c \exp[\lambda t]. \tag{3.82}$$

At intermediate times the exponential growth of $\delta(t)$ will be dominated by the fastest growing direction in phase space (which, in general, will change continuously with time). This equation defines the largest *Lyapunov exponent* λ for the system (λ is defined to be real, so any oscillating contribution is ignored). For the harmonic oscillator the phase separation is a constant of the motion and therefore the Lyapunov exponent λ is zero. Strictly, the exponential growth of trajectory separation is for initially infinitesimal displacements $\delta(0)$.

The largest Lyapunov exponent indicates the rate of growth of trajectory separation in phase space. If we consider a third phase point $\Gamma_2(t)$, which is constrained such that the vector between Γ_0 and Γ_1, $\delta_1 = \Gamma_1 - \Gamma_0$, is always orthogonal to the vector between Γ_0 and Γ_2, $\delta_2 = \Gamma_2 - \Gamma_0$, then we can follow the rate of change of an infinitesimal two-dimensional phase-space area, $V_2(t) = \delta_2(t) \cdot \delta_1(t)$. The rate of change of the volume element is given by:

$$V_2(t) = V_2(0) \exp[(\lambda_1 + \lambda_2)t]. \tag{3.83}$$

As the value of λ_1 is known, this defines the second largest Lyapunov exponent λ_2. In a similar way, if we construct a third phase-space vector $\delta_3(t)$ which is constrained to be orthogonal to both $\delta_1(t)$ and $\delta_2(t)$, then we can follow the rate of change of a three-dimensional volume element $V_3(t) = (\delta_1(t) \times \delta_2(t)) \cdot \delta_3(t)$ and calculate the third largest exponent λ_3:

$$V_3(t) = V_3(0) \exp((\lambda_1 + \lambda_2 + \lambda_3)t). \tag{3.84}$$

This construction can be generalized to calculate the full spectrum of Lyapunov exponents for an *N*-particle system. We consider the trajectory $\Gamma(t)$ of a dynamical system in phase space and study the convergence or divergence of neighboring trajectories by constructing a set of basis vectors (tangent vectors) which span phase space $\{\delta_1, \delta_2, \delta_3, \ldots\}$, where $\delta_i = \Gamma_i - \Gamma_0$. If the equation of motion for a trajectory is of the form:

$$\dot{\Gamma} = G(\Gamma), \tag{3.85}$$

then the equation of motion for the tangent vector δ_i is:

$$\dot{\delta}_i = \dot{\Gamma}_i - \dot{\Gamma}_0 = G(\Gamma_i) - G(\Gamma_0) = G(\Gamma_0 + \delta_i) - G(\Gamma_0)$$
$$= T(\Gamma) \cdot \delta_i + O(\delta_i^2). \tag{3.86}$$

Here $T(\Gamma)$ is the Jacobian matrix (or stability matrix $\partial G / \partial \Gamma$) for the system. If the magnitude of the tangent vector is small enough, the nonlinear terms in Equation (3.86) can be neglected. The formal solution of this equation is:

$$\delta_i(t) = \exp_L \left[\int_0^t ds\, T(s) \right] \delta_i(0) = L_\lambda(t)\delta_i(0), \tag{3.87}$$

which is closely related to the time evolution of a single phase point given by Equation (3.61) with B replaced by Γ, $\Gamma(t) = \exp(iLt)\Gamma(0)$. The subscript on the exponential means that it is a left-ordered exponential which we will meet later in Section 7.8. The time evolution operator $L_\lambda(t)$ gives the evolution of the set of basis vectors required for calculating the Lyapunov exponents. The mean exponential rate of growth of the ith tangent vector, gives the ith Lyapunov exponent:

$$\lambda_i(\Gamma(0), \delta_i(0)) = \lim_{t \to \infty} \frac{1}{t} \ln \frac{\|\delta_i(t)\|}{\|\delta_i(0)\|}. \tag{3.88}$$

The existence of the limit is ensured by the multiplicative ergodic theorem of - Oseledec (1968) (see also Eckmann and Ruelle, 1985). The Lyapunov exponents can be ordered $\lambda_1 > \lambda_2 > \ldots > \lambda_M$ and, if the system is ergodic, the exponents are independent of the initial phase $\Gamma(0)$ and the initial phase space separation $\delta_i(0)$. The full set of Lyapunov exponents for a system is called the Lyapunov spectrum.

If we consider the volume element V_N where N is the dimension of phase space, then we can show that the phase-space compression factor gives the rate of change of phase-space volume, and that this is simply related to the sum of the Lyapunov exponents by:

$$\dot{V}_N = \left\langle \frac{\partial}{\partial \Gamma} \cdot \dot{\Gamma} \right\rangle V_N = \left(\sum_{i=1}^{N} \lambda_i \right) V_N. \tag{3.89}$$

For a Hamiltonian system, the phase-space compression factor is identically zero, so the phase-space volume is conserved. This is a simple consequence of Liouville's theorem. From Equation (3.89) it follows that the sum of the Lyapunov exponents for a Hamiltonian is also equal to zero. If the system is time reversal symmetric, the Lyapunov exponents occur in pairs $(-\lambda_i, \lambda_i)$. This ensures that $\delta(t)$, $V_2(t)$, $V_3(t)$, etc. change at the same rate with both forward and backward time evolution. A system is said to be chaotic if it has at least one positive Lyapunov exponent. In Chapters 7 and 8 we will return to consider Lyapunov exponents in both equilibrium and nonequilibrium systems.

3.5 Equilibrium time-correlation functions

We shall often refer to averages over equilibrium distribution functions f_0 (we use the subscript zero to denote equilibrium, which should not be confused with $f(0)$, a distribution function at $t = 0$). Distribution functions are called equilibrium if they pertain to steady, unperturbed equations of motion and they have no explicit time dependence. An equilibrium distribution function satisfies a Liouville equation of the form:

$$\frac{\partial}{\partial t} f_0 = -i\mathscr{L} f_0 = 0. \tag{3.90}$$

This implies that the equilibrium average of any phase variable is a stationary quantity. That is, for an arbitrary phase variable B, using Equation (3.62)

$$\frac{d}{dt}\langle B(t)\rangle_0 = \frac{d}{dt}\int d\Gamma f_0(\Gamma)\exp[iLt]B(\Gamma) = \int d\Gamma f_0(\Gamma)\frac{\partial}{\partial t}\exp(iLt)B(\Gamma)$$

$$= \int d\Gamma f_0(\Gamma)iL\exp[iLt]B(\Gamma) = -\int d\Gamma (i\mathscr{L} f_0(\Gamma))\exp(iLt)B(\Gamma) = 0. \tag{3.91}$$

We will often need to calculate the equilibrium time-correlation function of a phase variable A with another phase variable B at some other time. We define the equilibrium time-correlation function of A and B by:

$$C_{AB}(t) \equiv \int d\Gamma f_0 B^* \exp[iLt]A = \langle A(t)B^*\rangle_0, \tag{3.92}$$

where B^* denotes the complex conjugate of the phase variable B. Sometimes we will refer to the autocorrelation function of a phase variable A. If this variable is real, one can form a simple graphical representation of how such functions are calculated (see Figure 3.4).

Because the averages are to be taken over a stationary equilibrium distribution function, time-correlation functions are only sensitive to time difference between which A and B are evaluated. $C_{AB}(t)$ is independent of the particular choice of the time origin. If $i\mathscr{L}$ *generates* the distribution function f_0, then the propagator

$\exp(-i\mathscr{L}t)$ preserves f_0. (The converse is not necessarily true.) To be more explicit $f_0(t_1) = \exp(-i\mathscr{L}t_1)f_0 = f_0$, so that $C_{AB}(t)$ becomes:

$$C_{AB}(t) = \int d\Gamma f_0 B^* \exp[iLt]A = \int d\Gamma f_0(t_1)B^* \exp[iLt]A$$

$$= \int d\Gamma(\exp[-i\mathscr{L}t_1]f_0)B^* \exp[iLt_1]A = \int d\Gamma f_0(\exp[iLt_1]B^*)\exp[iL(t_1 + t)]A$$

$$= \int d\Gamma f_0 A(t_1 + t)B^*(t_1). \tag{3.93}$$

In deriving the last form of Equation (3.93), we have used the important fact that since $iL \equiv \dot{\Gamma} \cdot (\partial/\partial\Gamma)$ and the equations of motion are real, it follows that L is pure imaginary. Thus, $(iL)^* = iL$ and $(\exp[iLt])^* = \exp[iLt]$. Comparing Equation (3.93) with the definition of $C_{AB}(t)$ above, we see that the equilibrium time-correlation function is independent of the choice of time origin. It is solely a function of the difference in time of the two arguments, A and B. A further identity follows from this result if we choose $t_1 = -t$. We find that:

$$C_{AB}(t) = \langle A(t)B^*(0)\rangle_0 = \langle A(0)B^*(-t)\rangle_0, \tag{3.94}$$

so that:

$$C^*_{AB}(t) = \langle A^*B(-t)\rangle = C_{BA}(-t), \tag{3.95}$$

or using the notation of Section 3.3:

$$\left(\int d\Gamma f_0 B^* \exp[iLt]A\right)^* = \left(\int d\Gamma A\exp[-i\mathscr{L}t](f_0 B^*)\right)^*$$

$$= \left(\int d\Gamma f_0 A\exp[-iLt]B^*\right)^* = \int d\Gamma f_0 A^*\exp[-iLt]B. \tag{3.96}$$

The second equality in Equation (3.96) follows by expanding the operator $\exp(-iLt)$ and repeatedly applying the identity:

$$i\mathscr{L}(f_0 B^*) = \frac{\partial}{\partial\Gamma} \cdot (\dot{\Gamma}f_0 B^*) = B^*\frac{\partial}{\partial\Gamma} \cdot (\dot{\Gamma}f_0) + f_0\dot{\Gamma} \cdot \frac{\partial}{\partial\Gamma}B^*$$

$$= B^*i\mathscr{L}f_0 + f_0 iLB^* = f_0 iLB^*.$$

The term $i\mathscr{L}f_0$ is zero from Equation (3.90).

Over the scalar product defined by Equation (3.92), L is an Hermitian operator. The *Hermitian* adjoint of L denoted L^* can be defined by the equation:

$$\left(\int d\Gamma f_0 B^* \exp[iLt]A\right)^* \equiv \int d\Gamma f_0 A^* \exp[-iL^\dagger t]B \tag{3.97}$$

Comparing Equation (3.97) with Equation (3.96) we see two things: we see that the Liouville operator L is self-adjoint or Hermitian ($L = L^\dagger$); and therefore the propagator $\exp[iLt]$, is *unitary*. This result stands in contrast to those of Section 3.3, for arbitrary distribution functions.

We can use the autocorrelation function of A to define a norm in Liouville space. This length or norm of a phase variable A, is defined by the equation:

$$\|A\|^2 = \int d\Gamma f_0 A(\Gamma) A^*(\Gamma) = \int d\Gamma f_0 |A(\Gamma)|^2 = \left\langle |A(\Gamma)|^2 \right\rangle_0 \geq 0 \qquad (3.98)$$

We can see immediately that the norm of any phase variable is time independent because:

$$\|A(t)\|^2 = \int d\Gamma f_0 A(t) A^*(t) = \int d\Gamma f_0 [\exp[iLt] A(\Gamma)][\exp[iLt] A^*(\Gamma)]$$

$$= \int d\Gamma f_0 \exp[iLt](A(\Gamma) A^*(\Gamma))$$

$$= \int d\Gamma (\exp[-i\mathscr{L}t] f_0) |A|^2 = \|A(0)\|^2. \qquad (3.99)$$

The propagator is said to be norm-preserving. This is a direct result of the fact that the propagator is a unitary operator. The propagator can be thought of as a rotation operator in Liouville space (Figure 3.5).

A phase variable whose norm is unity is said to be normalized. The scalar product, (A, B^*) of two phase variables A, B is simply the equilibrium average of A and B^* namely $\langle AB^* \rangle_0$. The norm of a phase variable is simply the scalar product of the variable with itself. The autocorrelation function $C_{AA}(t)$ has a zero time value which is equal to the norm of A. The propagator increases the angle

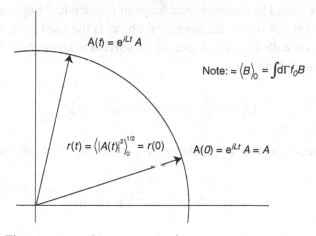

Figure 3.5 The propagator is norm-preserving

between A^* and $A(t)$, and the scalar product, which is the projection of $A(t)$ along A^*, therefore decreases. The autocorrelation function of a given phase variable therefore measures the rate at which the $6N$-dimensional rotation occurs.

We will now derive some relations for the time derivatives of time-correlation functions. It is easy to see that:

$$\frac{d}{dt} C_{AB}(t) = \frac{d}{dt} \int d\Gamma f_0 (\exp[iLt]A(\Gamma))B^*(\Gamma)$$

$$= \int d\Gamma f_0 (iL \exp[iLt]A(\Gamma))B^*(\Gamma) = C_{\dot{A}B}(t)$$

$$= -\int d\Gamma (\exp[iLt]A(\Gamma))i\mathscr{L}(f_0 B^*(\Gamma))$$

$$= -\int d\Gamma A(t)\frac{\partial}{\partial\Gamma}\left(\dot{\Gamma} f_0 B^*(\Gamma)\right)$$

$$= -\int d\Gamma f_0 (\exp[iLt]A(\Gamma))iLB^*(\Gamma)$$

$$= -\int d\Gamma f_0 (\exp[iLt]A(\Gamma))\dot{B}^*(\Gamma) = -C_{A\dot{B}}(t). \qquad (3.100)$$

3.6 Operator identities

In this section we develop the operator algebra that we will need to manipulate expressions containing Liouvilleans and their associated propagators. Most of the identities which we obtain are valid for arbitrary time-independent operators. Thus far we have been dealing with propagators in the time domain. For many problems it is more useful to consider their frequency dependent Laplace, or Fourier–Laplace, transforms. A useful mathematical object is the Laplace transform of the propagator. This is called the resolvent. The resolvent is an operator in the domain of the Laplace transform variable s:

$$G(s) = \int_0^\infty dt \exp[-st]\exp[-iLt]. \qquad (3.101)$$

Our first operator identity is obtained by considering two arbitrary operators A and B:

$$(A + B)^{-1} = A^{-1} - A^{-1}B(A + B)^{-1}. \qquad (3.102)$$

This identity is easily verified by operating from the right-hand side of this equation with $(A + B)$, so:

$$(A + B)^{-1}(A + B) = \left[A^{-1} - A^{-1}B(A + B)^{-1}\right](A + B)$$
$$= A^{-1}(A + B) - A^{-1}B$$
$$= A^{-1}A + A^{-1}B - A^{-1}B = I. \tag{3.103}$$

The operator expression $(A + B)^{-1}$ is the inverse of the operator $(A + B)$. To interpret an operator inverse of $(A + B)^{-1}$, we use the series expansion

$$(I + A)^{-1} = \sum_{n=0}^{\infty}(-A)^n. \tag{3.104}$$

First we prove that the right-hand side of this expression is indeed the inverse of the operator $(I + A)$. To do this consider:

$$\sum_{n=0}^{\infty}(-A)^n(I + A) = \sum_{n=0}^{\infty}(-A)^n - \sum_{n=1}^{\infty}(-A)^n = I, \tag{3.105}$$

so that this series expansion allows us to represent the inverse of $(I + A)$ in terms of an infinite series of products of the operator A.

The Dyson decomposition of propagators

Now we can investigate the Laplace transform (or resolvent) of the exponential of an operator in more detail. We use the expansion of the exponential to show that:

$$\int_0^{\infty} dt \exp[-st]\exp[-At] = \int_0^{\infty} dt \exp[-st]\sum_{n=0}^{\infty}\frac{(-)^n}{n!}(At)^n = \sum_{n=0}^{\infty}\frac{(-)^n}{n!}A^n\int_0^{\infty} dt \exp[-st]t^n$$
$$= \sum_{n=0}^{\infty}(-)^n\frac{A^n}{s^{n+1}} = \frac{1}{s}\left(1 + \frac{A}{s}\right)^{-1} = (s + A)^{-1}. \tag{3.106}$$

This means that the resolvent of the operator, $\exp[-At]$, is simply $(s + A)^{-1}$. We can now consider the resolvent derived from the operator $(A + B)$, and using the first identity above, relate this resolvent to the resolvent of A. We can write:

$$(s + A + B)^{-1} = (s + A)^{-1} - (s + A)^{-1}B(s + A + B)^{-1}. \tag{3.107}$$

Substituting the Laplace integrals for the operators $(s + A)^{-1}$ and $(s + A + B)^{-1}$ into this equation gives:

$$\int_0^\infty dt \exp[-st]\exp[-(A+B)] = \int_0^\infty dt \exp[-st]\exp[-At]$$

$$- \int_0^\infty dt_1 \exp[-st_1]\exp[-At_1]B$$

$$\times \int_0^\infty dt_2 \exp[-st_2]\exp[-(A+B)t_2]$$

$$= \int_0^\infty dt \exp[-st]\Big\{ \exp[-At]$$

$$- \int_0^t dt_1 \exp[-At_1]B\exp[-(A+B)(t-t_1)]\Big\} \quad (3.108)$$

As the equality holds for all values of s, the integrands must be equal, so:

$$\exp[-(A + B)] = \exp[-At] - \int_0^t dt_1 \exp[-At_1]B\exp[-(A+B)(t - t_1)] \quad (3.109)$$

This result is a very important step towards understanding the relationship between different propagators and is referred to as the *Dyson decomposition* when applied to propagators (Dyson, 1949 and Feynman, 1951). The derivation that we have used here is only valid if both of the operators A and B have no explicit time dependence. If we consider the propagators $\exp((A + B)t)$ and $\exp(At)$, then a second Dyson decomposition can be obtained:

$$\exp[(A + B)t] = \exp[At]$$

$$+ \int_0^t dt_1 \exp[At_1]B\exp[(A + B)(t - t_1)] \quad (3.110)$$

It is handy to use a graphical shorthand for the Dyson equation. Using this shorthand notation these two equations become:

$$\Leftarrow \quad = \quad \leftarrow \quad - \quad \Leftarrow (\otimes - \oplus) \leftarrow, \quad (3.111)$$

and

$$\Rightarrow \quad = \quad \rightarrow \quad + \quad \Rightarrow (\otimes - \oplus) \rightarrow . \quad (3.112)$$

The symbol $\otimes$ denotes the $(A + B)$-Liouvillean and $\oplus$ denotes the A-Liouvillean; the arrows $\Leftarrow$ and $\Rightarrow$ denote the propagators $\exp(-(A + B)t)$ and $\exp[(A + B)t]$ respectively, while $\leftarrow$ and $\rightarrow$ denote $\exp[-At]$ and $\exp[At]$ respectively. Any chain of arrows denotes a convolution over all intermediate times.

As an example of the application of this result, consider the case where B is a small perturbation to the operator A. In this case the Dyson decomposition gives the full $(A + B)$-propagator as the sum of the unperturbed A-propagator plus a correction term. One often faces the situation where we want to compare the operation of different propagators on either a phase variable or a distribution function. For example, one might like to know the difference between the value of a phase variable $A(\Gamma)$ propagated under the combined influence of the N-particle interactions *and* an applied external field F_e, with the value the phase variable might take at the same time in the absence of the external field. In that case (Evans and Morriss, 1984a):

$$A(t, F_e) ==\Rightarrow A(\Gamma)$$

$$= \{\rightarrow + \rightarrow (\otimes - \oplus) \rightarrow + \rightarrow (\otimes - \oplus) \rightarrow (\otimes - \oplus) \rightarrow + \cdots\} A(\Gamma). \tag{3.113}$$

Therefore we can write:

$$\Rightarrow A(\Gamma) = \sum_{n=0}^{\infty} \rightarrow [(\otimes - \oplus) \rightarrow]^n A(\Gamma). \tag{3.114}$$

This equation is of limited usefulness because, in general, $\otimes$ and $\rightarrow$ do not commute. This means that the Liouvillean $\otimes$ is locked inside a convolution of propagators with which it does not commute. A more useful expression can be derived from Equation (3.112) by realizing that $\otimes$ commutes with its own propagator $\otimes \Rightarrow = \Rightarrow \otimes$. Similarly $\oplus$ commutes with its own propagator $\oplus \rightarrow = \rightarrow \oplus$. We can *unlock* the respective Liouvilleans from the chain in Equation (3.112) by writing:

$$\Rightarrow = \rightarrow + \otimes \rightarrow \rightarrow - \rightarrow \rightarrow \oplus. \tag{3.115}$$

We can recursively substitute for $\Rightarrow$, yielding

$$\Rightarrow = \rightarrow + \otimes \rightarrow \rightarrow - \rightarrow \rightarrow \oplus$$
$$+ \otimes \otimes \rightarrow \rightarrow \rightarrow - 2\otimes \rightarrow \rightarrow \rightarrow \oplus + \rightarrow \rightarrow \rightarrow \oplus \oplus + \cdots \tag{3.116}$$

Now it is easy to show that, $(\rightarrow)^n = (t^n/n!) \rightarrow$. Thus Equation (3.116) can be written as:

$$\Rightarrow - \{1 + t(\otimes - \otimes) + \frac{t^2}{2!}(\otimes \otimes - 2 \otimes \oplus + \oplus \oplus)$$

$$+ \frac{t^3}{3!}(\otimes \otimes \otimes - 3 \otimes \otimes \oplus + 3 \otimes \oplus \oplus - \oplus \oplus \oplus) + \cdots\} \rightarrow \tag{3.117}$$

This equation was first derived by Evans and Morriss (1984a). Its utility arises from the fact that by *unrolling* the Liouville operators to the left and the propagator to the right, *explicit* formulae for the expansion can usually be derived. A limitation of the formula is that successive terms on the right-hand side do *not* constitute a power series expansion of the difference in the two propagators in powers of the difference between the respective Liouvilleans. To be more explicit, the term,

$\frac{t^3}{3!}(\otimes \otimes \otimes - 3 \otimes \otimes \oplus + 3 \otimes \oplus \oplus - \oplus \oplus \oplus)$ is *not*, in general, of order $(\otimes - \oplus)^3$.

Campbell–Baker–Hausdorff theorem

If A and B are noncommuting operators, then the operator, expression $\exp(A)$ $\exp(B)$ can be written in the form $\exp(C)$ where C is given by:

$$C = A + B + \frac{1}{2}[A, B] + \frac{1}{12}\{[[A, B], B] + [[B, A], A]\} + \dots . \qquad (3.118)$$

The notation [,] is the usual quantum mechanical commutator. A rearrangement of this expansion, known as the Magnus expansion is well known to quantum theorists (Magnus, 1954). Any finite truncation of the Magnus expansion for the time-displacement operator, gives a unitary time-displacement operator approximation (Pechukas and Light, 1966). This result has not proved as useful for nonequilibrium statistical mechanics as it has for quantum theory. We give it here mainly for the sake of completeness.

3.7 The Irving–Kirkwood procedure

In Chapter 2 we gave a brief outline of the structure of macroscopic hydrodynamics. We saw that, given appropriate boundary conditions, it is possible to use the Navier–Stokes equations to describe the resulting macroscopic flow patterns. In this chapter we began the microscopic description of nonequilibrium systems using the Liouville equation. We will now follow a procedure first outlined by Irving and Kirkwood (1950), to derive microscopic expressions for the thermodynamic forces and fluxes appearing in the phenomenological equations of hydrodynamics.

In our treatment of the macroscopic equations we stressed the role played by the densities of conserved quantities. Our first task here will be to define microscopic expressions for the local densities of mass, momentum, and energy. If the mass of the individual atoms in our system is m, then the mass per unit volume at a position $\mathbf{r}$ and time t can be obtained by taking an appropriate average over the normalised N-particle distribution function $f(\Gamma, t)$. To specify that particle i is at the position $\mathbf{r}_i$, we will use a delta function, $\delta(\mathbf{r} - \mathbf{r}_i)$. We will assume that particle dynamics are given by field-free Newtonian equations of motion.

The mass density $\rho(\mathbf{r}, t)$ can be calculated from the following average:

$$\rho(\mathbf{r}, t) = \int d\Gamma\, f(\Gamma, t) \sum_i m\delta(\mathbf{r} - \mathbf{r}_i)$$

$$= \int d\Gamma\, f(\Gamma, 0) \sum_i m\delta(\mathbf{r} - \mathbf{r}_i(t)) = \left\langle \sum_i m\big|_{\mathbf{r}_i(t)=\mathbf{r}} \right\rangle, \qquad (3.119)$$

where the time dependence in $\rho(\mathbf{r}, t)$ is initially due to the time dependence of the distribution function $f(\Gamma, t)$ (the $\mathbf{r}_i$s are integrated in the integral over Γ). Thus in the first line of Equation (3.119) we have the Schrödinger representation of the mass density, while the second and third equalities are Heisenberg representations, where the time dependence is transferred to the phase variable. We have already seen the equivalence of these two representations in Section 3.3. Note that the propagator advances particle positions, but has no effect on the position of the fluid element at $\mathbf{r}$.

The momentum density $\rho(\mathbf{r}, t)\mathbf{u}(\mathbf{r}, t)$, and total energy density $\rho(\mathbf{r}, t)\, e(\mathbf{r}, t)$, can be defined in an analogous manner:

$$\rho(\mathbf{r}, t)\mathbf{u}(\mathbf{r}, t) = \int d\Gamma f(\Gamma, t) \sum_i m\dot{\mathbf{r}}_i\delta(\mathbf{r} - \mathbf{r}_i)$$

$$= \int d\Gamma f(\Gamma, t) \sum_i \mathbf{p}_i\delta(\mathbf{r} - \mathbf{r}_i) = \left\langle \sum_i \mathbf{p}_i(t)\big|_{\mathbf{r}_i(t)=\mathbf{r}} \right\rangle, \qquad (3.120)$$

$$\rho(\mathbf{r}, t)e(\mathbf{r}, t) = \int d\Gamma f(\Gamma, t)e_i\, \delta(\mathbf{r} - \mathbf{r}_i) = \langle e_i|_{\mathbf{r}_i(t)=\mathbf{r}} \rangle. \qquad (3.121)$$

In these equations $\dot{\mathbf{r}}_i$ is the velocity of particle i, $\mathbf{p}_i$ is its momentum, $\mathbf{r}_{ij} \equiv \mathbf{r}_j - \mathbf{r}_i$, and we assume that the total potential energy of the system, Φ, is pair-wise additive and can be written as:

$$\Phi = \frac{1}{2}\sum_{i \neq j} \phi_{ij}. \qquad (3.122)$$

We arbitrarily assign one half of the potential energy to each of the two particles which contribute ϕ_{ij} to the total potential energy of the system, and $e_i = \frac{1}{2}m_i\dot{\mathbf{r}}_i^2 + \frac{1}{2}\sum_j \psi(\mathbf{r}_{ij})$ is the energy of particle i.

The conservation equations involve time derivatives of the averages of the densities of conserved quantities. We begin by calculating the time derivative of the mass density. Using Equations (3.56) and (3.62) it can be shown that:

$$\frac{\partial}{\partial t}\rho(\mathbf{r}, t) = \int d\Gamma \frac{\partial f(\Gamma, t)}{\partial t} \sum_i m\delta(\mathbf{r} - \mathbf{r}_i)$$

$$= -\int d\Gamma \sum_i m\delta(\mathbf{r} - \mathbf{r}_i) i\mathscr{L}f(\Gamma, t)$$

$$= \int d\Gamma f(\Gamma, t) iL \sum_i m\delta(\mathbf{r} - \mathbf{r}_i)$$

$$= \int d\Gamma f(\Gamma, t) \sum_i m\dot{\mathbf{r}}_i \cdot \frac{\partial \delta(\mathbf{r} - \mathbf{r}_i)}{\partial \mathbf{r}_i}$$

$$= -\int d\Gamma f(\Gamma, t) \sum_i m\dot{\mathbf{r}}_i \cdot \frac{\partial \delta(\mathbf{r} - \mathbf{r}_i)}{\partial \mathbf{r}}$$

$$= -\nabla \cdot \int d\Gamma f(\Gamma, t) \sum_i m\dot{\mathbf{r}}_i \delta(\mathbf{r} - \mathbf{r}_i)$$

$$= -\nabla \cdot [\rho(\mathbf{r}, t)\mathbf{u}(\mathbf{r}, t)]. \qquad (3.123)$$

The fifth equality follows from the important delta function identity:

$$\frac{\partial}{\partial \mathbf{r}_i} \delta(\mathbf{r} - \mathbf{r}_i) \equiv -\frac{\partial}{\partial \mathbf{r}} \delta(\mathbf{r} - \mathbf{r}_i). \qquad (3.124)$$

We have shown that the time derivative of the mass density yields the mass continuity Equation (2.4) as expected, and gives a microscopic representation for the momentum density. Strictly speaking therefore, we did not really need to define the momentum density in Equation (3.120), as the mass density definition, combined with the mass continuity equation, determines the microscopic expression for the momentum density.

3.8 Instantaneous microscopic representation of fluxes

The Irving–Kirkwood procedure has given us microscopic expressions for the thermodynamic fluxes in terms of ensemble averages. At equilibrium in a uniform fluid, the Irving–Kirkwood expression for the pressure tensor is the same expression as that derived using Gibbs' ensemble theory for equilibrium statistical mechanics.

In this section we derive *instantaneous* expressions for the fluxes (Todd and Evans, 1995; Todd *et al.*, 1995; Monaghan and Morriss, 1997) rather than the ensemble based, Irving–Kirkwood expressions. The reason for considering instantaneous expressions is two-fold. The fluxes are based upon conservation laws and these laws are valid instantaneously for every member of the ensemble – they do not require ensemble averaging. Second, most computer simulation involves calculating properties of a system from a single trajectory. Ensemble averaging is almost never used because it is relatively expensive in computer time. The ergodic hypothesis, that the result obtained by ensemble averaging is equal to that obtained by time averaging the same property along a single phase-space trajectory, implies that one should be able to develop expressions for the fluxes which do not require ensemble averaging. For this to be practically realizable it is clear that the mass, momentum, and energy densities must be definable at each instant along the trajectory.

We consider a single phase-space trajectory evolving in time and define the mass density for a fluid element at time t to be the mass of the element divided by its volume. In the quasi-microscopic picture we imagine that the mass of the element is the sum of the masses of the atoms it contains. However, in the microscopic picture the mass density at point $\mathbf{r}$ is zero if there is no particle at position $\mathbf{r}$, and infinite if there is a particle at $\mathbf{r}$. Therefore we write the mass density as:

$$\rho(\mathbf{r}, t) = \sum_i m_i \delta(\mathbf{r} - \mathbf{r}_i(t)). \tag{3.125}$$

Integrating this expression over a fluid element gives the mass of the fluid element. The LHS is a macroscopic hydrodynamic quantity for a single system and the RHS is its microscopic representation. It is important to realize that this definition is consistent with the Eulerian picture in that position $\mathbf{r}$ is fixed in space, and the only time dependence in the RHS is the time dependence of $\mathbf{r}_i(t)$ which arises through the motion of the particles. Substituting this mass density into the LHS of the mass conservation Equation (2.4) gives:

$$\frac{\partial}{\partial t} \rho(\mathbf{r}, t) = \frac{\partial}{\partial t} \sum_i m_i \delta(\mathbf{r} - \mathbf{r}_i(t)) = \sum_i m_i \frac{\partial \mathbf{r}_i}{\partial t} \frac{\partial}{\partial \mathbf{r}_i} \delta(\mathbf{r} - \mathbf{r}_i(t))$$

$$= -\sum_i m_i \dot{\mathbf{r}}_i \frac{\partial}{\partial \mathbf{r}} \delta(\mathbf{r} - \mathbf{r}_i(t)) = -\frac{\partial}{\partial \mathbf{r}} \sum_i m_i \dot{\mathbf{r}}_i \delta(\mathbf{r} - \mathbf{r}_i(t)), \tag{3.126}$$

and comparing with the RHS of Equation (2.4), we see that the instantaneous momentum density $\mathbf{J}(\mathbf{r}, t)$ is:

$$\rho(\mathbf{r}, t) \mathbf{u}(\mathbf{r}, t) = \mathbf{J}(\mathbf{r}, t) = \sum_i m_i \dot{\mathbf{r}}_i \delta(\mathbf{r} - \mathbf{r}_i). \tag{3.127}$$

There is no instantaneous representation for the streaming velocity $\mathbf{u}(\mathbf{r}, t)$ at the particle level that can be constructed from the instantaneous representations for $\rho(\mathbf{r}, t)$ and $\mathbf{J}(\mathbf{r}, t)$. Taking the ratio $\mathbf{J}(\mathbf{r}, t)/\rho(\mathbf{r}, t)$ would give the streaming velocity to be the particle velocity at the position of each particle, and undefined elsewhere. This is not a useful definition of the streaming velocity. Any realistic representation for streaming velocity $\mathbf{u}(\mathbf{r}, t)$ of a fluid element necessarily involves some form of coarse graining, either in space or time. Once a streaming velocity has been determined (by whatever means), we can divide the laboratory velocity of each particle into a thermal part $\mathbf{v}_i$ and a streaming part $\mathbf{u}(\mathbf{r}_i, t)$. That is, the laboratory velocity is:

$$\dot{\mathbf{r}}_i = \mathbf{v}_i + \mathbf{u}(\mathbf{r}_i, t). \tag{3.128}$$

Using this representation, we can write the instantaneous momentum density as:

$$\mathbf{J}(\mathbf{r}, t) = \sum_i m_i \mathbf{v}_i \delta(\mathbf{r} - \mathbf{r}_i) + \rho(\mathbf{r}, t)\mathbf{u}(\mathbf{r}, t). \tag{3.129}$$

We see immediately, from the definition of $\mathbf{J}(\mathbf{r}, t)$, that the thermal velocities do not contribute to the momentum current and:

$$\sum_i m_i \mathbf{v}_i \delta(\mathbf{r} - \mathbf{r}_i) = 0. \tag{3.130}$$

A key step in all the microscopic derivations is that if $f(\mathbf{r})$ is a simple function of $\mathbf{r}$ (that is, not an operator) then $f(\mathbf{r})\delta(\mathbf{r} - \mathbf{r}_i) \equiv f(\mathbf{r}_i)\delta(\mathbf{r} - \mathbf{r}_i)$. There are some subtle points with regard to the interpretation of Equation (3.129). Both $\mathbf{J}(\mathbf{r}, t)$ and $\rho(\mathbf{r}, t)\mathbf{u}(\mathbf{r}, t)$ are macroscopic quantities, defined for a fluid element, and are numerically equal. This implies that at the fluid-element level Equation (3.130) is equal to zero. Clearly this separation between thermal and streaming parts is physically correct. We can also interpret this equation as a condition that the streaming velocity of a fluid element must satisfy. Thus:

$$\mathbf{u}(\mathbf{r}) = \frac{\sum_{i \in E(\mathbf{r})} m_i \dot{\mathbf{r}}_i \delta(\mathbf{r} - \mathbf{r}_i)}{\sum_{i \in E(\mathbf{r})} m_i \delta(\mathbf{r} - \mathbf{r}_i)}, \tag{3.131}$$

where the summation is over particles within fluid element $E(\mathbf{r})$, but this form for $\mathbf{u}(\mathbf{r})$ will change discontinuously as particles enter or leave the fluid element. The approach that we will adopt in the formal derivations that follow is to derive the microscopic representations using the full particle velocities and afterwards make the separation into thermal and streaming components.

We will now use exactly the same procedure to differentiate the instantaneous momentum density:

$$\frac{\partial}{\partial t}[\rho(\mathbf{r}, t)\mathbf{u}(\mathbf{r}, t)] = \frac{\partial}{\partial t} \sum_i m_i \dot{\mathbf{r}}_i \delta(\mathbf{r} - \mathbf{r}_i)$$

$$= -\frac{\partial}{\partial \mathbf{r}} \cdot \sum_i m_i \dot{\mathbf{r}}_i \dot{\mathbf{r}}_i \delta(\mathbf{r} - \mathbf{r}_i) + \sum_i m_i \ddot{\mathbf{r}}_i \delta(\mathbf{r} - \mathbf{r}_i). \tag{3.132}$$

If we consider the first term on the right-hand side then:

$$\sum_i m_i \dot{\mathbf{r}}_i \dot{\mathbf{r}}_i \delta(\mathbf{r} - \mathbf{r}_i) = \sum_i m_i (\mathbf{v}_i + \mathbf{u}(\mathbf{r}_i, t))(\mathbf{v}_i + \mathbf{u}(\mathbf{r}_i, t))\delta(\mathbf{r} - \mathbf{r}_i)$$

$$= \sum_i m_i \mathbf{v}_i \mathbf{v}_i \delta(\mathbf{r} - \mathbf{r}_i) + \sum_i m_i \mathbf{v}_i \mathbf{u}(\mathbf{r}_i, t)\delta(\mathbf{r} - \mathbf{r}_i)$$

$$+ \sum_i m_i \mathbf{u}(\mathbf{r}_i, t)\mathbf{v}_i \delta(\mathbf{r} - \mathbf{r}_i) + \sum_i m_i \mathbf{u}(\mathbf{r}_i, t)\mathbf{u}(\mathbf{r}_i, t)\delta(\mathbf{r} - \mathbf{r}_i)$$

$$= \sum_i m_i \mathbf{v}_i \mathbf{v}_i \delta(\mathbf{r} - \mathbf{r}_i) + \mathbf{u}(\mathbf{r}, t)\sum_i m_i \mathbf{v}_i \delta(\mathbf{r} - \mathbf{r}_i)$$

$$+ \mathbf{u}(\mathbf{r}, t)\sum_i m_i \mathbf{v}_i \delta(\mathbf{r} - \mathbf{r}_i) + \mathbf{u}(\mathbf{r}, t)\mathbf{u}(\mathbf{r}, t)\sum_i m_i \delta(\mathbf{r} - \mathbf{r}_i)$$

$$= \sum_i m_i \mathbf{v}_i \mathbf{v}_i \delta(\mathbf{r} - \mathbf{r}_i) + \rho(\mathbf{r}, t)\mathbf{u}(\mathbf{r}, t)\mathbf{u}(\mathbf{r}, t). \tag{3.133}$$

We have used $\mathbf{u}(\mathbf{r}_i, t)\delta(\mathbf{r} - \mathbf{r}_i) \equiv \mathbf{u}(\mathbf{r}, t)\delta(\mathbf{r} - \mathbf{r}_i)$ to remove the particle index from the streaming velocity so it can be factored out of the summations. The term $\sum m_i \mathbf{v}_i \delta(\mathbf{r} - \mathbf{r}_i) = 0$ from Equation (3.130) and the remaining summation is equal to the mass density. Combining these results it follows that:

$$\frac{\partial}{\partial t}[\rho(\mathbf{r}, t)\mathbf{u}(\mathbf{r}, t)] = -\frac{\partial}{\partial \mathbf{r}} \cdot \left(\sum_i m_i \mathbf{v}_i \mathbf{v}_i \delta(\mathbf{r} - \mathbf{r}_i) + \rho(\mathbf{r}, t)\mathbf{u}(\mathbf{r}, t)\mathbf{u}(\mathbf{r}, t) \right)$$

$$+ \sum_i \mathbf{F}_i \delta(\mathbf{r} - \mathbf{r}_i). \tag{3.134}$$

We will now consider the second term on the right-hand side of this equation in some detail. A physical macroscopic property cannot depend upon the labels attached to individual particles, so we symmetrize summations by replacing $\sum_{i,j} a_i b_j$ by $\frac{1}{2}\sum_{i,j}(a_i b_j + a_j b_i)$. As the force on particle i due to particle j, $\mathbf{F}_{ij}$, is equal and opposite to the force on particle j due to particle i, $\mathbf{F}_{ji}$, we can write:

$$\sum_i \delta(\mathbf{r} - \mathbf{r}_i)\mathbf{F}_i = \sum_{i,j} \delta(\mathbf{r} - \mathbf{r}_i)\mathbf{F}_{ij} = \frac{1}{2}\sum_{i,j}\left[\delta(\mathbf{r} - \mathbf{r}_i)\mathbf{F}_{ij} + \delta(\mathbf{r} - \mathbf{r}_j)\mathbf{F}_{ji}\right]$$

$$= \frac{1}{2}\sum_{i,j}\left[\delta(\mathbf{r} - \mathbf{r}_i) - \delta(\mathbf{r} - \mathbf{r}_j)\right]\mathbf{F}_{ij}. \tag{3.135}$$

The first equality introduces j, the second symmetrizes the sum, and the third uses the fact that the forces are equal and opposite. The difference between two delta functions can be written in a very convenient integral form (Noll, 1955):

$$\delta(\mathbf{r} - \mathbf{r}_i) - \delta(\mathbf{r} - \mathbf{r}_j) = \frac{\partial}{\partial \mathbf{r}} \cdot \mathbf{r}_{ij} \int_0^1 d\lambda \delta(\mathbf{r} - \mathbf{r}_i - \lambda \mathbf{r}_{ij}). \tag{3.136}$$

Using this equation for the difference of the two delta functions $\delta(\mathbf{r} - \mathbf{r}_i)$ and $\delta(\mathbf{r} - \mathbf{r}_j)$ leads to:

$$\frac{\partial}{\partial t}[\rho(\mathbf{r}, t)\mathbf{u}(\mathbf{r}, t)]$$

$$= -\frac{\partial}{\partial \mathbf{r}} \cdot \left(\sum_i m_i \mathbf{v}_i \mathbf{v}_i \delta(\mathbf{r} - \mathbf{r}_i) + \rho(\mathbf{r}, t)\mathbf{u}(\mathbf{r}, t)\mathbf{u}(\mathbf{r}, t) \right)$$

$$+ \frac{1}{2}\frac{\partial}{\partial \mathbf{r}} \cdot \sum_{i,j} \mathbf{r}_{ij}\mathbf{F}_{ij} \int_0^1 d\lambda \delta(\mathbf{r} - \mathbf{r}_i - \lambda \mathbf{r}_{ij}). \tag{3.137}$$

Comparing this equation with the momentum conservation Equation (2.12) (without external forces) we see that the microscopic representation of the pressure tensor is:

$$\mathbf{P}(\mathbf{r}, t) = \sum_i m_i \mathbf{v}_i \mathbf{v}_i \delta(\mathbf{r} - \mathbf{r}_i) - \frac{1}{2}\sum_{i,j} \mathbf{r}_{ij}\mathbf{F}_{ij} \int_0^1 d\lambda \delta(\mathbf{r} - \mathbf{r}_i - \lambda \mathbf{r}_{ij}). \tag{3.138}$$

This expression implies that if $\mathbf{r}$ is on the straight line from $\mathbf{r}_i$ to $\mathbf{r}_j$, then that term is included in the sum, otherwise it is not included.

We will now use the same technique to calculate the instantaneous microscopic expression for the heat flux vector from the energy density:

$$\rho(\mathbf{r}, t)e(\mathbf{r}, t) = \sum_i e_i \delta(\mathbf{r} - \mathbf{r}_i(t)),$$

where $e_i = \frac{1}{2}m_i\dot{\mathbf{r}}_i^2 + \frac{1}{2}\sum_j \phi(\mathbf{r}_{ij})$ is the energy of particle i. The partial time derivative of the energy density is then:

$$\frac{\partial}{\partial t}[\rho(\mathbf{r}, t)e(\mathbf{r}, t)] = \frac{\partial}{\partial t}\left(\sum_i e_i \delta(\mathbf{r} - \mathbf{r}_i) \right)$$

$$= \sum_i \dot{e}_i \delta(\mathbf{r} - \mathbf{r}_i) + \sum_i e_i \frac{\partial \mathbf{r}_i}{\partial t} \cdot \frac{\partial}{\partial \mathbf{r}_i}\delta(\mathbf{r} - \mathbf{r}_i)$$

$$= \sum_i \dot{e}_i \delta(\mathbf{r} - \mathbf{r}_i) - \frac{\partial}{\partial \mathbf{r}} \cdot \sum_i e_i\dot{\mathbf{r}}_i \delta(\mathbf{r} - \mathbf{r}_i). \tag{3.139}$$

In the second line of Equation (3.139), the gradient operator $\partial/\partial \mathbf{r}$ is contracted into the laboratory velocity $\dot{\mathbf{r}}_i$. The term:

$$\dot{e}_i = \sum_j \dot{\mathbf{r}}_i \cdot \mathbf{F}_{ij} - \frac{1}{2}\sum_j \left(\dot{\mathbf{r}}_i \cdot \mathbf{F}_{ij} + \dot{\mathbf{r}}_j \cdot \mathbf{F}_{ji}\right)$$

$$= \sum_j \left\{ \dot{\mathbf{r}}_i \cdot \mathbf{F}_{ij} - \frac{1}{2}(\dot{\mathbf{r}}_i \cdot \mathbf{F}_{ij} - \dot{\mathbf{r}}_j \cdot \mathbf{F}_{ij}) \right\} = \frac{1}{2}\sum_j \left\{ \dot{\mathbf{r}}_i \cdot \mathbf{F}_{ij} + \dot{\mathbf{r}}_j \cdot \mathbf{F}_{ij} \right\}.$$

Using our previous result for the difference of two delta functions, Equation (3.136), and symmetrizing gives:

$$\sum_i \dot{e}_i \delta(\mathbf{r} - \mathbf{r}_i) = \frac{1}{2}\sum_{i,j} \left(\dot{\mathbf{r}}_i \cdot \mathbf{F}_{ij} + \dot{\mathbf{r}}_j \cdot \mathbf{F}_{ij}\right)\delta(\mathbf{r} - \mathbf{r}_i)$$

$$= \frac{1}{2}\sum_{i,j} \left(\dot{\mathbf{r}}_i \cdot \mathbf{F}_{ij}\delta(\mathbf{r} - \mathbf{r}_i) + \dot{\mathbf{r}}_j \cdot \mathbf{F}_{ij}\delta(\mathbf{r} - \mathbf{r}_i)\right)$$

$$= \frac{1}{2}\sum_{i,j} \left(\dot{\mathbf{r}}_i \cdot \mathbf{F}_{ij}\delta(\mathbf{r} - \mathbf{r}_i) + \dot{\mathbf{r}}_i \cdot \mathbf{F}_{ji}\delta(\mathbf{r} - \mathbf{r}_j)\right)$$

$$= \frac{1}{2}\sum_{i,j} \dot{\mathbf{r}}_i \cdot \mathbf{F}_{ij}\left(\delta(\mathbf{r} - \mathbf{r}_i) - \delta(\mathbf{r} - \mathbf{r}_j)\right)$$

$$= \frac{1}{2}\frac{\partial}{\partial \mathbf{r}} \cdot \sum_{i,j} \mathbf{r}_{ij}(\mathbf{F}_{ij} \cdot \dot{\mathbf{r}}_i) \int_0^1 d\lambda \delta(\mathbf{r} - \mathbf{r}_i - \lambda \mathbf{r}_{ij}). \qquad (3.140)$$

From the energy conservation Equation (2.24) we conclude that:

$$\rho(\mathbf{r}, t)e(\mathbf{r}, t)\mathbf{u}(\mathbf{r}, t) + \mathbf{J}_Q(\mathbf{r}, t) + \mathbf{P}(\mathbf{r}, t) \cdot \mathbf{u}(\mathbf{r}, t)$$

$$= \sum_i e_i \dot{\mathbf{r}}_i \delta(\mathbf{r} - \mathbf{r}_i) - \frac{1}{2}\sum_{i,j} \mathbf{r}_{ij}(\mathbf{F}_{ij} \cdot \dot{\mathbf{r}}_i) \int_0^1 d\lambda \delta(\mathbf{r} - \mathbf{r}_i - \lambda \mathbf{r}_{ij}). \qquad (3.141)$$

Both sides of this equation contain components due to the streaming motion of the fluid. We can remove these by considering:

$$\sum_i e_i \dot{\mathbf{r}}_i \delta(\mathbf{r} - \mathbf{r}_i) = \sum_i e_i(\mathbf{v}_i + \mathbf{u}(\mathbf{r}_i))\delta(\mathbf{r} - \mathbf{r}_i)$$

$$= \sum_i e_i \mathbf{v}_i \delta(\mathbf{r} - \mathbf{r}_i) + \sum_i e_i \mathbf{u}(\mathbf{r}_i)\delta(\mathbf{r} - \mathbf{r}_i)$$

$$= \sum_i e_i \mathbf{v}_i \delta(\mathbf{r} - \mathbf{r}_i) + \rho(\mathbf{r}, t)e(\mathbf{r}, t)\mathbf{u}(\mathbf{r}, t). \qquad (3.142)$$

However, there are still kinetic streaming components remaining in e_i. Removing these by defining the internal energy of particle i to be $U_i = \frac{1}{2}m_i v_i^2 + \frac{1}{2}\sum_j \phi_{ij}$ it can be shown that:

$$\sum_i e_i \mathbf{v}_i \delta(\mathbf{r} - \mathbf{r}_i(t)) = \sum_i \left(\frac{1}{2}m\dot{\mathbf{r}}_i^2 + \frac{1}{2}\sum_j \phi_{ij}\right) \mathbf{v}_i \delta(\mathbf{r} - \mathbf{r}_i(t))$$

$$= \sum_i \left(\frac{1}{2}m(\mathbf{v}_i + \mathbf{u}(\mathbf{r}_i))^2 + \frac{1}{2}\sum_j \phi_{ij}\right) \mathbf{v}_i \delta(\mathbf{r} - \mathbf{r}_i(t))$$

$$= \sum_i \left(\frac{1}{2}m\mathbf{v}_i^2 + \frac{1}{2}\sum_j \phi_{ij}\right) \mathbf{v}_i \delta(\mathbf{r} - \mathbf{r}_i(t))$$

$$+ \sum_i m\mathbf{v}_i \mathbf{v}_i \cdot \mathbf{u}(\mathbf{r}_i) \delta(\mathbf{r} - \mathbf{r}_i(t))$$

$$= \sum_i U_i \mathbf{v}_i \delta(\mathbf{r} - \mathbf{r}_i(t)) + \left(\sum_i m\mathbf{v}_i \mathbf{v}_i \delta(\mathbf{r} - \mathbf{r}_i(t))\right) \cdot \mathbf{u}(\mathbf{r}).$$

$$(3.143)$$

To obtain a microscopic representation for the heat-flux vector in the comoving frame we need to remove the $\mathbf{P} \cdot \mathbf{u}$ term. From the microscopic representation of $\mathbf{P}$ (Equation 3.138), we have:

$$\mathbf{P}(\mathbf{r}, t) \cdot \mathbf{u}(\mathbf{r}) = \sum_i m_i \mathbf{v}_i \mathbf{v}_i \cdot \mathbf{u}(\mathbf{r}) \delta(\mathbf{r} - \mathbf{r}_i)$$

$$(3.144)$$

$$- \frac{1}{2}\sum_{i,j} \mathbf{r}_{ij} \mathbf{F}_{ij} \cdot \mathbf{u}(\mathbf{r}) \int_0^1 d\lambda \delta(\mathbf{r} - \mathbf{r}_i - \lambda \mathbf{r}_{ij}).$$

Combining Equations (3.139–3.144) gives the final result for the heat flux vector:

$$\mathbf{J}_Q(\mathbf{r}, t) = \sum_i U_i \mathbf{v}_i \delta(\mathbf{r} - \mathbf{r}_i) - \frac{1}{2}\sum_{i,j} \mathbf{r}_{ij} \mathbf{F}_{ij} \cdot (\mathbf{v}_i + \mathbf{u}(\mathbf{r}_i) - \mathbf{u}(\mathbf{r}))$$

$$(3.145)$$

$$\times \int_0^1 d\lambda \, \delta(\mathbf{r} - \mathbf{r}_i - \lambda \mathbf{r}_{ij}).$$

Here the streaming term $\mathbf{u}(\mathbf{r}_i) - \mathbf{u}(\mathbf{r})$ is the difference between the streaming velocity at the position of particle i, $\mathbf{u}(\mathbf{r}_i)$, and the streaming velocity of the comoving frame $\mathbf{u}(\mathbf{r})$. It is easy to show that the integral term is invariant under interchange of i and j, and the sum is over all i and j so that no further symmetrizing is necessary.

Our procedure for calculating microscopic expressions for the hydrodynamic densities and fluxes relies upon establishing a correspondence between the microscopic and macroscopic forms of the continuity equations. These equations refer only to the divergence of the pressure tensor and heat flux. Strictly speaking, therefore, we can only determine the divergences of the flux tensors. We can add any divergence free quantity to our expressions for the flux tensors without affecting the identification process.

k-Space representations

It is sometimes easier to see if we calculate microscopic expressions for the fluxes in **k**-space rather than real space. If we define the Fourier transform and inverse in three-dimensions by:

$$f(\mathbf{k}) = \int d\mathbf{r} \exp[i\mathbf{k} \cdot \mathbf{r}] f(\mathbf{r}) \quad f(\mathbf{r}) = \frac{1}{(2\pi)^3} \int d\mathbf{k} \exp[-i\mathbf{k} \cdot \mathbf{r}] f(\mathbf{k}), \quad (3.146)$$

then the fact that transform followed by inverse transform returns the original function, means that the delta function is given by:

$$\delta(\mathbf{r} - \mathbf{r}') = \frac{1}{(2\pi)^3} \int d\mathbf{k} \exp[-i\mathbf{k} \cdot (\mathbf{r} - \mathbf{r}')]. \quad (3.147)$$

The Fourier transform of the instantaneous **r**-space mass density (Equation 3.125) is then:

$$\rho(\mathbf{k}, t) = \int d\mathbf{r} \sum_{i=1}^{N} m\delta(\mathbf{r} - \mathbf{r}_i(t)) \exp[i\mathbf{k} \cdot \mathbf{r}] = \sum_{i=1}^{N} m \exp[i\mathbf{k} \cdot \mathbf{r}_i(t)]. \quad (3.148)$$

The explicit time dependence of the macroscopic quantity $\rho(\mathbf{r}, t)$ (the time dependence differentiated by the derivative $\partial/\partial t$, at **r** fixed) in microscopic representation is the time dependence of $\mathbf{r}_i(t)$. Therefore:

$$\frac{\partial}{\partial t} \rho(\mathbf{k}, t) = i\mathbf{k} \cdot \sum_{i=1}^{N} m\dot{\mathbf{r}}_i(t) \exp[i\mathbf{k} \cdot \mathbf{r}_j(t)]. \quad (3.149)$$

Comparing this with the Fourier transform of Equation (2.4) (noting that $\partial/\partial t|_{\mathbf{k}}$ in Equation (3.149) corresponds to $\partial/\partial t|_{\mathbf{r}}$ in Equation (2.4)), we see that if we let $\mathbf{J}(\mathbf{r}, t) = \rho(\mathbf{r}, t)\mathbf{u}(\mathbf{r}, t)$ then:

$$\mathbf{J}(\mathbf{k}, t) = \sum_{i=1}^{N} m\dot{\mathbf{r}}_i \exp[i\mathbf{k} \cdot \mathbf{r}_i(t)]. \quad (3.150)$$

This equation is clearly the Fourier transform of the instantaneous momentum density, Equation (3.127). To look at the *instantaneous pressure tensor* we only need to differentiate Equation (3.150) with respect to time. Thus:

$$\frac{\partial}{\partial t} \mathbf{J}(\mathbf{k}, t) = \sum_{i=1}^{N} (i\mathbf{k} \cdot m\mathbf{v}_i(t)\mathbf{v}_i(t) \exp[i\mathbf{k} \cdot \mathbf{r}_i(t)] + \mathbf{F}_i \exp[i\mathbf{k} \cdot \mathbf{r}_i(t)]) \quad (3.151)$$

We can write the second term on the right-hand side of this equation in the form of the Fourier transform of a divergence by noting that:

$$\sum_{i=1}^{N} \mathbf{F}_i \exp[i\mathbf{k} \cdot \mathbf{r}_i] = \frac{1}{2} \sum_{i,j=1}^{N} \left(\mathbf{F}_{ij} \exp[i\mathbf{k} \cdot \mathbf{r}_i] + \mathbf{F}_{ji} \exp[i\mathbf{k} \cdot \mathbf{r}_j] \right)$$

$$= \frac{1}{2} \sum_{i,j=1}^{N} \mathbf{F}_{ij} \left(\exp[i\mathbf{k} \cdot \mathbf{r}_i] - \exp[i\mathbf{k} \cdot \mathbf{r}_j] \right)$$

$$= -\frac{1}{2} \sum_{i,j=1}^{N} \mathbf{F}_{ij} \left(\exp[i\mathbf{k} \cdot \mathbf{r}_{ij}] - 1 \right) \exp[i\mathbf{k} \cdot \mathbf{r}_i]$$

$$= -i\mathbf{k} \cdot \frac{1}{2} \sum_{i,j=1}^{N} \mathbf{r}_{ij} \mathbf{F}_{ij} \frac{\exp[i\mathbf{k} \cdot \mathbf{r}_{ij}] - 1}{i\mathbf{k} \cdot \mathbf{r}_{ij}} \exp[i\mathbf{k} \cdot \mathbf{r}_i]. \quad (3.152)$$

Combining Equations (3.151) and (3.152) and performing an inverse Fourier transform we obtain Equation (3.134). We could, of course, continue this analysis to remove the streaming contribution from the pressure tensor but we will obtain the transform of Equation (3.138). In transforming to **k**-space it is sufficient to Fourier transform the microscopic representations for the pressure tensor and the heat-flux vector. The microscopic representation for the **k**-dependent pressure tensor is given by:

$$\mathsf{P}(\mathbf{k}, t) = \sum_{i=1}^{N} m_i \mathbf{v}_i \mathbf{v}_i \exp[i\mathbf{k} \cdot \mathbf{r}_i] - \frac{1}{2} \sum_{i,j}^{N} \mathbf{r}_{ij} \mathbf{F}_{ij} g(i\mathbf{k} \cdot \mathbf{r}_{ij}) \exp[i\mathbf{k} \cdot \mathbf{r}_i], \quad (3.153)$$

where the function $g(x) = (\exp[x] - 1)/x$. We can obtain the same result by Fourier transforming the **r**-space result directly, as the Fourier transform of:

$$\int d\mathbf{r} \exp[i\mathbf{k} \cdot \mathbf{r}] \int_0^1 d\lambda \, \delta(\mathbf{r} - \mathbf{r}_i - \lambda \mathbf{r}_{ij}) = \frac{\exp[i\mathbf{k} \cdot \mathbf{r}_{ij}] - 1}{i\mathbf{k} \cdot \mathbf{r}_{ij}} \exp[i\mathbf{k} \cdot \mathbf{r}_i]$$

$$= g(i\mathbf{k} \cdot \mathbf{r}_{ij}) \exp[i\mathbf{k} \cdot \mathbf{r}_i]. \quad (3.154)$$

In the limit as $\mathbf{k} \cdot \mathbf{r}_{ij} \to 0$, the function $g_{ij}(\mathbf{k}) \to 1$. For an atomic system where the force between two atoms acts in the direction of their separation, $\mathbf{r}_{ij} \mathbf{F}_{ij} = \mathbf{r}_{ij} \mathbf{r}_{ij} \phi'(r_{ij})$ and it follows that the pressure tensor is symmetric. We can use our instantaneous expression for the pressure tensor to describe fluctuations in an equilibrium system.

The **k**-space representation for the heat-flux vector is the Fourier transform of the **r**-space representation Equation (3.145), that is:

$$\mathbf{J}_Q(\mathbf{k}, t) = \sum_i u_i \mathbf{v}_i \exp[i\mathbf{k} \cdot \mathbf{r}_i]$$

$$-\frac{1}{2} \sum_{i,j} \mathbf{r}_{ij} \mathbf{F}_{ij} \cdot \left\{ \mathbf{v}_i g_{ij}(\mathbf{k}) - \int_0^1 d\lambda \exp[i\mathbf{k} \cdot \mathbf{r}_{ij}\lambda](\mathbf{u}(\mathbf{r}_i + \lambda\mathbf{r}_{ij}) - \mathbf{u}(\mathbf{r}_i)) \right\} \exp[i\mathbf{k} \cdot \mathbf{r}_i]$$

$$(3.155)$$

The heat-flux vector depends on the explicit form of the streaming velocity $\mathbf{u}(\mathbf{r})$, and its integral along the $\mathbf{r}_{ij}$ vector. Clearly if the streaming term is constant, the integral term is zero.

The full generality of the expressions derived here for the **r**-space and **k**-space instantaneous fluxes have not been fully exploited. However, a nice application of the definition of the mass density in a nonisotropic system is the method of Daivis *et al.* (1996). For a system such as Poiseuille flow we expect the density to vary as a function of a single variable y, the distance from the surface. If we consider the time average of the mass density Equation (3.125) and average over the other directions then:

$$\rho(y) = \frac{1}{TA} \int_0^T dt \int_A dx dz \sum_i m_i \delta(\mathbf{r} - \mathbf{r}_i(t)), \qquad (3.156)$$

where A is the area of the surface in the x, z-directions. The three-dimensional delta function reduces to a one-dimensional delta function and gives a contribution $\delta(y - y_i(t)) = \delta(t - t_\alpha)/|\dot{y}_i(t_\alpha)|$ each time a particle intersects the plane. The time average then becomes the sum over all intersection times for each particle, thus:

$$\rho(y) = \frac{1}{TA} \sum_i \sum_\alpha \frac{m_i}{|\dot{y}_i(t_\alpha)|}. \qquad (3.157)$$

An interpolation is required to find $\dot{y}_i(t_\alpha)$ as t_α generally does not correspond to an integer number of time steps. The same procedure can be used for the time average of any density in a Poiseulle flow so the momentum, energy, and temperature can all be calculated as functions of y.

3.9 Microscopic representation of the temperature

At equilibrium, we obtain an instantaneous expression for the temperature by analyzing the expression for the pressure tensor (Equation 3.138). Thus if $n(\mathbf{r}, t)$ is the local instantaneous number density, then:

$$\frac{3}{2}n(\mathbf{r}, t)k_B T(\mathbf{r}, t) = \sum_{i=1}^N \frac{1}{2}m_i \mathbf{v}_i(t)^2 \delta(\mathbf{r}_i(t) - \mathbf{r}). \qquad (3.158)$$

We will call this expression for the temperature, the kinetic temperature. In using this expression for the temperature we are employing a number of approximations. Firstly we are ignoring the number of degrees of freedom, which are frozen by the instantaneous determination of the local streaming velocity $\mathbf{u}(\mathbf{r}, t)$. Secondly, and more importantly, we are assuming that in a nonequilibrium system the kinetic temperature is identical to the thermodynamic temperature T_{th}:

$$T_{th} = \frac{\partial E}{\partial S}\bigg|_{N,V}. \tag{3.159}$$

This is undoubtedly an approximation. It would be true if the postulate of local thermodynamic equilibrium was exact. However, we know that the energy, pressure, enthalpy etc. are all functions of the thermodynamic forces driving the system away from equilibrium. These are nonlinear effects which vanish for Newtonian fluids. Presumably the entropy is also a function of these driving forces. It is extremely unlikely that the field dependence of the entropy and the energy are precisely those required for the exact equivalence of the kinetic and thermodynamic temperatures for all nonequilibrium systems. Recent calculations of the entropy of systems very far from equilibrium support the hypothesis that the kinetic and thermodynamic temperatures are in fact different (Evans, 1989). Outside the linear (Newtonian) regime, the kinetic temperature is a convenient operational (as opposed to thermodynamic) state variable. If a nonequilibrium system is in a steady state, both the kinetic and the thermodynamic temperatures must be constant in time. Furthermore we expect that outside the linear regime in systems with a unique nonequilibrium steady state, that the thermodynamic temperature should be a monotonic function of the kinetic temperature.

In a recent paper by Rugh (1997), a method of determining the temperature in a Hamiltonian dynamical system was proposed. It begins with the definition of the entropy S as the (canonically invariant) weighted area of the energy surface Ω (the level set of the Hamiltonian $H(\mathbf{q}, \mathbf{p})$), under the assumption that the dynamical system is ergodic on the energy surface Ω. Defining the temperature $T(E)$ in the usual thermodynamic way, using Equation (3.159), we have a phase variable:

$$\Psi(\Gamma) = \frac{\partial}{\partial\Gamma} \cdot \left(\frac{\frac{\partial H}{\partial\Gamma}}{\frac{\partial H}{\partial\Gamma} \cdot \frac{\partial H}{\partial\Gamma}}\right), \tag{3.160}$$

whose time average is the inverse of the temperature for a system with total energy E. We will return to the concept of temperature, for both equilibrium and nonequilibrium systems, in Chapter 10.

4

The Green–Kubo relations

4.1 The Langevin equation

In 1828 the botanist Robert Brown (1828a, 1828b) observed the motion of pollen grains suspended in a fluid. Although the system was allowed to come to equilibrium, he observed that the grains seemed to undergo a kind of unending irregular motion. This motion is now known as Brownian motion. The motion of large pollen grains suspended in a fluid composed of much lighter particles can be modeled by dividing the accelerating force into two components: a slowly varying *drag* force, and a rapidly varying *random* force, due to the thermal fluctuations in the velocities of the solvent molecules. The Langevin equation, as it is known, is conventionally written in the form:

$$\frac{d\mathbf{v}}{dt} = -\zeta \mathbf{v} + \mathbf{F}_R. \tag{4.1}$$

Using the Navier–Stokes equations to model the flow around a sphere, it is known that the friction coefficient $\zeta = 6\pi\eta d/m$, where η is the shear viscosity of the fluid, d is the diameter of the sphere and m is its mass. The random force per unit mass $\mathbf{F}_R$ is used to model the force on the sphere due to the bombardment of solvent molecules. This force is called random because it is assumed that $\langle \mathbf{v}(0) \cdot \mathbf{F}_R(t) \rangle = 0, \forall t$. A more detailed investigation of the drag on a sphere which is forced to oscillate in a fluid shows that a nonMarkovian generalization (see Section 2.4), of the Langevin equation (Langevin, 1908) is required to describe the time-dependent drag on a rapidly oscillating sphere:

$$\frac{d\mathbf{v}(t)}{dt} = -\int_0^t dt' \, \zeta(t - t')\mathbf{v}(t') + \mathbf{F}_R(t). \tag{4.2}$$

In this case, the viscous drag on the sphere is not simply linearly proportional to the instantaneous velocity of the sphere as in Equation (4.1). Instead it is linearly proportional to the velocity at all previous times in the past. As we will see, there are many transport processes which can be described by an equation of this form. We will refer to the equation:

$$\frac{\mathrm{d}\,A(t)}{\mathrm{d}t} = -\int_0^t \mathrm{d}t' K(t-t')A(t') + F(t),$$
(4.3)

as the generalized Langevin equation for the phase variable $A(\Gamma)$. $K(t)$ is the time-dependent transport coefficient that we seek to evaluate. We assume that the equilibrium canonical ensemble average of the random force and the phase variable A, vanishes for all times, so:

$$\langle A(0)F(t)\rangle = \langle A(t_0)F(t_0+t)\rangle = 0, \quad \forall\, t \text{ and } t_0.$$
(4.4)

The time displacement by t_0 is allowed because the equilibrium time-correlation function is independent of the time origin. Multiplying both sides of Equation (4.3) by the complex conjugate of $A(0)$ and taking a canonical average we see that:

$$\frac{\mathrm{d}C(t)}{\mathrm{d}t} = -\int_0^t \mathrm{d}t' K(t-t')C(t'),$$
(4.5)

where $C(t)$ is defined to be the equilibrium autocorrelation function:

$$C(t) \equiv \langle A(t)A^*(0)\rangle.$$
(4.6)

Another function we will find useful is the flux autocorrelation function $\phi(t)$:

$$\phi(t) = \langle \dot{A}(t)\dot{A}^*(0)\rangle.$$
(4.7)

Taking a Laplace transform of Equation (4.5) we see that there is an intimate relationship between the transport memory kernel $K(t)$ and the equilibrium fluctuations in A. The left-hand side of (4.5) becomes:

$$\int_0^\infty \mathrm{d}t\, e^{-st}\frac{\mathrm{d}C(t)}{\mathrm{d}t} = \left[e^{-st}C(t)\right]_0^\infty - \int_0^\infty \mathrm{d}t\left(-se^{-st}\right)C(t) = s\tilde{C}(s) - C(0),$$

and, as the right-hand side is a Laplace transform convolution:

$$s\tilde{C}(s) - C(0) = -\tilde{K}(s)\tilde{C}(s),$$
(4.8)

so that:

$$\tilde{C}(s) = \frac{C(0)}{s + \tilde{K}(s)}.$$
(4.9)

One can convert the A autocorrelation function into a flux autocorrelation function by realizing that:

$$\frac{d^2}{dt^2}C(t) = \frac{d}{dt}\left\langle\frac{dA(t)}{dt}A^*(0)\right\rangle = \frac{d}{dt}\langle(iLA(t))A^*(0)\rangle$$

$$= \frac{d}{dt}\langle A(t)(-iLA^*(0))\rangle = -\langle(iLA(t))(-iLA^*(0))\rangle = -\phi(t). \quad (4.10)$$

Then we take the Laplace transform of a second derivative to find:

$$-\tilde{\phi}(s) = \int_0^\infty dt\,\exp[-st]\frac{d^2}{dt^2}C(t) = \left[\exp[-st]\frac{d}{dt}C(t)\right]_0^\infty + s\int_0^\infty dt\,\exp[-st]\frac{d}{dt}C(t)$$

$$= s[\exp[-st]C(t)]_0^\infty + s^2\int_0^\infty dt\,\exp[-st]C(t) = s^2\tilde{C}(s) - sC(0). \quad (4.11)$$

Here we have used the result that $C^Y(0) = 0$. Eliminating $\tilde{C}(s)$ between Equations (4.9) and (4.11) gives:

$$\tilde{K}(s) = \frac{\tilde{\phi}(s)}{C(0) - \dfrac{\tilde{\phi}(s)}{s}}. \quad (4.12)$$

Rather than try to give a general interpretation of this equation, it may prove more useful to apply it to the Brownian motion problem. $C(0)$ is the time zero value of an equilibrium time-correlation function, and can be easily evaluated as k_BT/m, and $\dot{v} = F/m$ where $\mathbf{F}$ is the total force on the Brownian particle:

$$\tilde{\zeta}(s) = \frac{\tilde{C}^F(s)}{mk_BT - \dfrac{\tilde{C}^F(s)}{s}}, \quad (4.13)$$

where

$$\tilde{C}^F(s) = \frac{1}{3}\langle\mathbf{F}(0)\cdot\tilde{\mathbf{F}}(s)\rangle, \quad (4.14)$$

is the Laplace transform of the total force autocorrelation function. In writing (4.14) we have used the fact that the equilibrium ensemble average, denoted $\langle\ldots\rangle$, must be isotropic. The average of any second-rank tensor, say $\langle\mathbf{F}(0)\mathbf{F}(t)\rangle$, must therefore be a scalar multiple of the second-rank identity tensor. That scalar must, of course, be $\frac{1}{3}\text{Tr}\{\langle\mathbf{F}(0)\mathbf{F}(t)\rangle\} = \frac{1}{3}\langle\mathbf{F}(0)\cdot\mathbf{F}(t)\rangle$.

In the so-called Brownian limit, where the ratio of the Brownian particle mass to the mean square of the force becomes infinite:

$$\tilde{\zeta}(s) = \frac{\beta}{3m} \int_0^\infty dt \, \exp[-st] \langle \mathbf{F}(t) \cdot \mathbf{F}(0) \rangle. \tag{4.15}$$

For any finite value of the Brownian ratio, Equation (4.13) shows that the integral of the force autocorrelation function is zero. This is seen most easily by solving Equation (4.13) for C^F and taking the limit as $s \to 0$.

Equation (4.9), which gives the relationship between the memory kernel and the force autocorrelation function, implies that the velocity autocorrelation function $Z(t) \equiv \frac{1}{3} \langle \mathbf{v}(0) \cdot \mathbf{v}(t) \rangle$ is related to the friction coefficient by the equation:

$$\tilde{Z}(s) = \frac{k_B T / m}{s + \tilde{\zeta}(s)}. \tag{4.16}$$

This equation is valid outside the Brownian limit. The integral of the velocity autocorrelation function, is related to the growth of the mean-square displacement, giving yet another expression for the friction coefficient:

$$\tilde{Z}(0) = \lim_{t \to \infty} \int_0^t dt' \frac{1}{3} \langle \mathbf{v}(0) \cdot \mathbf{v}(t') \rangle = \lim_{t \to \infty} \int_0^t dt' \frac{1}{3} \langle \mathbf{v}(t) \cdot \mathbf{v}(t') \rangle$$

$$= \lim_{t \to \infty} \frac{1}{3} \langle \mathbf{v}(0) \cdot \Delta \mathbf{r}(t) \rangle = \lim_{t \to \infty} \frac{1}{6} \frac{d}{dt} \langle \Delta \mathbf{r}(t)^2 \rangle. \tag{4.17}$$

Here the displacement vector $\Delta \mathbf{r}(t)$ is defined by:

$$\Delta \mathbf{r}(t) = \mathbf{r}(t) - \mathbf{r}(0) = \int_0^t dt' \mathbf{v}(t'). \tag{4.18}$$

Assuming that the mean-square displacement is linear in time, in the long time limit, it follows from Equation (4.16) that the friction coefficient can be calculated from:

$$\frac{k_B T}{m \tilde{\zeta}(0)} \equiv D = \frac{1}{6} \lim_{t \to \infty} \frac{d}{dt} \langle \Delta \mathbf{r}(t)^2 \rangle = \frac{1}{6} \lim_{t \to \infty} \frac{\langle \Delta \mathbf{r}(t)^2 \rangle}{t}. \tag{4.19}$$

This is the Einstein (1905) relation for the diffusion coefficient D. The relationship between the diffusion coefficient and the integral of the velocity autocorrelation function Equation (4.17) is an example of a Green–Kubo relation (Green, 1954; Kubo, 1957).

It should be pointed out that the transport properties we have just evaluated are properties of systems at equilibrium. The Langevin equation describes the irregular

Brownian motion of particles *in an equilibrium system*. Similarly the self-diffusion coefficient characterizes the random walk executed by a particle in an equilibrium system. The identification of the zero-frequency friction coefficient $6\pi\eta d/m$, with the viscous drag on a sphere, which is forced to move with constant velocity through a fluid, implies that equilibrium fluctuations can be modeled by nonequilibrium transport coefficients, in this case the shear viscosity of the fluid. This hypothesis is known as the Onsager regression hypothesis (Onsager, 1931). The hypothesis can be inverted: one can calculate transport coefficients from a knowledge of the equilibrium fluctuations. We will now discuss these relations in more detail.

4.2 Mori–Zwanzig theory

We will show that for an arbitrary phase variable $A(\Gamma)$, evolving under equations of motion which preserve the equilibrium distribution function, one can always write down a Langevin equation. Such an equation is an exact consequence of the equations of motion. We will use the symbol iL, to denote the Liouvillean associated with these equations of motion. These *equilibrium* equations of motion could be field-free Newtonian equations of motion or they could be field-free thermostatted equations of motion, such as Gaussian isokinetic or Nosé–Hoover equations. The equilibrium distribution could be microcanonical, canonical, or even isothermal-isobaric provided that if the latter is the case, suitable distribution-preserving dynamics are employed. For simplicity we will compute equilibrium time-correlation functions over the canonical distribution function, f_c:

$$f_c(\Gamma) = \frac{\exp[\beta H_0(\Gamma)]}{\int d\Gamma \exp[\beta H_0(\Gamma)]}. \tag{4.20}$$

We saw in the previous section that a key element of the derivation was that the correlation of the random force, $\mathbf{F}_R(t)$ with the Langevin variable A, vanished for all time. We will now use the notation first developed in Section 3.5, which treats phase variables, $A(\Gamma)$, $B(\Gamma)$, as vectors in $6N$-dimensional phase space with a scalar product defined by $\int d\Gamma f_0(\Gamma) B(\Gamma) A^*(\Gamma)$, and denoted as (B, A^*). We will define a projection operator which will transform any phase variable B into a vector which has no correlation with the Langevin variable A. The component of B parallel to A is just:

$$PB(\Gamma, t) = \frac{(B(\Gamma, t), A^*(\Gamma))}{(A(\Gamma), A^*(\Gamma))} A(\Gamma). \tag{4.21}$$

This equation defines the projection operator P.

The operator $Q = 1 - P$, is the complement of P and computes the component of B orthogonal to A:

$$(QB(t), A^*) = \left(B(t) - \frac{(B(t), A^*)}{(A, A^*)} A, A^* \right)$$

$$= (B(t), A^*) - \frac{(B(t), A^*)}{(A, A^*)} (A, A^*) = 0.$$

(4.22)

In more physical terms, the projection operator Q computes that part of any phase variable which is random with respect to a Langevin variable, A (Figure 4.1).

Other properties of the projection operators are that:

$$PP = P,$$

$$QQ = Q,$$

$$QP = PQ.$$

(4.23)

Secondly, P and Q are Hermitian operators (like the Liouville operator itself). To prove this we note that:

$$(PB, C^*)^* = \frac{((B, A^*)A, C^*)^*}{(A, A^*)^*} = \frac{(B, A^*)^*(A, C^*)^*}{(A, A^*)^*} = \frac{(B^*, A)(A^*, C)}{(A, A^*)}$$

$$= \frac{(A, B^*)(C, A^*)}{(A, A^*)} = \frac{((C, A^*)A, B^*)}{(A, A^*)} = (PC, B^*).$$

(4.24)

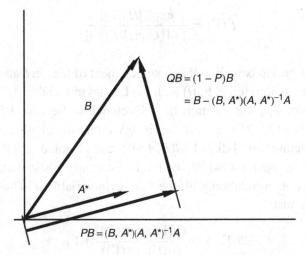

Figure 4.1 The projection operator P, operating on B produces a vector which is the component of B parallel to A

Furthermore, since $Q = 1 - P$ where 1 is the identity operator, and since both the identity operator and P are Hermitian, so is Q.

We will wish to compute the random and direct components of the propagator $\exp[iLt]$. The random and direct parts of the Liouvillean iL are iQL and iPL respectively. These Liouvilleans define the corresponding random and direct propagators, $\exp[iQLT]$ and $\exp[iPLt]$. We can use the Dyson equation to relate these two propagators. If we take $\exp[iQLt]$ as the reference propagator in Equation (3.110) and $\exp[iLt]$ as the test propagator then:

$$\exp[iLt] = \exp[iQLt] + \int_0^t d\tau \exp[iL(t - \tau)]iPL \exp[iQL\tau]. \qquad (4.25)$$

The rate of change of $A(t)$, the Langevin variable at time t is:

$$\frac{dA(t)}{dt} = \exp[iLt]iLA = \exp[iLt]i(Q + P)LA, \qquad (4.26)$$

but:

$$\exp[iLt]iPLA = \exp[iLt]\frac{(iLA, A^*)}{(A, A^*)}A = \frac{(iLA, A^*)}{(A, A^*)}\exp[iLt]A \equiv i\Omega A(t). \qquad (4.27)$$

This defines the frequency $i\Omega$ which is an equilibrium property of the system. It only involves equal time averages. Substituting this equation into Equation (4.26) gives:

$$\frac{dA(t)}{dt} = i\Omega A(t) + \exp[iLt]iQLA. \qquad (4.28)$$

Using the Dyson decomposition of the propagator given in Equation (4.25), this leads to:

$$\frac{dA(t)}{dt} = i\Omega A(t) + \int_0^t d\tau \exp[iL(t - \tau)] iPL \exp[iQL\tau]iQLA$$
$$+ \exp[iQLt]iQLA. \qquad (4.29)$$

We identify $\exp[iQLt] iQLA$ as the random force $F(t)$ because:

$$(F(t), A^*) = (\exp[iQLt]iQLA, A^*) = (QF(t), A^*) = 0, \qquad (4.30)$$

where we have used Equation (4.23). It is very important to remember that the propagator which generates $F(t)$ from $F(0)$ is not the propagator e^{iLt}; rather it is the random propagator e^{iQLt}. The integral in Equation (4.29) involves the term:

$$iPL \exp[iQLt]iQLA = iPLF(t) = iPLQF(t) = \frac{(iLQF(t), A^*)}{(A, A^*)}A$$

$$= -\frac{(QF(t), (iLA)^*)}{(A, A^*)}A, \qquad (4.31)$$

as L is Hermitian and i is anti-Hermitian, $(iL)^* = (d/dt)^* = (\dot{\Gamma} \cdot \partial/\partial\Gamma)^* = d/dt = iL$, (since the equations of motion are real). Since Q is Hermitian:

$$iPL \exp[iQLt]iQLA = -\frac{(F(t), (iQLA)^*)}{(A, A^*)}A = -\frac{(F(t), F(0)^*)}{(A, A^*)}A,$$

$$\equiv -K(t)A \qquad (4.32)$$

where we have defined a memory kernel $K(t)$. It is basically the autocorrelation function of the random force. Substituting this definition into Equation (4.29) gives:

$$\frac{dA(t)}{dt} = i\Omega A(t) - \int_0^t d\tau \, \exp[iL(t - \tau)]K(\tau)A + F(t)$$

$$= i\Omega A(t) - \int_0^t d\tau \, K(\tau)A(t - \tau) + F(t). \qquad (4.33)$$

This shows that the *generalized Langevin equation* is an exact consequence of the equations of motion for the system (Zwanzig, 1961; Mori, 1965a, 1965b). Since the random force is random with respect to A, multiplying both sides of Equation (4.33) by $A^*(0)$ and taking a canonical average gives the memory function equation:

$$\frac{dC(t)}{dt} = i\Omega C(t) - \int_0^t d\tau K(\tau)C(t - \tau) \qquad (4.34)$$

This is essentially the same as Equation (4.5).

As we mentioned in the introduction to this section, the generalized Langevin equation and the memory function equation are exact consequences of any dynamics which preserve the equilibrium distribution function. As such, the equations therefore describe equilibrium fluctuations in the phase variable A, and the equilibrium autocorrelation function for A, namely $C(t)$.

However, the generalized Langevin equation bears a striking resemblance to a nonequilibrium constitutive relation. The memory kernel $K(t)$ plays the role of a transport coefficient. Onsager's regression hypothesis (Onsager, 1931) states that the equilibrium fluctuations in a phase variable are governed by the same transport coefficients as is the relaxation of that same phase variable to equilibrium. This hypothesis implies that the generalized Langevin equation can be interpreted as a linear, *nonequilibrium* constitutive relation with the memory function $K(t)$, given by the *equilibrium* autocorrelation function of the random force.

Onsager's hypothesis can be justified by the fact that in observing an *equilibrium* system for a time which is of the order of the relaxation time for the memory kernel, it is impossible to tell whether the system is at equilibrium or not. We could be observing the final stages of a relaxation towards equilibrium, or we could be simply observing the small time-dependent fluctuations in an equilibrium system. On a short timescale there is simply no way of telling the difference between these two possibilities. When we interpret the generalized Langevin equation as a nonequilibrium constitutive relation, it is clear that it can only be expected to be valid close to equilibrium. This is because it is a linear constitutive equation.

4.3 Shear viscosity

It is relatively straightforward to apply the Mori–Zwanzig formalism to the calculation of fluctuation expressions for linear transport coefficients. Our first application of the method will be the calculation of shear viscosity. Before we do this we will say a little more about constitutive relations for shear viscosity. The Mori–Zwanzig formalism leads naturally to a non-Markovian expression for the viscosity. Equation (4.33) refers to a memory function rather than a simple Markovian transport coefficient such as the Newtonian shear viscosity. We will thus be led to a discussion of viscoelasticity (see Section 2.4).

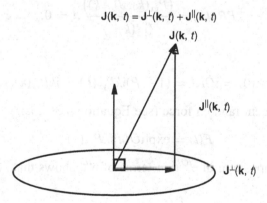

$$\mathbf{J}(\mathbf{k},\,t) = \mathbf{J}^{\perp}(\mathbf{k},\,t) + \mathbf{J}^{\parallel}(\mathbf{k},\,t)$$

$\mathbf{J}(\mathbf{k},\,t)$

$\mathbf{J}^{\parallel}(\mathbf{k},\,t)$

$\mathbf{J}^{\perp}(\mathbf{k},\,t)$

Figure 4.2 We can resolve the wave vector-dependent momentum density into components which are parallel and orthogonal to the wave vector, **k**

We choose our test variable A, to be the x-component of the wave-vector-dependent transverse momentum current $\mathbf{J}^\perp(\mathbf{k}, t)$.

For simplicity, we define the coordinate system so that $\mathbf{k}$ is in the y direction and $\mathbf{J}^\perp$ is in the x direction (Figure 4.2):

$$J_x(k_y, t) = \sum_i m v_{xi}(t) \exp[ik_y y_i(t)].\qquad(4.35)$$

In Section 3.8 we saw that:

$$\dot{J} = ikP_{yx}(k, t),\qquad(4.36)$$

where for simplicity we have dropped the Cartesian indices for J and k. We note that at zero wave vector the transverse momentum current is a constant of the motion, $\dot{J} = 0$. The quantities we need in order to apply the Mori–Zwanzig formalism are easily computed.

The frequency matrix $i\Omega$, defined in Equation (4.27), is identically zero. This is always so in the single variable case as $\langle A^* \dot{A} \rangle = 0$, for any phase variable A. The norm of the transverse current is calculated:

$$\langle J(k)J^*(k)\rangle = \left\langle \sum_{i=1}^N p_{xi} \exp[iky_i] \sum_{j=1}^N p_{xj} \exp[-iky_j]\right\rangle$$

$$= N\langle p_{x1}^2\rangle + N(N-1)\langle p_{x1}p_{x2} \exp[ik(y_1 - y_2)]\rangle = Nmk_BT. \quad(4.37)$$

At equilibrium p_{x1} is independent of p_{x2} and $(y_1 - y_2)$, so the correlation function factors into the product of three equilibrium averages. The values of $\langle p_{x1}\rangle$ and $\langle p_{x2}\rangle$ are identically zero. The random force, F, can also easily be calculated since, if we use Equation (4.36):

$$PP_{yx}(k) = \frac{\left(P_{yx}(k), J(-k)\right)}{\langle |J(k)|^2\rangle} J = 0,\qquad(4.38)$$

we can write:

$$F(0) = iQLJ = (1 - P)ikP_{yx}(k) = ikP_{yx}(k).\qquad(4.39)$$

The time-dependent random force (see Equation 4.30), is:

$$F(t) = \exp[iQLt]ikP_{yx}(k).\qquad(4.40)$$

A Dyson decomposition of e^{iQLt} in terms of e^{iLt} shows that,

$$\exp[iLt] = \exp[iQLt] + \int_0^t ds \, \exp[iL(t - s)]iPL \exp[iQLs].\qquad(4.41)$$

Now for any phase variable B:

$$iPLB = \langle J^* iLB \rangle \frac{J}{Nmk_B T} = -\langle B(iLJ)^* \rangle \frac{J}{Nmk_B T}$$

$$= -ik\langle BP_{yx}(-k) \rangle \frac{J}{Nmk_B T}. \tag{4.42}$$

Substituting this observation into Equation (4.41) shows that the difference between the propagators e^{iQLt} and e^{iLt} is of order k, and can therefore be ignored in the zero wave vector limit.

From Equation (4.32) the memory kernel $K(t)$ is $\langle F(t)F^*(0) \rangle / \langle AA^* \rangle$. Using Equation (4.40), the small wave vector form for $K(t)$ becomes:

$$K(t) = k^2 \frac{\langle P_{yx}(k, t)P_{yx}(-k, 0) \rangle}{Nmk_B T}. \tag{4.43}$$

The generalized Langevin equation (the analog of Equation 4.33) is

$$\lim_{k \to 0} \frac{dJ_x(k_y, t)}{dt} = -\frac{k^2}{Nmk_B T} \int_0^t ds \langle P_{yx}(k_y, s)P_{yx}(-k_y, 0) \rangle_0 J_x(k_y, t - s)$$
$$+ ik_y P_{yx}(k_y, t), \tag{4.44}$$

where we have taken explicit note of the Cartesian components of the relevant functions. Now, we know that the rate of change of the transverse current is $ikP_{yx}(k, t)$. This means that the left-hand side of Equation (4.44) is related to equilibrium fluctuations in the shear stress. We also know that $J(k) = \int dk' \rho(k' - k)u(k')$, so, close to equilibrium, the transverse momentum current (our Langevin variable A), is closely related to the wave-vector-dependent shear rate $\gamma(k)$. In fact the wave-vector-dependent shear rate $\gamma(k)$ is $-ikJ(k)/\rho(k = 0)$. Putting these two observations together we see that the generalized Langevin equation for the transverse momentum current is essentially a relation between fluctuations in the shear stress and the shear rate – a constitutive relation. Ignoring the random force (constitutive relations are deterministic), we find that Equation (4.44) can be written in the form of the constitutive relation (2.76):

$$\lim_{k \to 0} P_{yx}(t) = -\int_0^t ds \, \eta(k = 0, t - s)\gamma(k = 0, s). \tag{4.45}$$

If we use the fact that, $P_{yx}V = \lim_{k \to 0} P_{yx}(k)$, $\eta(t)$ is easily seen to be:

$$\eta(t) = \beta V \langle P_{yx}(t)P_{yx}(0) \rangle. \tag{4.46}$$

Equation (4.45) is identical to the viscoelastic generalization of Newton's law of viscosity Equation (2.76).

The Mori–Zwanzig procedure has derived a viscoelastic constitutive relation. No mention has been made of the shearing boundary conditions required for shear flow. Neither is there any mention of viscous heating or possible nonlinearities in the viscosity coefficient. Equation (4.44) is a description of equilibrium fluctuations. However, unlike the case for the Brownian friction coefficient or the self diffusion coefficient, the viscosity coefficient refers to nonequilibrium rather than equilibrium systems.

The zero wave vector limit is subtle. We can imagine longer and longer wavelength fluctuations in the shear rate $\gamma(k)$. For an equilibrium system, however, $\gamma(k = 0) \equiv 0$ and $\langle \gamma(k = 0)\gamma^*(k = 0)\rangle \equiv 0$. There are *no* equilibrium fluctuations in the shear rate at $k = 0$. The zero wave vector shear rate is completely specified by the boundary conditions.

If we invoke Onsager's regression hypothesis, we can obviously identify the memory kernel $\eta(t)$ as the memory function for planar (i.e. $k = 0$) Couette flow. We might observe that there is no fundamental way of knowing whether we are watching small equilibrium fluctuations at small but nonzero wave vector, or the *last* stages of relaxation toward equilibrium of a finite k, nonequilibrium disturbance. Provided the nonequilibrium system is sufficiently close to equilibrium, the Langevin memory function will be the nonequilibrium memory kernel. However, the Onsager regression hypothesis is additional to, and not part of, the Mori–Zwanzig theory. In Section 6.3 we prove that the nonequilibrium linear viscosity coefficient is given exactly by the infinite time integral of the stress fluctuations. In Section 6.3 we will not use the Onsager regression hypothesis.

At this stage one might legitimately ask the question: what happens to these equations if we do *not* take the zero wave vector limit? After all, we have already defined a wave-vector-dependent shear viscosity in Equation (2.77). It is not a simple matter to apply the Mori–Zwanzig formalism to the finite wave-vector case. We will instead use a method which makes a direct appeal to the Onsager regression hypothesis.

Provided the time and spatially dependent shear rate is of sufficiently small amplitude, the generalized viscosity can be defined as Equation (2.77):

$$P_{yx}(k, t) = - \int_0^t ds\, \eta(k, t - s)\gamma(k, s). \qquad (4.47)$$

Using the fact that $\gamma(k, t) = -iku_x(k, t) = -ikJ(k, t)/\rho$, and Equation (4.36), we can rewrite Equation (4.47) as:

$$\dot{J}(k, t) = -\frac{k^2}{\rho} \int_0^t ds\, \eta(k, t - s)J(k, s). \qquad (4.48)$$

If we Fourier–Laplace transform both sides of this equation in time, and using Onsager's hypothesis, multiply both sides by $J(-k, 0)$ and average with respect to the equilibrium canonical ensemble, we obtain:

$$\tilde{C}(k, \omega) = \frac{C(k, 0)}{i\omega + \dfrac{k^2\tilde{\eta}(k, \omega)}{\rho}}, \tag{4.49}$$

where $C(k, t)$ is the equilibrium transverse current autocorrelation function $\langle J(k, t) J(-k, 0)\rangle$ and the tilde notation denotes a Fourier–Laplace transform in time:

$$\tilde{C}(\omega) = \int_0^\infty dt\, C(t) e^{-i\omega t}. \tag{4.50}$$

We call the autocorrelation function of the wave-vector-dependent shear stress:

$$N(k, t) \equiv \frac{1}{V k_B T} \langle P_{yx}(k, t) P_{yx}(-k, 0)\rangle. \tag{4.51}$$

We can use Equation (4.36) to transform from the transverse current autocorrelation function $C(k, t)$ to the stress autocorrelation function $N(k, t)$, since:

$$\frac{d^2}{dt^2} \langle J(k, t) J(-k, 0)\rangle = -\langle \dot{J}(k, t) \dot{J}(-k, 0)\rangle = -k^2 \langle P_{yx}(k, t) P_{yx}(-k, 0)\rangle. \tag{4.52}$$

This derivation closely parallels that for Equations (4.11) and (4.12) in Section 4.1. The reader should refer to that section for more details. Using the fact that, $\rho = Nm/V$, we see that:

$$k^2 V k_B T \tilde{N}(k, \omega) = \omega^2 \tilde{C}(k, \omega) + i\omega C(k, 0). \tag{4.53}$$

The equilibrium average $C(k, 0)$ is given by Equation (4.37). Substituting this equation into Equation (4.49) gives us an equation for the frequency and wave-vector-dependent shear viscosity in terms of the stress autocorrelation function:

$$\tilde{\eta}(k, \omega) = \frac{\tilde{N}(k, \omega)}{1 - \dfrac{k^2\tilde{N}(k, \omega)}{i\omega\rho}}. \tag{4.54}$$

This equation is *not* of the Green–Kubo form (Evans, 1981). Green–Kubo relations are exceptional, being only valid for infinitely slow processes. Momentum relaxation is only infinitely slow at *zero* wave vector. At finite wave vectors,

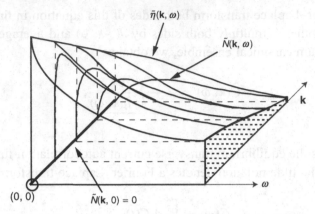

Figure 4.3 The relationship between the viscosity, $\tilde{\eta}(k, \omega)$, and the stress autocorrelation function, $\tilde{N}(k, \omega)$. At $k = 0$, both functions are identical. At $\alpha = 0$, but $k \neq 0$, the stress autocorrelation function is identically zero. The stress autocorrelation function is discontinuous at the origin. The viscosity is continuous everywhere, but nonanalytic at the origin

momentum relaxation is a *fast* process. We can obtain the usual Green–Kubo form by taking the zero k limit of Equation (4.54). In that case:

$$\tilde{\eta}(0, \omega) = \lim_{k \to 0} \tilde{N}(k, \omega). \tag{4.55}$$

Because there are no fluctuations in the zero wave vector shear rate, the function $\tilde{N}(k, \omega)$ is discontinuous at the origin. For all nonzero values of k, $\tilde{N}(k, 0) = 0$! Over the years many errors have been made as a result of this fact. Figure 4.3 above illustrates these points schematically. The results for shear viscosity precisely parallel those for the friction constant of a Brownian particle. Only in the Brownian limit is the friction constant given by the autocorrelation function of the Brownian force.

An immediate conclusion from the theory we have outlined is that *all* fluids are viscoelastic. Viscoelasticity is a direct result of the generalized Langevin equation, which is, in turn, an exact consequence of the microscopic equations of motion.

4.4 Green–Kubo relations for Navier–Stokes transport coefficients

It is relatively straightforward to derive Green–Kubo relations for the other Navier–Stokes transport coefficients, namely bulk viscosity and thermal conductivity. In Section 6.3, when we describe the SLLOD equations of motion for viscous flow, we will find a simpler way of deriving Green–Kubo relations for both viscosity coefficients. For now we simply state the Green–Kubo relation for bulk viscosity as (Zwanzig, 1965):

$$\eta_V = \frac{1}{V k_B T} \int_0^\infty dt \langle (p(t)V(t) - \langle pV \rangle)(p(0)V(0) - \langle pV \rangle) \rangle. \qquad (4.56)$$

The Green–Kubo relation for thermal conductivity can be derived by similar arguments to those used in the viscosity derivation. Firstly we note from Equation (2.26), that, in the absence of a velocity gradient, the internal energy per unit volume ρU obeys a continuity equation, $\rho dU/dt = -\nabla \cdot \mathbf{J}_Q$. Secondly, we note that Fourier's definition of the thermal conductivity coefficient λ, from Equation (2.56a), is $\mathbf{J}_Q = -\lambda \nabla T$. Combining these two results we obtain:

$$\rho \frac{dU}{dt} = \lambda \nabla^2 T. \qquad (4.57)$$

Unlike the previous examples, both U and T have nonzero equilibrium values, namely, $\langle U \rangle$ and $\langle T \rangle$. A small change in the left-hand side of Equation (4.57) can be written as $(\rho + \Delta\rho)d(\langle U \rangle + \Delta U)/dt$. By definition, $d\langle U \rangle/dt = 0$, so to first order in Δ, we have $\rho d\Delta U/dt$. Similarly, the spatial gradient of $\langle T \rangle$ does not contribute, so we can write:

$$\rho \frac{d\Delta U}{dt} = \lambda \nabla^2 \Delta T. \qquad (4.58)$$

The next step is to relate the variation in temperature ΔT to the variation in energy per unit volume $\Delta(\rho U)$. To do this we use the thermodynamic definition:

$$\frac{1}{V} \frac{\partial E}{\partial T}\bigg|_P = \frac{\partial(\rho U)}{\partial T}\bigg|_P = \rho c_P, \qquad (4.59)$$

where c_P is the specific heat per unit mass. We see from the second equality, that a small variation in the temperature ΔT is equal to $\Delta(\rho U)/\rho c_P$. Therefore:

$$\rho \Delta \dot{U} = \frac{\lambda}{\rho c_P} \nabla^2 \rho \Delta U. \qquad (4.60)$$

If $D_T \equiv \lambda/\rho c_P$ is the thermal diffusivity, then in terms of the wave-vector-dependent internal energy density Equation (4.60) becomes:

$$\rho \Delta \dot{U}(\mathbf{k}, t) = -k^2 D_T \rho \Delta U(\mathbf{k}, t). \qquad (4.61)$$

If $C(k, t)$ is the wave-vector-dependent internal energy density autocorrelation function:

$$C(k, t) \equiv \langle \rho \Delta U(\mathbf{k}, t) \rho \Delta U(-\mathbf{k}, 0) \rangle, \qquad (4.62)$$

then the frequency and wave-vector-dependent diffusivity is the memory function of the energy-density autocorrelation function:

$$\tilde{C}(k, \omega) = \frac{C(k, 0)}{i\omega + k^2 \tilde{D}_T(k, \omega)}. \tag{4.63}$$

Using exactly the same procedures as in Section 4.1 we can convert Equation (4.63) to an expression for the diffusivity in terms of a current correlation function. From Equations (4.7 and 4.11), if $\phi = \dot{C}$ then:

$$\phi(k, t) = k^2 \langle J_{Qx}(k, t) J_{Qx}(-k, 0) \rangle. \tag{4.64}$$

Using Equation (4.11), we obtain the analog of Equation (4.12):

$$k^2 \tilde{D}_T(k, \omega) = \frac{C(k, 0) - i\omega \tilde{C}(k, \omega)}{\tilde{C}(k, \omega)} = \frac{\tilde{\phi}(k, \omega)}{C(k, 0) - \dfrac{\tilde{\phi}(k, \omega)}{i\omega}}. \tag{4.65}$$

If we define the analog of Equation (4.51), that is $\phi(k, t) = k^2 N_Q(k, t)$, then Equation (4.65), for the thermal diffusivity, can be written in the same form as the wave-vector-dependent shear viscosity equation (4.54). That is:

$$\tilde{D}_T(k, \omega) = \frac{\tilde{N}_Q(k, \omega)}{C(k, 0) - \dfrac{k^2}{i\omega} \tilde{N}_Q(k, \omega)}. \tag{4.66}$$

Again we see that we must take the zero wave vector limit *before* we take the zero frequency limit, and using the canonical ensemble fluctuation formula for the specific heat:

$$\rho c_P = \frac{V}{k_B T^2} \lim_{k \to 0} C(k, 0), \tag{4.67}$$

we obtain the Green–Kubo expression for the thermal conductivity:

$$\lambda = \frac{V}{k_B T^2} \int_0^\infty dt \langle J_{Qx}(t) J_{Qx}(0) \rangle. \tag{4.68}$$

This completes the derivation of the Green–Kubo formulae for thermal transport coefficients. These formulae relate thermal transport coefficients to equilibrium properties. In the next chapter we will develop nonequilibrium routes to the thermal transport coefficients.

5

Linear-response theory

5.1 Adiabatic linear response theory

In this chapter we will discuss how an external field F_e, perturbs an N-particle system. We assume that the field is sufficiently weak that only the linear response of the system need be considered. These considerations will lead us to equilibrium fluctuation expressions for mechanical transport coefficients such as electrical conductivity. These expressions are formally identical to the Green–Kubo formulae that were derived in the last chapter. The difference is that the Green–Kubo formulae pertain to thermal transport processes where boundary conditions perturb the system away from equilibrium – all Navier–Stokes processes fall into this category. Mechanical transport coefficients, on the other hand, refer to systems where mechanical fields which appear explicitly in the equations of motion for the system, drive the system away from equilibrium. As we will see, it is no coincidence that there is such a close similarity between the fluctuation expressions for thermal and mechanical transport coefficients. In fact one can often mathematically transform the nonequilibrium boundary conditions for a thermal transport process into a mechanical field. The two representations of the system are then said to be *congruent*.

A major difference between the derivations of the equilibrium fluctuation expressions for the two representations is that in the mechanical case one does not need to invoke Onsager's regression hypothesis. The linear mechanical response of a nonequilibrium system is analyzed mathematically with resultant expressions for the response that involve equilibrium time-correlation functions. In the thermal case – Chapter 4 – equilibrium fluctuations were studied and after invoking Onsager's hypothesis, the connection with nonequilibrium transport coefficients was made. Given a congruent mechanical representation of a thermal-transport process, one can, in fact, prove the validity of Onsager's hypothesis.

The mechanical field F_e performs work on the system, preventing relaxation to equilibrium. This work is converted into heat. It is easy to show that the rate at which the field performs work on the system is, for small fields, proportional to F_e^2. As such this is, at least formally, a nonlinear effect. This is why, in the complete absence of any thermostatting mechanism, Kubo (1957) was able to derive correct expressions for the linear response. However, in spite of heating being a nonlinear effect, a *thermostatted* treatment of linear response theory leads to a considerably more satisfying discussion. We will therefore include in this chapter, a description of thermostats and isothermal linear response theory.

Consider a system of N atoms to which a time-dependent external field $F_e(t)$ is applied at $t = 0$ (Evans and Morriss, 1984b). The generalization to vector or tensor fields is straightforward. For simplicity we will assume that the particles move in a three-dimensional Cartesian space. For times greater than zero the system obeys the dynamics given below:

$$\dot{\mathbf{q}}_i = \frac{1}{m}\mathbf{p}_i + \mathbf{C}_i F_e(t)$$

$$\dot{\mathbf{p}}_i = \mathbf{F}_i + \mathbf{D}_i F_e(t). \tag{5.1}$$

The phase variables $\mathbf{C}_i(\Gamma)$ and $\mathbf{D}_i(\Gamma)$ describe the coupling of the field to the system. We assume that the equations have been written in such a way that at equilibrium, in the absence of the external field, the canonical kinetic energy K satisfies the equipartition relation:

$$\frac{3N}{2}k_B T = \left\langle \sum_{i=1}^{N} \frac{p_i^2}{2m} \right\rangle \equiv \langle K \rangle. \tag{5.2}$$

This implies that the canonical momenta give the peculiar velocities of each of the particles and that therefore:

$$\sum_{i=1}^{N} \mathbf{p}_i = \mathbf{0}. \tag{5.3}$$

In this case, the instantaneous expression for the internal energy H_0 is:

$$H_0(\Gamma) = \sum_{i=1}^{N} \frac{p_i^2}{2m} + \Phi(\mathbf{q}). \tag{5.4}$$

We do not assume that a Hamiltonian exists which will generate the field-dependent equations of motion. In the absence of the external field and the thermostat, H_0 is the total energy, and is therefore a constant of the motion. The rate of change of internal energy due to the field is:

$$\dot{H}_0(\Gamma, t) = \sum_{i=1}^{N} \left[\frac{1}{m} \dot{\mathbf{p}}_i(t) \cdot \mathbf{p}_i - \dot{\mathbf{q}}_i(t) \cdot \mathbf{F}_i \right]$$

$$= - \sum_{i=1}^{N} \left[-\frac{1}{m} \mathbf{D}_i \cdot \mathbf{p}_i + \mathbf{C}_i \cdot \mathbf{F}_i \right] F_e(t) \equiv -J(\Gamma) F_e(t), \qquad (5.5)$$

where $J(\Gamma)$ is called the dissipative flux generated by Equation (5.1).

Monitoring the average response of an arbitrary phase variable $B(\Gamma)$ at some later time can assess the response of the system to the external field. The average response is obtained by perturbing an ensemble of initial phases. As a typical example we will consider a canonical ensemble as the initial distribution, thus:

$$f(\Gamma, 0) = f_c(\Gamma) = \frac{\exp[-\beta H_0(\Gamma)]}{\int d\Gamma' \exp[-\beta H_0(\Gamma')]} = \frac{\exp[-\beta H_0(\Gamma)]}{Z(\beta)}. \qquad (5.6)$$

The average response $\langle B(t) \rangle$ can be calculated from the Schrödinger representation:

$$\langle B(t) \rangle = \int d\Gamma B(\Gamma) f(\Gamma, t), \qquad (5.7)$$

where the distribution function carries the time dependence. Thus, determining the response reduces to determining the perturbed distribution function $f(t)$. The rate of change of the distribution function is given by the Liouville equation:

$$\frac{\partial}{\partial t} f(\Gamma, t) = -iL f(\Gamma, t) = -\left(\frac{\partial}{\partial \Gamma} \cdot \dot{\Gamma}(t) + \dot{\Gamma}(t) \cdot \frac{\partial}{\partial \Gamma} \right) f(\Gamma, t). \qquad (5.8)$$

The $\dot{\Gamma}(t)$ in Equation (5.8) is given by the equations of motion (5.1) with the external field evaluated at the current time, t.

If the equations of motion are derivable from a Hamiltonian, it is easy to show that $\left(\frac{\partial}{\partial \Gamma} \right) \cdot \dot{\Gamma} = 0$ (Section 3.3). In general, we will assume that even in the case where no Hamiltonian exists that $\left(\frac{\partial}{\partial \Gamma} \right) \cdot \dot{\Gamma} = 0$. We refer to this condition as the adiabatic incompressibility of phase space (AIΓ). A sufficient, but not necessary, condition for this to hold is that the unthermostatted or adiabatic equations of motion are derivable from a Hamiltonian. It is, of course, possible to pursue the theory without this condition, but in practice it is rarely necessary to do so (the only known exception is discussed: Macgowan and Evans (1986a)).

Thus in the adiabatic case, if AIΓ holds, we know that the Liouville operator is Hermitian (see Sections 3.3 and 3.5) and therefore:

$$i\mathscr{L}A = \frac{\partial}{\partial \Gamma} \cdot \left(\dot{\Gamma}A \right) = \dot{\Gamma} \cdot \frac{\partial}{\partial \Gamma} A + A \frac{\partial}{\partial \Gamma} \cdot \dot{\Gamma} = \dot{\Gamma} \cdot \frac{\partial}{\partial \Gamma} A = iLA. \qquad (5.9)$$

If we denote the Liouvillean for the field free equations of motion as $i\mathcal{L}_0$, and break up the total Liouvillean into its field-free and field-dependent parts, Equation (5.8) becomes:

$$\frac{\partial}{\partial t}(f_0 + \Delta f(\Gamma, t)) = -(i\mathcal{L}_0 + i\Delta\mathcal{L}(t))(f_0 + \Delta f(\Gamma, t)), \qquad (5.10)$$

where the distribution function $f(\Gamma, t)$, is separated into two parts $f_0 + \Delta f(\Gamma, t)$, where f_0 is the solution of the field-free Liouvillean $i\mathcal{L}_0$. That is:

$$\frac{\partial f_0}{\partial t} = -i\mathcal{L}_0 f_0 = 0. \qquad (5.11)$$

Substituting Equation (5.11) into Equation (5.10) and equating first-order terms we see that:

$$\frac{\partial}{\partial t}\Delta f(\Gamma, t) + i\mathcal{L}_0\Delta f(\Gamma, t) = -i\Delta\mathcal{L}(t)f_0(\Gamma) + O(\Delta^2). \qquad (5.12)$$

In Equation (5.12) we are ignoring perturbations to the distribution function which are second order in the field. (The Schrödinger–Heisenberg equivalence [Section 3.3] proves that these second-order terms for the distribution are identical to the second-order trajectory perturbations.) In Section 7.1 we discuss the nature of this linearization procedure in some detail. The solution of Equation (5.12) is:

$$\Delta f(\Gamma, t) = -\int\limits_0^t ds \exp[-i\mathcal{L}_0(t - s)]i\Delta\mathcal{L}(s)f_0(\Gamma) + O(\Delta^2). \qquad (5.13)$$

We will now operate on the equilibrium distribution function with the operator, $i\Delta\mathcal{L}(t)$. We again use the fact that $i\mathcal{L}_0$ preserves the equilibrium distribution function f_0, so:

$$i\Delta\mathcal{L}(t)f_0 = i\mathcal{L}(t)f_0 - i\mathcal{L}_0 f_0 = i\mathcal{L}(t)f_0 \equiv \beta J(\Gamma)F_e(t)f_0. \qquad (5.14)$$

The last equality in Equation (5.14) defines the dissipation function $J(\Gamma)$ and its character depends on both the form of f_0 and the field-dependent time evolution implicit in $i\mathcal{L}(t)$. Substituting Equation (5.14) into Equation (5.13) and in turn into Equation (5.7), the linear response of the phase variable B is given by:

$$\langle B(t)\rangle = \langle B(0)\rangle + \int d\Gamma B(\Gamma)\Delta f(\Gamma, t)$$

$$= \langle B(0)\rangle - \int_0^t ds \int d\Gamma B(\Gamma)\exp[-i\mathcal{L}_0(t - s)]\beta J(\Gamma)F_e(s)f_0$$

$$= \langle B(0)\rangle - \beta\int_0^t ds F_e(s)\int d\Gamma B(\Gamma(t - s))J(\Gamma)f_0(\Gamma). \qquad (5.15)$$

In deriving the third line of this equation from the second we have *unrolled* the propagator from the dissipative flux onto the response variable B.

It is useful to explore a particular example where f_0 is the canonical distribution. Then the dissipative flux $J(\Gamma)$ can be easily calculated as:

$$i\Delta \mathcal{L}(s)f_c = i\mathcal{L}(s)f_c = \dot{\Gamma}(s) \cdot \frac{\partial}{\partial \Gamma} \frac{\exp[-\beta H_0(\Gamma)]}{Z(\beta)}$$

$$= -\beta\dot{\Gamma}(s) \cdot \frac{\partial H_0(\Gamma) \exp[-\beta H_0(\Gamma)]}{\partial \Gamma \quad Z(\beta)} = -\beta \dot{H}_0(s)f_c(\Gamma), \quad (5.16)$$

and

$$\dot{H}_0(s) = \sum_{i=1}^{N} \left(\dot{\mathbf{q}}_i \cdot \frac{\partial H_0}{\partial \mathbf{q}_i} + \dot{\mathbf{p}}_i \cdot \frac{\partial H_0}{\partial \mathbf{p}_i} \right)$$

$$= \sum_{i=1}^{N} \left(\left(\frac{1}{m}\mathbf{p}_i + \mathbf{C}_iF_e(s) \right) \cdot (-\mathbf{F}_i) + (\mathbf{F}_i + \mathbf{D}_iF_e(s)) \cdot \frac{1}{m}\mathbf{p}_i \right)$$

$$= -F_e(s) \sum_{i=1}^{N} \left(\mathbf{C}_i \cdot \mathbf{F}_i - \mathbf{D}_i \cdot \frac{1}{m}\mathbf{p}_i \right) = -F_e(s)J(\Gamma). \quad (5.17)$$

Here we see explicitly that the coupling of the field in the equations of motion, together with the form of the canonical distribution function, determine the dissipation function $J(\Gamma)$. The linear response of an arbitrary phase variable B with zero mean at equilibrium (that is $\langle B(0) \rangle = 0$) is then:

$$\langle B(t) \rangle = -\beta \int_0^t ds \int d\Gamma B(t-s)J(\Gamma)f_c(\Gamma)F_e(s)$$

$$= -\beta \int_0^t ds\chi_{BJ}(t-s)F_e(s), \quad (5.18)$$

where χ_{BJ} is a linear susceptibility in terms of the equilibrium time-correlation function of B and J, $\chi_{BJ}(t) \equiv \beta\langle B(t)J(0)\rangle_c$.

To linear order, the canonical ensemble averaged linear response for $B(t)$ is:

$$\langle B(t) \rangle = \langle B(0) \rangle - \lim_{F_e \to 0} \int_0^t ds\chi_{BJ}(t-s)F_e(s). \quad (5.19)$$

This equation is very similar to the response functions we met in Chapter 4 when we discussed viscoelasticity and constitutive relations for thermal-transport coefficients. The equation shows that the linear response is non-Markovian. All systems have memory. All N-body systems remember the field history over the decay time of the relevant time-correlation function, $\langle B(t)J(0)\rangle$. Markovian behavior is only an idealization brought about by a lack of sensitivity in our measurements of the time resolved many-body response.

There are a number of deficiencies in the derivation we have just given. Suppose that by monitoring $\langle B(t)\rangle$ for a family of external fields F_e, we wish to deduce the susceptibility $\chi(t)$. One cannot blindly use Equation (5.19). This is because as the system heats up through the conversion of work into heat, the system temperature will change in time. This effect is quadratic with respect to the magnitude of the external field. If χ increases with temperature, the long time limiting value of $\langle B(t)\rangle$ will be infinite. If χ decreases with increasing temperature the limiting value of $\langle B(t)\rangle$ could well be zero. This is simply a reflection of the fact that in the absence of a thermostat there is no steady state. The linear steady-state value for the response can only be obtained if we take the field strength to zero before we let time go to infinity. This procedure will inevitably lead to difficulties in both the experimental and numerical determination of the linear susceptibilities.

Another difficulty with the derivation is that if adiabatic linear response theory were applied to computer simulation, one would prefer not to use canonical averaging. This is because a single Newtonian equilibrium trajectory cannot generate or span the canonical ensemble. A single Newtonian trajectory can, at most, span a microcanonical subset of the canonical ensemble of states. A canonical evaluation of the susceptibility therefore requires an ensemble of trajectories if one is using Newtonian dynamics. This is inconvenient and very expensive in terms of computer time.

One cannot simply extend this adiabatic theory to the microcanonical ensemble. Kubo (1982) recently showed that if one subjects a cumulative microcanonical ensemble (all states less than a specified energy have the same probability) to a mechanical perturbation, then the linear susceptibility is given by the equilibrium correlation of the test variable B and the dissipative flux J, averaged over the microcanonical ensemble (all states with a precisely specified energy have the same probability). When the equilibrium ensemble of starting states is not identical to the equilibrium ensemble used to compute the susceptibilities, we say that the theory is *ergodically inconsistent*. We will now show how both of these difficulties can be resolved.

5.2 Thermostats and equilibrium distribution functions

The Gaussian isokinetic thermostat

Thermostats were first introduced as an aid to performing nonequilibrium computer simulations. Only later was it realized that these devices have a fundamental role in the statistical mechanics of many-body systems. The first deterministic method for thermostatting molecular dynamics simulations was proposed simultaneously and independently by Hoover and Evans (Hoover *et al.*, 1982; Evans, 1983a). Their method employs a damping or friction term in the equations of motion. Initially the use of such damping terms had no theoretical justification. Later it was realized (Evans *et al.*, 1983) that these equations of motion could be derived using Gauss' principle of least constraint (Section 3.1). This systematized the extension of the method to other constraint functions.

Using Gauss' principle (Chapter 3), the isokinetic equations of motion for a system subject to an external field can be written as:

$$\dot{\mathbf{q}}_i = \frac{\mathbf{p}_i}{m} + \mathbf{C}_i F_e(t)$$

$$\dot{\mathbf{p}}_i = \mathbf{F}_i + \mathbf{D}_i F_e(t) - \alpha \mathbf{p}_i. \tag{5.20}$$

This is the thermostatted generalization of Equation (5.1) where the thermostatting term $\alpha \mathbf{p}_i$ has been added. In writing these equations we are assuming:

(1) that the equations have been written in a form in which the canonical momenta are peculiar with respect to the local streaming velocity;
(2) that $\sum \mathbf{p}_i = \mathbf{0}$;
(3) and that H_0 is the phase variable which corresponds to the internal energy.

In order to know that these three conditions are valid, we must know quite a lot about the possible flows induced in the system by the external field. This means that if we are considering shear flow, for example, the Reynolds number must be small enough for laminar flow to be stable. Otherwise we cannot specify the streaming component of a particles motion ($\mathbf{C}_i$ must contain the local hydrodynamic flow field $\mathbf{u}(\mathbf{r}, t)$) and we cannot expect condition (1) to be valid.

The isokinetic expression for the multiplier is easily seen to be:

$$\alpha = \alpha_0 + \alpha_1 F_e(t) = \frac{\sum_i \frac{1}{m_i} \mathbf{F}_i \cdot \mathbf{p}_i}{\sum_i \frac{1}{m_i} \mathbf{p}_i^2} + \frac{\sum_i \frac{1}{m_i} \mathbf{D}_i \cdot \mathbf{p}_i}{\sum_i \frac{1}{m_i} \mathbf{p}_i^2} F_e(t). \tag{5.21}$$

It is instructive to compare this result with the corresponding field-free multiplier given in Equation (3.40). It is important to keep in mind that the expression for the multiplier depends explicitly on the external field and therefore on time. This is why we define the time and field-independent phase variables α_0, α_1.

It is easy to show that if Gauss' principle is used to fix the internal energy H_0, then the equations of motion take on exactly the same form (Evans, 1983a), except that the multiplier is:

$$\alpha = \frac{\sum\limits_{i=1}^{N} \mathbf{D}_i \cdot \mathbf{p}_i - \mathbf{C}_i \cdot \mathbf{F}_i}{\sum\limits_{i=1}^{N} \frac{1}{m_i} \mathbf{p}_i^2} F_e(t). \tag{5.22}$$

It may seem odd that the form of the field-dependent equations of motion is independent of whether we are constraining the kinetic or the total energy. This occurs because the vector character of the constraint force is the same for both forms of constraint. In the isoenergetic case, it is clear that the multiplier vanishes when the external field is zero. This is as expected, since in the absence of an external field, Newton's equations conserve the total energy.

Gaussian thermostats remove heat from the system at a rate:

$$\dot{Q}(t) = \left(\frac{\mathrm{d}H_0}{\mathrm{d}t}\right)^{\mathrm{therm}} = \alpha(t) \sum_{i=1}^{N} \frac{\mathbf{p}_i^2}{m_i}, \tag{5.23}$$

by applying a force of constraint which is parallel to the peculiar velocity of each particle in the system.

We will now discuss the equilibrium properties of Gaussian isokinetic systems in more detail. At equilibrium, the Gaussian isokinetic equations become:

$$\dot{\mathbf{q}}_i = \frac{\mathbf{p}_i}{m}$$

$$\dot{\mathbf{p}}_i = \mathbf{F}_i - \alpha\mathbf{p}_i, \tag{5.24}$$

with the multiplier given by Equation (5.21) with $F_e = 0$. Clearly the average value of the multiplier is zero at equilibrium, with fluctuations in its value being precisely those required to keep the kinetic energy constant. Following our assumption that the initial value of the total linear momentum is zero, it is trivial to see that, like the kinetic energy, it is a constant of the motion.

The ergodically generated equilibrium distribution function $f_K(\Gamma)$, can be obtained by solving the Liouville equation for these equations of motion. Consider the total time derivative of f. From the Liouville equation (3.42), we see that:

$$\frac{\mathrm{d}f}{\mathrm{d}t} = -f \frac{\partial}{\partial \Gamma} \cdot \dot{\Gamma} = -f \sum_{i=1}^{N} \frac{\partial}{\partial \mathbf{p}_i} \cdot \dot{\mathbf{p}}_i = f \sum_{i=1}^{N} \frac{\partial}{\partial \mathbf{p}_i} \cdot (\alpha\mathbf{p}_i). \tag{5.25}$$

In computing the final derivative in this equation we get $3N$ identical intensive terms from the $3N$ derivatives, $\alpha\left(\frac{\partial}{\partial \mathbf{p}_i} \cdot \mathbf{p}_i\right)$. We also get $3N$ terms from $\mathbf{p}_i \cdot \partial\alpha/\partial\mathbf{p}_i$ which sum to give α. Since we are interested in statistical mechanical systems, we will ignore terms of relative order $1/N$, in the remaining discussion. It is certainly possible to retain these terms, but this would add considerably to the algebraic complexity, without revealing any new physics. This being the case, Equation (5.25) above becomes:

$$\frac{df}{dt} = 3N\alpha f + O(1)f. \tag{5.26}$$

From Equation (5.25) it is can be shown that:

$$\frac{df}{dt} = -\frac{3N}{2K}f\dot{\Phi}, \qquad \text{or} \qquad \frac{d\ln f}{dt} = -\frac{3N}{2K}\frac{d\Phi}{dt}. \tag{5.27}$$

Integrating both sides with respect to time enables us to evaluate the time independent equilibrium distribution function:

$$f_K(\Gamma) = \frac{\exp[-\beta\Phi(\Gamma)]\delta(K(\Gamma) - K_0)}{\int d\Gamma \exp[-\beta\Phi(\Gamma)]\delta(K(\Gamma) - K_0)}, \tag{5.28}$$

where the constant, $\beta = 3N/2K_0$. We call this distribution function the isokinetic distribution f_K (Evans and Morriss, 1983a; 1983c). It has a very simple form: the kinetic degrees of freedom are distributed microcanonically, and the configurational degrees of freedom are distributed canonically. The thermodynamic temperatures $(\partial E/\partial S)_{N,V} = T$ of these two subsystems are, of course, identical.

If one retains terms of order $1/N$ in the above derivation, the result is the same except that $\beta = (3N - 4)/2K_0$. Such a result could have been anticipated in advance because, in our Gaussian isokinetic system, four degrees of freedom are frozen, one by the kinetic energy constraint, and three because the linear momentum is fixed.

One can check that the isokinetic distribution is an equilibrium solution of the equilibrium Liouville equation. Clearly $df_K/dt \neq 0$. As one follows the streaming motion of an evolving point in phase space $\Gamma(t)$, the streaming derivative of the comoving local density is:

$$\frac{df_K}{dt} = \frac{3N}{2K(\Gamma)}\dot{\Phi}(\Gamma)f_K(\Gamma) \neq 0. \tag{5.29}$$

This is a direct consequence of the fact that, for a Gaussian isokinetic system, phase space is compressible. It is clear, however, that in the absence of external fields $\langle df_K/dt \rangle = 0$, because the mean value of $\dot{\Phi}$ must be zero. If we sit at a fixed point in phase space and ask whether, under Gaussian isokinetic dynamics, the isokinetic distribution function changes, then the answer is no. The isokinetic distribution is the equilibrium distribution function. It is preserved by the dynamics.

Substitution into the Liouville equation gives:

$$\frac{\partial f_K}{\partial t} = -\dot{\Gamma} \cdot \frac{\partial f_K}{\partial \Gamma} - f_K \frac{\partial}{\partial \Gamma} \cdot \dot{\Gamma} = (\beta \dot{\Phi} + 3N\alpha) f_K = 0. \qquad (5.30)$$

The proof that the last two terms sum to zero is easily given using the fact that, $\beta = 3N/2K_0$ and that $K = \sum \frac{1}{2m} \mathbf{p}^2$ is a constant of the motion. Hence

$$\beta\dot{\Phi} + 3N\alpha = -\beta \sum_i \frac{1}{m_i} \mathbf{p}_i \cdot \mathbf{F}_i + 3N \frac{\sum_i \frac{1}{m_i} \mathbf{p}_i \cdot \mathbf{F}_i}{\sum_i \frac{1}{m_i} \mathbf{p}_i^2} = 0. \qquad (5.31)$$

If the equilibrium isokinetic system is ergodic, a single trajectory in phase space will eventually generate the isokinetic distribution. On the other hand, a single iso-kinetic trajectory cannot ergodically generate a canonical distribution. We can, however, ask whether isokinetic dynamics will preserve the canonical distribution. If we integrate the equations of motion for an ensemble of systems which are initially distributed canonically, will that distribution be preserved by isokinetic dynamics? Clearly:

$$\frac{\partial f_c}{\partial t} = f_c(3N\alpha + \beta\dot{K} + \beta\dot{\Phi}) = f_c(\beta - \frac{3N}{2K(\Gamma)})\dot{\Phi}(\Gamma) = f_c\Delta(\beta)\dot{\Phi}(\Gamma), \qquad (5.32)$$

is not identically zero. In this expression K is a phase variable and not a constant, and $\dot{\Phi}$ is only equal to zero on average. K would only be a constant if all members of the ensemble had identical kinetic energies. The mean value of $3N/2K$ is, of course, β.

Consider the time derivative of the canonical average of an arbitrary extensive phase variable B where the dynamics are Gaussian isokinetic:

$$\frac{d}{dt}\langle B(t)\rangle = \int d\Gamma B \frac{\partial f_c}{\partial t} = \int d\Gamma B \Delta(\beta)\dot{\Phi} f_c. \qquad (5.33)$$

The time derivative of the ensemble average is:

$$\frac{d}{dt}\langle B(t)\rangle = \left\langle B\left(\frac{3N}{2K} - \beta\right)\dot{\Phi}\right\rangle = \frac{\beta}{K_0}\langle B\Delta K\dot{\Phi}\rangle + O(\Delta^2), \qquad (5.34)$$

where $\Delta K \equiv K - \langle K\rangle = K - K_0$. Equation (5.34) can be written as the time derivative of a product of three extensive, zero-mean variables:

$$\frac{d}{dt}\langle B(t)\rangle = \frac{\beta}{K_0}[\langle B\rangle\langle\Delta K\dot{\Phi}\rangle + \langle\Delta B\Delta K\dot{\Phi}\rangle] = \frac{\beta}{K_0}\langle\Delta B\Delta K\dot{\Phi}\rangle = O(1). \qquad (5.35)$$

In deriving these equations we have used the fact that $\langle \Delta K \dot{\Phi} \rangle = 0$, and that the ensemble average of the product of three extensive, zero-mean phase variables is of order N, while $K_0 = \langle K \rangle$ is extensive.

The above equation shows that although B is extensive, the change in $\langle B(t) \rangle$ with time, (as the ensemble changes from canonical at $t = 0$, to whatever for the Gaussian isokinetic equations generate as $t \to \infty$) is of order 1 and therefore can be ignored relative to the average of B itself. In the thermodynamic limit the canonical distribution is *preserved* by Gaussian isokinetic dynamics.

Hamiltonian formulation of the GIK thermostat

One of the most surprising results is that there is a Hamiltonian formulation of the Gaussian isokinetic (GIK) thermostat (Dettmann and Morriss, 1996a). Related to this is a variational formulation in terms of geodesic motion in a curved space. The existence of a Hamiltonian allows a correspondence to be made between Gaussian thermostatted systems and other Hamiltonian dynamical systems, so that both can be treated on an equal footing. The implied conservation of phase volume and the associated symplectic structure of phase space give a new understanding of thermostatted systems. Later we will see that this Hamiltonian structure survives away from equilibrium for some systems. For the moment we emphasize that the seemingly nonHamiltonian equations of motion become Hamiltonian after a particular change of variable involving the momenta and the time. The apparent contradiction is resolved by distinguishing between physical and canonical momenta, (in analogy with a charged particle in a magnetic field).

Setting $m = 1$ and scaling the magnitude of the momentum vector $|(\mathbf{p}_1, \ldots, \mathbf{p}_N)|$ to be equal to one, the Gaussian isokinetic equations of motion can be rewritten as:

$$\frac{d\mathbf{q}_i}{dt} = \mathbf{p}_i, \qquad \frac{d\mathbf{p}_i}{dt} = \mathbf{F}_i - \alpha \mathbf{p}_i, \qquad \alpha = \sum_{i=1}^{N} \mathbf{F}_i \cdot \mathbf{p}_i. \tag{5.36}$$

In these variables the total kinetic energy is equal to $1/2$. The central result is a one-parameter family of Hamiltonians which generate the GIK equations of motion (Dettmann and Morriss, 1996a):

$$H_\beta(\mathbf{q}, \boldsymbol{\pi}; \lambda) = \frac{1}{2} \exp[(\beta + 1)\Phi(\mathbf{q})] \sum_{i=1}^{N} \pi_i^2 - \frac{1}{2} \exp[(\beta + 1)\Phi(\mathbf{q})]. \tag{5.37}$$

Here π_i is the *Hamiltonian* momentum of particle i and λ is the *Hamiltonian* time, so the Hamiltonian equations of motion are:

$$\frac{d}{d\lambda}\mathbf{q}_i = \frac{\partial H}{\partial \boldsymbol{\pi}_i} = \exp[(\beta + 1)\Phi]\boldsymbol{\pi}_i = (\exp[\Phi]\boldsymbol{\pi}_i)\exp[\beta\Phi] \tag{5.38}$$

$$\frac{d}{d\lambda}\boldsymbol{\pi}_i = -\frac{\partial H}{\partial \mathbf{q}_i} = -\frac{1}{2}\frac{\partial \Phi}{\partial \mathbf{q}_i}\exp[(\beta - 1)\Phi]$$

$$\times \left((\beta + 1)\exp[2\Phi]\sum_i \boldsymbol{\pi}_i^2 - (\beta - 1)\right). \tag{5.39}$$

We are now free to choose the connection between Hamiltonian variables $(\mathbf{q}, \boldsymbol{\pi}, \lambda)$ and physical variables $(\mathbf{q}, \mathbf{p}, t)$. Equation (5.38) suggests a particular choice:

$$\frac{dt}{d\lambda} = \exp[\beta\Phi] \qquad \text{and} \qquad \mathbf{p}_i = \exp[\Phi]\boldsymbol{\pi}_i. \tag{5.40}$$

As the physical kinetic energy is equal to $1/2$, Equation (5.40) implies that $\exp[2\Phi]\sum_i \boldsymbol{\pi}_i^2 = 1$, and hence the conserved value of the Hamiltonian (5.37) is zero.

We now demonstrate that the equations of motion (5.38) and (5.39) together with the connection between Hamiltonian and physical variables, leads to the GIK equations of motion. Combining Equations (5.38) and (5.40) the left-hand side of Equation (5.38) becomes:

$$\frac{d}{d\lambda}\mathbf{q}_i = \frac{dt}{d\lambda}\frac{d}{dt}\mathbf{q}_i = \exp[\beta\Phi]\frac{d}{dt}\mathbf{q}_i = \exp[\beta\Phi]\dot{\mathbf{q}}_i,$$

and the right-hand side is $(\exp[\Phi]\boldsymbol{\pi}_i)\exp[\beta\Phi] = \mathbf{p}_i\exp[\beta\Phi]$ so we obtain $\dot{\mathbf{q}}_i = \mathbf{p}_i$, that is the first equation of (5.36). Similarly for Equation (5.39), the left-hand side is:

$$\frac{d}{d\lambda}\boldsymbol{\pi}_i = \frac{dt}{d\lambda}\frac{d}{dt}(\exp[-\Phi]\mathbf{p}_i) = \exp[\beta\Phi]\left(\exp[-\Phi]\dot{\mathbf{p}}_i - \mathbf{p}_i\exp[-\Phi]\frac{d\Phi}{dt}\right)$$

$$= \exp[(\beta - 1)\Phi]\left(\dot{\mathbf{p}}_i - \mathbf{p}_i\sum_{i=1}^{N}\dot{\mathbf{q}}_i \cdot \frac{d\Phi}{d\mathbf{q}_i}\right)$$

$$= \exp[(\beta - 1)\Phi]\left(\dot{\mathbf{p}}_i + \mathbf{p}_i\sum_{i=1}^{N}\mathbf{p}_i \cdot \mathbf{F}_i\right) = \exp[(\beta - 1)\Phi](\dot{\mathbf{p}}_i + \alpha\mathbf{p}_i),$$

and the right-hand side is:

$$\frac{1}{2}\mathbf{F}_i\exp[(\beta - 1)\Phi]\left((\beta + 1)\exp[2\Phi]\sum_i \boldsymbol{\pi}_i^2 - (\beta - 1)\right)$$

$$= \frac{1}{2}\mathbf{F}_i\exp[(\beta - 1)\Phi]\left((\beta + 1)\sum_i \mathbf{p}_i^2 - (\beta - 1)\right)$$

$$= \frac{1}{2}\mathbf{F}_i\exp[(\beta - 1)\Phi](2) = \mathbf{F}_i\exp[(\beta - 1)\Phi].$$

Combining these two equations gives the second equation in (5.36) $\dot{\mathbf{p}}_i = \mathbf{F}_i - \alpha\mathbf{p}_i$, where α is also given by Equation (5.36). This completes the proof that the Hamiltonian (5.37) generates the GIK equations of motion.

Notice that the value of β is completely arbitrary, and does not affect the equations of motion, however, there are three particular values of β in which the canonical variables have a simple interpretation. Hamiltonians with different β are not related by any of the usual types of canonical transformation, since the time variables differ in each case.

(1) For $\beta = -1$, the Hamiltonian reduces to kinetic plus potential energy. The thermostatted equations are thus equivalent to a potential problem with zero total energy. Alternatively, any system of particles with purely attractive forces and zero total energy can be represented in terms of GIK thermostatted dynamics.
(2) For $\beta = 0$, the canonical and physical times are the same, so this is the most natural form in which to derive the thermostatted dynamics.
(3) For $\beta = 1$, the Hamiltonian is a quadratic form:

$$H_{\beta=1}(\mathbf{q}, \boldsymbol{\pi}, \lambda) = \frac{1}{2}\exp[2\Phi(\mathbf{q})]\sum_{i=1}^{N}\pi_i^2 - \frac{1}{2}. \tag{5.41}$$

Ignoring the constant, we see that GIK dynamics are equivalent to a geodesic in a curved space with metric $g^{\mu\nu}$, that is $H_g(\mathbf{q}, \boldsymbol{\pi}) = \frac{1}{2}g^{\mu\nu}(\mathbf{q})\pi_\mu\pi_\nu$. $H_{\beta=1}$ is equivalent to geodesic motion on configuration space with a metric given by:

$$ds^2 = \exp[-2\Phi]\sum_{i=1}^{N}d\mathbf{q}_i^2. \tag{5.42}$$

The trajectory followed between two points in configuration space has extremal length with respect to the metric. That is, for any two points in configuration space A and B, the trajectory followed by the system has a minimum value of $\int ds$ for all paths between A and B. Occasionally it may be only a local minimum, or even (for sufficiently pathological Φ) a maximum. We have also incidentally proved that the dynamics is time reversible, as there is no preferred direction along a geodesic. We have shown that the Gaussian isokinetic thermostat is intimately related to more conventional dynamical systems, augmenting the link which has already been made between quadratic Hamiltonians and geodesic motion in a curved manifold (Szydlowski and Biesiada, 1991).

Nosé–Hoover thermostat – canonical ensemble

The Gaussian thermostat generates the isokinetic ensemble by a differential feedback mechanism. The change in the kinetic temperature is constrained to be precisely zero. Control theory provides a range of alternative feedback processes: proportional, differential, and integral. After the Gaussian thermostat was developed (which uses differential feedback), (Nosé, 1984a; 1984b) utilized an integral feedback mechanism to generate the canonical ensemble.

The original Nosé method considers an extended system with an additional coordinate s and its conjugate momentum p_s. The extra two variables s and p_s can be considered to be the simplest realization of an external reservoir, interacting with the system by scaling the velocities $\mathbf{v}_i = s\dot{\mathbf{q}}_i$. The potential energy associated with s is $(g+1)k_BT \ln s$, where g is the number of degrees of freedom of the system and T is the target value of the temperature. It is the particular choice of the potential for s and the interaction with the rest of the system which leads to dynamics which generate the canonical ensemble. Subsequently, Hoover (1985) rewrote the Nosé equations of motion in a form that is similar to the Gaussian equations. The only difference is that the multiplier α has its own equation of motion.

The Nosé Hamiltonian for the extended system of N particles in a potential Φ is:

$$H_N(\mathbf{q}, \boldsymbol{\pi}, s, p_s) = \sum_{i=1}^{N} \frac{\pi_i^2}{2\,m_i s^2} + \Phi(\mathbf{q}) + \frac{p_s^2}{2Q} + (g+1)k_BT \ln s, \qquad (5.43)$$

where Q is an arbitrary constant corresponding to the mass of the reservoir (s is dimensionless so Q does not have the dimensions of mass). The equations of motion generated by this Hamiltonian are:

$$\frac{d\mathbf{q}_i}{d\lambda} = \frac{\boldsymbol{\pi}_i}{ms^2}, \quad \frac{d\boldsymbol{\pi}_i}{d\lambda} = \mathbf{F}_i,$$

$$\frac{ds}{d\lambda} = \frac{p_s}{Q}, \quad \frac{dp_s}{d\lambda} = \sum_{i=1}^{N} \frac{\pi_i^2}{m_i s^3} - \frac{(g+1)k_BT}{s}, \qquad (5.44)$$

where we choose λ to be the *Hamiltonian* time variable. We can eliminate the variable p_s from the equations of motion by combining the last two equations to obtain a single second-order differential equation for s. For an equilibrium system, the average force on the s variable must be zero, so:

$$\left\langle \frac{d^2 s}{d\lambda^2} \right\rangle = \frac{1}{Q} \left\langle \sum_{i=1}^{N} \frac{\pi_i^2}{m_i s^3} - \frac{(g+1)k_BT}{s} \right\rangle = 0. \qquad (5.45)$$

If we choose the Hamiltonian momenta to be related to the physical momenta by $\boldsymbol{\pi}_i = s\mathbf{p}_i$ then:

$$\left\langle \sum_{i=1}^{N} \frac{\pi_i^2}{m_i s^3} \right\rangle = \left\langle \sum_{i=1}^{N} \frac{1}{s} \frac{\mathbf{p}_i^2}{m_i} \right\rangle = (g+1)k_BT \left\langle \frac{1}{s} \right\rangle. \qquad (5.46)$$

This suggests that the dynamical average of k_BT and an arbitrary phase variable A has the general form:

$$k_BT = \frac{\left\langle \sum_{i=1}^{N} \frac{1}{s}\frac{\mathbf{p}_i^2}{m_i}\right\rangle}{(g+1)\left\langle \frac{1}{s}\right\rangle} \Rightarrow \langle A \rangle = \frac{\int_0^T \left(\frac{A(\lambda)}{s}\right)\mathrm{d}\lambda}{\int_0^T \left(\frac{1}{s}\right)\mathrm{d}\lambda}. \tag{5.47}$$

Comparing this with the usual physical time t we take the relation between the Hamiltonian time λ and physical time t to be $\mathrm{d}t = \mathrm{d}\lambda/s$.

To calculate the equilibrium distribution function for the Nosé Hamiltonian we use the fact that for an ergodic system, the extended system is microcanonical. Hence from Equation (5.43):

$$Z = \frac{1}{N!}\int \mathrm{d}\mathbf{q}\,\mathrm{d}\boldsymbol{\pi}\,\mathrm{d}s\,\mathrm{d}p_s\,\delta\left(\sum_{i=1}^{N}\frac{\boldsymbol{\pi}_i^2}{2\,m_i s^2} + \Phi(\mathbf{q}) + \frac{p_s^2}{2Q} + (g+1)k_BT\ln s - E\right), \tag{5.48}$$

where $\mathbf{q}$ and $\boldsymbol{\pi}$ are $3N$-dimensional vectors, $\mathbf{q} \equiv (\mathbf{q}_1,\ldots,\mathbf{q}_N)$ and $\boldsymbol{\pi} \equiv (\boldsymbol{\pi}_1,\ldots,\boldsymbol{\pi}_N)$. If we change variables from $\boldsymbol{\pi}$ to $\mathbf{p}$, where $\mathbf{p} \equiv (\mathbf{p}_1,\ldots,\mathbf{p}_N)$ and $\mathbf{p}_i = \boldsymbol{\pi}_i/s$ for each i, then:

$$Z = \frac{1}{N!}\int \mathrm{d}\mathbf{q}\,\mathrm{d}\mathbf{p}\,\mathrm{d}s\,\mathrm{d}p_s s^{3N}\delta\left(H_0(\mathbf{q},\mathbf{p}) + \frac{p_s^2}{2Q} + (g+1)k_BT\ln s - E\right), \tag{5.49}$$

where $H_0(\mathbf{q},\mathbf{p}) = \sum_i \mathbf{p}_i^2/2\,m_i + \Phi(\mathbf{q})$ is the usual N-particle Hamiltonian. The integral over s can be performed as the only contributions come from the zeros of the argument of the delta function. If $G(s) = H_0(\mathbf{q},\mathbf{p}) + p_s^2/2Q + (g+1)k_BT\ln s - E$, then G has only one zero, that is:

$$s_0 = \exp\left[-\frac{H_0(\mathbf{q},\mathbf{p}) + p_s^2/2Q - E}{(g+1)k_BT}\right]. \tag{5.50}$$

Using the identity $\delta(G(s)) = \delta(s - s_0)/G'(s)$ it is easy to show that the integral over s gives:

$$Z = \frac{1}{N!}\int \mathrm{d}\mathbf{q}\,\mathrm{d}\mathbf{p}\,\mathrm{d}p_s\frac{1}{(g+1)k_BT}\exp\left[-\frac{1}{k_BT}\frac{3N+1}{(g+1)}\left(H_0(\mathbf{q},\mathbf{p}) + \frac{p_s^2}{2Q} - E\right)\right]. \tag{5.51}$$

Here the choice $g = 3N$ cancels the factor in the exponent. The integral over p_s is the infinite integral of a Gaussian and the result is:

$$Z = \frac{1}{(3N+1)} \left(\frac{2\pi Q}{k_B T}\right)^{1/2} \frac{1}{N!} \int d\mathbf{q} d\mathbf{p} \, \exp\left[-\frac{H_0(\mathbf{q}, \mathbf{p}) - E}{k_B T}\right]. \tag{5.52}$$

If the variables $\mathbf{q}, \pi, s, p_s$ are distributed microcanonically, then the variables $\mathbf{p}$ and $\mathbf{q}$ are distributed canonically.

To compute averages in real physical time the equations of motion can be written in terms of the physical variables $(\mathbf{q}, \mathbf{p}, t)$, eliminating the Hamiltonian variables π, s, p_s, λ, entirely. We use the same transformations as before $\pi_i = s\mathbf{p}_i$ and $dt = d\lambda/s$ to rewrite the Nosé equations of motion as:

$$\frac{d\mathbf{q}_i}{dt} = \frac{\mathbf{p}_i}{m_i}, \qquad \frac{d\mathbf{p}_i}{dt} = \mathbf{F}_i - \left(\frac{1}{s}\frac{ds}{dt}\right)\mathbf{p}_i,$$

$$\frac{ds}{dt} = \frac{p_s}{Q}s, \qquad \frac{dp_s}{dt} = \sum_{i=1}^{N} \frac{\mathbf{p}_i^2}{m_i} - (g+1)k_B T. \tag{5.53}$$

Introducing a new variable $\zeta = p_s/Q$ to replace p_s, the equations of motion become:

$$\dot{\mathbf{q}}_i = \frac{\mathbf{p}_i}{m_i}, \quad \dot{\mathbf{p}}_i = \mathbf{F}_i - \zeta\mathbf{p}_i,$$

$$\dot{s} = \zeta s, \quad \dot{\zeta} = \frac{1}{Q}\left(\sum_{i=1}^{N} \frac{\mathbf{p}_i^2}{m_i} - (g+1)k_B T\right) = \frac{1}{\tau^2}\left(\frac{K(\mathbf{p})}{K_0} - 1\right), \tag{5.54}$$

where the dot represents the derivative with respect to t. The term $K_0 = \frac{1}{2}(g+1)k_B T$ is the target value of the kinetic energy, $K(\mathbf{p})$ is the instantaneous value of the kinetic energy, and τ is a relaxation time which is related to the *mass* of the s degree of freedom $(\tau^2 = \frac{1}{2}Q/K_0)$. The motion of the system of interest can now be determined without reference to s. It is an irrelevant variable! The variable $\dot{\zeta}$ is a function of $\mathbf{p}$ only, so the complete description of the system can be given in terms of the variables $(\mathbf{q}, \mathbf{p}, \zeta)$. A time average in the physical variables now takes the usual form:

$$\langle A \rangle_t = \frac{1}{T'} \int_0^{T'} dt' A(\mathbf{q}, \mathbf{p}). \tag{5.55}$$

Notice that a Nosé isoenergetic (NIE) thermostat can be constructed from Equations (5.54) by replacing $K(\mathbf{p})$ by $E(\mathbf{q}, \mathbf{p})$, and K_0 by E_0.

We have dispensed with the original form of Nosé's equations entirely. The N particle distribution function $f(\Gamma, \zeta)$ generated by the Nosé–Hoover equations of motion can be obtained by solving the Liouville equation for the equations of motion for the physical variables $(\mathbf{q}, \mathbf{p}, \zeta)$. The total time derivative of $f(\Gamma, \zeta)$ from the Liouville equation is:

$$\frac{\mathrm{d}f}{\mathrm{d}t} = -f\left(\frac{\partial}{\partial \Gamma} \cdot \dot{\Gamma} + \frac{\partial}{\partial \zeta}\dot{\zeta}\right). \tag{5.56}$$

From the equations of motion (5.54) it is easy to see that $\dot{\zeta}$ is a function of $\mathbf{p}$, and hence independent of ζ. The only nonzero contribution to the right-hand side comes from the $\mathbf{p}$ dependence of $\dot{\mathbf{p}}$, so that:

$$\frac{\mathrm{d}}{\mathrm{d}t}\ln f = 3N\zeta. \tag{5.57}$$

Considering the time derivative of the quantity $H_0 + \frac{1}{2}Q\zeta^2$:

$$\frac{\mathrm{d}}{\mathrm{d}t}\left(H_0 + \frac{1}{2}Q\zeta^2\right) = \frac{\mathrm{d}}{\mathrm{d}t}H_0 + Q\zeta\dot{\zeta} = -(g+1)\zeta k_B T. \tag{5.58}$$

If we take $g + 1 = 3N$, then from Equations (5.57) and (5.58) we find that the equilibrium distribution function is the extended canonical distribution f_c, where:

$$f_c(\Gamma, \zeta) = \frac{\exp\left[-\beta\left(H_0 + \frac{1}{2}Q\zeta^2\right)\right]}{\int \mathrm{d}\Gamma \mathrm{d}\zeta \exp\left[-\beta\left(H_0 + \frac{1}{2}Q\zeta^2\right)\right]}. \tag{5.59}$$

In the Hoover representation of the equations of motion the scaling variable s has been eliminated so the number of degrees of freedom of the system changes from $3N + 1$ to $3N$.

5.3 Isothermal linear response theory

In Section 5.2 we considered two forms of thermostatted dynamics – the Gaussian isokinetic dynamics and the Nosé–Hoover canonical ensemble dynamics. Both of these thermostatted equations of motion can add or remove energy from the system to control its temperature. It is particularly important to incorporate thermostatted dynamics when an external field perturbs the system. This allows the irreversibly produced heat to be removed continuously, and the system maintained in a steady, nonequilibrium state. We now generalize the adiabatic linear response theory of Section 5.1, to treat perturbed thermostatted systems we have developed in Section 5.2. Following Morriss and Evans (1985), we consider an N-particle system evolving under the Gaussian isokinetic dynamics for $t < 0$, but subject to an external field F_e, for all times $t > 0$. The equations of motion are given by:

$$\dot{\mathbf{q}}_i = \frac{\mathbf{p}_i}{m} + \mathbf{C}_i F_e(t)$$

$$\dot{\mathbf{p}}_i = \mathbf{F}_i + \mathbf{D}_i F_e(t) - \alpha \mathbf{p}_i. \tag{5.60}$$

The term $\alpha\mathbf{p}_i$ couples the system to a thermostat and we shall take:

$$\alpha = \alpha_0 + \alpha_1 F_e(t) = \frac{\sum_i \frac{1}{m_i}\mathbf{F}_i \cdot \mathbf{p}_i}{\sum_i \frac{1}{m_i}\mathbf{p}_i^2} + \frac{\sum_i \frac{1}{m_i}\mathbf{D}_i \cdot \mathbf{p}_i}{\sum_i \frac{1}{m_i}\mathbf{p}_i^2} F_e(t), \tag{5.61}$$

so that the peculiar kinetic energy, $K(\Gamma) = \sum_i p_i^2/2m = K_0$, is a constant of the motion. In the absence of the field, these equations of motion ergodically generate the isokinetic distribution function, f_K, Equation (5.28), with $\beta = 3N/2K_0$. As we have seen, the isokinetic distribution function f_K, is preserved by the field-free isokinetic equations of motion and that:

$$\frac{\partial f_K}{\partial t} = -i\mathscr{L}_K f_K = 0. \tag{5.62}$$

We use $i\mathscr{L}_K$ for the zero-field isokinetic Liouvillean.

To calculate the linear thermostatted response we need to solve the linearized Liouville equation for thermostatted systems. Following the same arguments used in the adiabatic case (Equations (5.8–5.12)), the linearized Liouville equation is:

$$\frac{\partial}{\partial t}\Delta f(\Gamma, t) + i\mathscr{L}_K \Delta f(\Gamma, t) = -\Delta i\mathscr{L}(t)f_K(\Gamma) + O(\Delta^2), \tag{5.63}$$

where $i\mathscr{L}(t)$ is the external field dependent, isokinetic Liouvillean and $\Delta i\mathscr{L}(t) = i\mathscr{L}(t) - i\mathscr{L}_K$. Its solution is the analog of (5.13), namely:

$$\Delta f(\Gamma, t) = -\int_0^t \mathrm{d}s \exp(-i\mathscr{L}_K(t-s))\Delta i\mathscr{L}(s)f_K(\Gamma) + O(\Delta^2), \tag{5.64}$$

Using Equations (5.8), (5.28), and (5.60), and the fact that $\beta = 3N/2K_0$, it is easy to show that:

$$\Delta i\mathscr{L}(t)f_K(\Gamma) = i\mathscr{L}(t)f_K(\Gamma) - i\mathscr{L}_K f_K(\Gamma) = i\mathscr{L}(t)f_K(\Gamma)$$

$$= \left(\dot{\Gamma}(t) \cdot \frac{\partial}{\partial\Gamma} + \left(\frac{\partial}{\partial\Gamma} \cdot \dot{\Gamma}(t)\right)\right)f_K(\Gamma) \tag{5.65}$$

$$= -\beta\dot{\Phi}f_K(\Gamma) - f_K(\Gamma)\sum_i \frac{\partial}{\partial\mathbf{p}_i} \cdot (\alpha\mathbf{p}_i).$$

There is one subtle point in deriving the last line of Equation (5.65):

$$\sum_{i=1}^N \dot{\mathbf{p}}_i \cdot \frac{\partial f_K}{\partial\mathbf{p}_i} = \sum_{i=1}^N \dot{\mathbf{p}}_i \cdot \frac{\partial}{\partial\mathbf{p}_i}\frac{\delta(K(\mathbf{p}) - K_0)\exp[-\beta\Phi(\mathbf{q})]}{Z_K(\beta)}$$

$$= \frac{\exp[-\beta\Phi(\mathbf{q})]}{Z_K(\beta)}\sum_{i=1}^N \dot{\mathbf{p}}_i \cdot \frac{\partial K(\mathbf{p})}{\partial\mathbf{p}_i}\frac{\partial\delta(K(\mathbf{p}) - K_0)}{\partial K(\mathbf{p})} \tag{5.66}$$

$$= \frac{\exp[-\beta\Phi(\mathbf{q})]}{Z_K(\beta)}\dot{K}(\mathbf{p})\frac{\partial\delta(K(\mathbf{p}) - K_0)}{\partial K(\mathbf{p})} = 0.$$

The last line follows because $K(p)$ is a constant of the motion for the Gaussian isokinetic equations of motion. We have also assumed that the only contribution to the phase-space compression factor comes from the thermostatting term $\alpha \mathbf{p}_i$. This means that, in the absence of a thermostat, that is the adiabatic case, the phase space is incompressible and:

$$\frac{\partial}{\partial \Gamma} \cdot \dot{\Gamma}(t)^{ad} = \sum_{i=1}^{N} \left(\frac{\partial}{\partial \mathbf{q}_i} \cdot \mathbf{C}_i + \frac{\partial}{\partial \mathbf{p}_i} \cdot \mathbf{D}_i \right) = 0. \tag{5.67}$$

This assumption or condition is known as the *adiabatic incompressibility of phase space* (AIΓ). A sufficient, but not necessary, condition for it to hold is that the adiabatic equations of motion should be derivable from a Hamiltonian. It is important to note that AIΓ does not imply that the phase space for the thermostatted system should be incompressible. Rather it states that if the thermostat is removed from the field-dependent equations of motion, the phase space is incompressible. It is essentially a condition on the external field coupling terms $\mathbf{C}_i(\mathbf{q}, \mathbf{p})$ and $\mathbf{D}_i(\mathbf{q}, \mathbf{p})$. It is not necessary that $\mathbf{C}_i$ be independent of $\mathbf{q}$, and $\mathbf{D}_i$ be independent of $\mathbf{p}$. Indeed, in Section 6.3 we find that this is not the case for planar Couette flow, but the combination of partial derivatives in Equation (5.67) is zero. It is possible to generalize the theory to treat systems where AIΓ does not hold, but this generalization has proved to be unnecessary.

Using Equation (5.61) for the multiplier α, to first order in N we have:

$$\Delta i \mathscr{L}(t) f_K(\Gamma) = -(\beta \dot{\Phi}(t) + 3N\alpha) f_K(\Gamma)$$

$$= \beta \sum_{i=1}^{N} \left(\mathbf{C}_i \cdot \mathbf{F}_i - \mathbf{D}_i \cdot \frac{\mathbf{p}_i}{m} \right) F_e(t) f_K(\Gamma)$$

$$= \beta J(\Gamma) F_e(t) f_K(\Gamma). \tag{5.68}$$

This equation shows that $\Delta i \mathscr{L}(t) f(\Gamma)$ is independent of thermostatting. Equations (5.68) and (5.15) are essentially identical. This is why the dissipative flux J is defined in terms of the adiabatic derivative of the internal energy. Interestingly, the kinetic part of the dissipative flux, $J(\Gamma)$, comes from the multiplier α, while the potential part comes from the time derivative of Φ.

Substituting Equation (5.68) into Equation (5.64), the change in the isokinetic distribution function is given by:

$$\Delta f(\Gamma, t) = -\beta \int_0^t ds \, \exp[-i\mathscr{L}_K(t-s)] J(\Gamma) F_e(s) f_K(\Gamma). \tag{5.69}$$

Using this result to calculate the mean value of $B(t)$, the isothermal linear response formula corresponding to Equation (5.16), is:

$$\langle B(t) \rangle_K - \langle B(0) \rangle_K = \int d\Gamma B(\Gamma) \Delta f(\Gamma, t)$$

$$= -\beta \int_0^t ds \int d\Gamma B(\Gamma) \exp[-i\mathscr{L}_K(t-s)] J(\Gamma) F_e(s) f_K(\Gamma)$$

$$= -\beta \int_0^t ds \int d\Gamma f_K(\Gamma) J(\Gamma) \exp[i\mathscr{L}_K(t-s)] B(\Gamma) F_e(s)$$

$$= -\beta \int_0^t ds \int d\Gamma f_K(\Gamma) J(\Gamma) B(t-s) F_e(s)$$

$$= -\beta \int_0^t ds \langle B(t-s) J(0) \rangle_{K,0} F_e(s). \tag{5.70}$$

Equation (5.70) is very similar in form to the adiabatic linear response formula derived in Section 5.1. The notation $\langle \ \rangle_{K,0}$ signifies that a field-free (0), isokinetic (K) ensemble average should be taken. Differences from the adiabatic formula are that:

(1) the field-free Gaussian isokinetic propagator governs the time evolution in the equilibrium time correlation function $\langle B(t-s) J(0) \rangle_{K,0}$;
(2) the ensemble averaging is Gaussian isokinetic rather than canonical;
(3) because both the equilibrium and nonequilibrium motions are thermostatted, the long time limit of $\langle B(t) \rangle_K$ on the left-hand side of Equation (5.70), is finite;
(4) and the formula is ergodically consistent. There is only one ensemble referred to in the expression, the Gaussian isokinetic distribution. The dynamics used to calculate the time evolution of the phase variable B in the equilibrium time-correlation function; ergodically generates the ensemble of time-zero starting states $f_K(\Gamma)$. We refer to this as *ergodically consistent linear response theory*.

The last point means that time averaging rather than ensemble averaging can be used to generate the time-zero starting states for the equilibrium time-correlation function on the right-hand side of Equation (5.70).

It can be useful, especially for theoretical treatments, to use ergodically inconsistent formulations of linear response theory. It may be convenient to employ canonical rather than isokinetic averaging, for example. For the canonical ensemble, assuming $A\Gamma$, we have, in place of Equation (5.65):

$$\Delta i L(t) f_c(\Gamma) = -(\beta \Delta \dot{\Phi} + 3N \Delta \alpha) f_c(\Gamma)$$

$$= \left(\beta \sum \mathbf{F}_i \cdot \mathbf{C}_i - \frac{3N}{2K} \sum \frac{\mathbf{p}_i}{m} \cdot \mathbf{D}_i \right) F_e f_c(\Gamma) \tag{5.71}$$

$$= \beta J(\Gamma) F_e f_c(\Gamma) + \beta \frac{\Delta K}{\langle K \rangle_{c,0}} \sum \frac{\mathbf{p}_i}{m} \cdot \mathbf{D}_i F_e f_c(\Gamma) + O(\Delta^2),$$

where $\Delta\dot{\Phi}$ is the difference between the rate of change of Φ with the external field turned on and with the field turned off $(\dot{\Phi}(F_e) - \dot{\Phi}(F_e = 0))$. Similarly $\Delta\alpha = \alpha(F_e) - \alpha(F_e = 0) = \alpha_1 F_e$ (see Equation 5.61). The response of a phase variable B, is therefore:

$$\langle B(t) \rangle_c = \langle B(0) \rangle_c - \beta \int_0^t ds \langle B(t-s)J(0) \rangle_{c,0} F(s)$$

$$- \beta \int_0^t ds \left\langle B(t-s) \frac{\Delta K}{\langle K \rangle_{c,0}} \sum \frac{\mathbf{p}_i}{m} \cdot \mathbf{D}_i(0) \right\rangle_{c,0} F(s). \tag{5.72}$$

Using the same methods as those used in deriving Equation (5.35), we can show that if B is extensive, the second integral in Equation (5.72) is of order one and can therefore be ignored.

Thus for a canonical ensemble of starting states and thermostatted Gaussian isokinetic dynamics, the response of an extensive variable B, is given by:

$$\langle B(t_K) \rangle_c = \langle B(0) \rangle_c - \beta \int_0^t ds \langle B(t-s)_K J(0) \rangle_{c,0} F(s). \tag{5.73}$$

Like the isokinetic ensemble formula, the response, $\langle B(t_K) \rangle_c$, possesses well-defined steady-state limits.

It is straightforward to apply the linear response formalism to a wide variety of combinations of statistical mechanical ensembles, and equilibrium dynamics. The resultant susceptibilities are shown in the Table 5.1 below. It is important toappreciate that the dissipative flux $J(\Gamma)$ is determined by both the choice of equilibrium ensemble of starting states and the choice of the equilibrium dynamics.

Table 5.1 *Linear susceptibilities expressed as equilibrium time-correlation functions*[a,b]

Adiabatic response of canonical ensemble	$\chi = \beta \langle B(t_N)J \rangle_c$	(T.5.1)
Isothermal response of canonical or isothermal ensemble	$\chi = \beta \langle B(t_K)J \rangle_{c,K}$	(T.5.2)
Isoenergetic response of canonical or microcanonical ensembles	$\chi = \beta \langle B(t_N)J \rangle_{c,E}$	(T.5.3)
Isoenthalpic response of isoenthalpic ensemble		
$-JF_e \equiv dI/dt$	$\chi = \beta \langle B(t_I)J \rangle_I$	(T.5.4)
Nosé dynamics of the canonical ensemble	$\chi = \beta \langle B(t_c)J \rangle_c$	(T.5.5)[c]

[a] Equilibrium dynamics: t_N = Newtonian; t_K = Gaussian Isokinetic; t_I = Gaussian isoenthalpic; t_c = Nosé–Hoover.
[b] Ensemble averaging: $\langle \ \rangle_c$ canonical; $\langle \ \rangle_K$ isokinetic; $\langle \ \rangle_E$ microcanonical; $\langle \ \rangle_I$ isoenthalpic.
[c] Proof of (T.5.5) can be found in a paper by Holian and Evans (1983).

5.4 The equivalence of thermostatted linear responses

We shall now address the important question of how the various linear susceptibilities described in Table 5.1, relate to one another. For simplicity, let us assume
that the initial unperturbed ensemble is canonical. In this case, the only difference
between the adiabatic, the isothermal, the isoenergetic, and the Nosé susceptibilities
is in the respective field-free propagators used to generate the equilibrium time-
correlation functions. We will now discuss the differences between the adiabatic
and isothermal responses, however, the analyses of the other cases involve
similar arguments. Without loss of generality we shall assume that the dissipative
flux J and the response phase variable B are both extensive and have mean values
which vanish at equilibrium. The susceptibility is of order N.

The only difference between (T.5.1) and (T.5.2) is in the time propagation of the
phase variable B:

$$B(t_K) = U_K(t)B(\Gamma) = \exp[iL_K t]B(\Gamma) \qquad (5.74)$$

and

$$B(t_N) = U_N(t)B(\Gamma) = \exp[iL_N t]B(\Gamma). \qquad (5.75)$$

In Equations (5.74) and (5.75) the Liouvillean iL_N is the Newtonian Liouvillean,
and iL_K is the Gaussian isokinetic Liouvillean obtained from the equations of
motion (5.23), with α given by the $F_e \to 0$ limit of Equation (5.20). In both
cases there is no explicit time dependence in the Liouvillean. We note that the
multiplier α is intensive.

We can now use the Dyson Equation (3.102), to calculate the difference between
the isothermal and adiabatic susceptibilities for the canonical ensemble.
If $\Rightarrow$ denotes the isothermal propagator and $\to$ the Newtonian, the difference
between the two relevant equilibrium time-correlation functions is:

$$\langle J \Rightarrow B \rangle - \langle J \to B \rangle = \langle J \Rightarrow \Delta \to B \rangle \equiv \delta \langle J \Rightarrow B \rangle, \qquad (5.76)$$

where we have used the Dyson Equation (3.106). Now the difference between the
isothermal and Newtonian Liouvillean is:

$$\Delta = iL_K - iL_N = \Delta\dot{\Gamma} \cdot \frac{\partial}{\partial\Gamma} = -\alpha \sum_{i=1}^{N} \mathbf{p}_i \cdot \frac{\partial}{\partial\mathbf{p}_i}. \qquad (5.77)$$

Thus

$$\delta \langle J \Rightarrow B \rangle = -\int_0^t ds \left\langle J \exp[iL_N s]\alpha \sum_{i=1}^{N} \mathbf{p}_i \cdot \frac{\partial}{\partial\mathbf{p}_i} \exp[iL_N(t-s)]B \right\rangle, \qquad (5.78)$$

where α is the field-free Gaussian multiplier appearing in the isothermal equation of motion. We assume that it is possible to define a new phase variable B' by:

$$\exp[iL_N t]B' = \sum_{i=1}^{N} \mathbf{p}_i \cdot \frac{\partial}{\partial \mathbf{p}_i} \exp[iL_N t]B. \tag{5.79}$$

This is a rather unusual definition of a phase variable, but if B is an analytic function of the momenta, and then an extensive phase variable B' always exists. First we calculate the average value of $B'(t)$:

$$\langle B'(t_N)\rangle = \left\langle \sum_{i=1}^{N} \mathbf{p}_i \cdot \frac{\partial}{\partial \mathbf{p}_i} B(t_N) \right\rangle = \int d\Gamma f_c(\Gamma) \sum_{i=1}^{N} \mathbf{p}_i \cdot \frac{\partial}{\partial \mathbf{p}_i} B(t_N)$$

$$= -\int d\Gamma B(t_N) \sum_{i=1}^{N} \cdot \frac{\partial}{\partial \mathbf{p}_i} \cdot (\mathbf{p}_i f_c(\Gamma)) \tag{5.80}$$

$$= -3N\langle B(t_N)\rangle + 2\beta\langle B(t_N)K(0)\rangle$$

$$= 2\beta\langle B(t_N)[K(0) - <K(0)>]\rangle.$$

Unless B is trivially related to the kinetic energy K, $\langle B'(t_N)\rangle = 0$. Typically B will be a thermodynamic flux, such as the heat-flux, vector or the symmetric traceless part of the pressure tensor. In these cases $\langle B'(t_N)\rangle$ vanishes because of Curie's principle (Section 2.3).

Assuming, without loss of generality, that $\langle B(t_N)\rangle = 0$, we can show that:

$$\delta\langle J \Rightarrow B\rangle = -\int_0^t ds \langle J \exp[iL_K s]\alpha \exp[iL_N(t-s)]B'\rangle$$

$$= -\int_0^t ds \langle J(-s_K)\alpha(0)B'(t_N - s_N)\rangle. \tag{5.81}$$

This is because $\langle J\rangle = \langle \alpha\rangle = 0$. Because J, B, and B' are extensive and α is intensive, Equation (5.81) can be expressed as the product of three zero-mean extensive quantities divided by N. The average of three local, zero-mean quantities is extensive, and thus the quotient is intensive. Therefore, except in the case where B is a scalar function of the kinetic energy, the difference between the susceptibilities computed under Newton's equations and under Gaussian isokinetic equations, is of order $1/N$ compared to the magnitude of the susceptibilities themselves. This means that in the large system limit the adiabatic and isokinetic susceptibilities are equivalent. Similar arguments can be used to show the thermodynamic equivalence of the adiabatic and Nosé susceptibilities. It is pleasing to be able to prove that the mechanical response is independent of the thermostatting mechanism and so only depends upon the thermodynamic state of the system.

Two further comments can be made at this stage: firstly, there is a simple reason why the difference in the respective susceptibilities is significant in the case where B is a scalar function of the kinetic energy. This is simply a reflection of the fact that in this case, B is intimately related to a constant of the motion for Gaussian isokinetic dynamics. One would expect to see a difference in the susceptibilities in this case. Secondly, in particular cases one can use Dyson decomposition techniques, (in particular Equation 3.111), to examine systematically the differences between the adiabatic and isokinetic susceptibilities. Evans and Morriss (1984a) used this approach to calculate the differences, evaluated using Newtonian and isokinetic dynamics, between the correlation functions for each of the Navier–Stokes transport coefficients. The results showed that the equilibrium time correlation functions for the shear viscosity, for the self-diffusion coefficient and for the thermal conductivity are independent of thermostatting in the large system limit.

6

Computer simulation algorithms

6.1 Introduction

Linear response theory can be used to design computer simulation algorithms for the calculation of transport coefficients. There are two types of transport coefficients: mechanical and thermal, and we will show how thermal transport coefficients can be calculated using mechanical methods.

In Nature nonequilibrium systems may respond essentially adiabatically, or depending upon circumstances, they may respond approximately isothermally – the quasi-isothermal response. No natural systems can be precisely adiabatic or isothermal. There will always be some transfer of the dissipative heat produced in nonequilibrium systems towards thermal boundaries. This heat may be radiated, convected, or conducted to the boundary reservoir. Provided this heat transfer is slow on a microscopic timescale and provided that the temperature gradients implicit in the transfer process lead to negligible temperature differences on a microscopic length scale, we call the system *quasi-isothermal*. We assume that quasi-isothermal systems can be modelled microscopically in computer simulations, as isothermal systems.

In view of the robustness of the susceptibilities and equilibrium time-correlation functions to various thermostatting procedures (see Sections 5.2 and 5.4), we expect that quasi-isothermal systems may be modeled using Gaussian or Nosé–Hoover thermostats or enostats. Furthermore, since heating effects are *quadratic* functions of the thermodynamic forces, the *linear* response of nonequilibrium systems can always be calculated by analyzing the adiabatic, isothermal, or isoenergetic response.

The fundamental relations between the linear nonequilibrium response and time-dependent equilibrium fluctuations (Table 6.1) give two ways of calculating the susceptibilities. We can perform an equilibrium simulation and calculate the appropriate equilibrium time-correlation functions. The practical advantage of this method is that all possible transport coefficients can, in principle, be calculated

Table 6.1 *Green–Kubo relations for Navier–Stokes transport coefficients*

Self diffusion	$D = \frac{1}{3} \int_0^\infty dt \langle \mathbf{v}_i(t) \cdot \mathbf{v}_i(0) \rangle$	(T.6.1)
Thermal conductivity	$\lambda = \dfrac{V}{3\,k_B T^2} \int_0^\infty dt \langle \mathbf{J}_Q(t) \cdot \mathbf{J}_Q(0) \rangle$	(T.6.2)
Shear viscosity	$\eta = \dfrac{V}{k_B T} \int_0^\infty dt \langle P_{xy}(t) P_{xy}(0) \rangle$	(T.6.3)
Bulk viscosity	$\eta_V = \dfrac{1}{V k_{BT}} \int_0^\infty dt \langle (p(t)V(t) - \langle pV \rangle)(p(0)V(0) - \langle pV \rangle) \rangle$	(T.6.4)

from a single molecular-dynamics run. This approach is however, very expensive in computer time with poor signal-to-noise ratios, and provides results that often depend strongly and non-monotonically upon the size of the system. Frequently a more useful approach is to perform a direct non-equilibrium simulation of the transport process. For mechanical transport processes we apply an external field, F_e, and calculate the transport coefficient L, from a linear constitutive relation:

$$L = \int_0^\infty dt\, \chi(t) = \lim_{F_e \to 0} \lim_{t \to \infty} \frac{\langle B(t) \rangle}{F_e}. \tag{6.1}$$

The use of Equation (6.1) necessitates a thermostat, since otherwise, the work done on the system would be transformed continuously into heat and no steady state could be achieved (the limit, $t \to \infty$, would not exist). This method, known as nonequilibrium molecular dynamics (NEMD), has the added advantage that it can, in principle, be used to calculate nonlinear as well as linear transport coefficients. They can be calculated as a function of external field strength, frequency, or wave vector. The most efficient, number-independent way to calculate *mechanical* transport coefficients is to ignore the beautiful results of response theory and to duplicate the transport process, essentially as it occurs in Nature.

Thermal transport processes are much more difficult to simulate on the computer. A thermal transport process is one which is driven by boundary conditions rather than mechanical fields. For thermal processes we cannot perform time-dependent perturbation theory because there is no external field appearing in the Hamiltonian which could be used as a perturbation variable. In spite of this difference, susceptibilities for thermal processes show many similarities to their mechanical counterparts (compare Equation (5.73) with the results of Chapter 4). If J is the flux of some conserved quantity (mass, momentum, or energy), and if X is a gradient in the density of that conserved quantity, then a linear Navier–Stokes transport coefficient is defined by a constitutive relation of the form:

$$J = LX. \tag{6.2}$$

A Green–Kubo relation relates each of the Navier–Stokes transport coefficients L to equilibrium fluctuations. These relations are set out in Table 6.1. Remarkably, Navier–Stokes thermal transport coefficients are related to equilibrium time-correlation functions in essentially the same way as mechanical transport coefficients. We must stress, however, that this close formal similarity between thermal and mechanical transport coefficients only applies to Navier–Stokes thermal transport processes. If fluxes of non-conserved variables are involved, then Green–Kubo relations must be generalized (see Equation (4.12) and Section 4.3).

The ensemble averages employed in Table 6.1 are usually taken to be canonical while the time dependence of the correlation functions is generated by field-free Newtonian equations of motion. In Section 5.4, we proved that, except for bulk viscosity, thermostatted equations of motion can also be used to generate the equilibrium time-correlation functions. For bulk viscosity the correlation function involves functions of the kinetic energy of the system. We *cannot* therefore use Gaussian isokinetic equations of motion (see Equations (5.86) and (5.87)). This is because, for these equations, the kinetic energy is a constant of the motion.

To calculate thermal transport coefficients using computer simulation, we have the same two options that were available to us in the mechanical case. We could use equilibrium molecular dynamics to calculate the appropriate equilibrium time-correlation functions, or we could mimic experiment as closely as possible and calculate the transport coefficients from their defining constitutive relations. Perhaps surprisingly the first technique to be used was equilibrium molecular dynamics (Alder and Wainwright, 1956). Much later, the more efficient nonequilibrium approach was pioneered by Hoover and Ashurst (1975). Although the realistic nonequilibrium approach proved more efficient than equilibrium simulations it was still far from ideal. This was because, for thermal transport processes, appropriate boundary conditions are needed to drive the system away from equilibrium – moving walls or walls maintained at different temperatures. These boundary conditions necessarily make the system inhomogeneous. In dense fluids particles pack against these walls, giving rise to significant number dependence and interpretative difficulties.

The most effective way to calculate thermal transport coefficients exploits the formal similarities between susceptibilities for thermal and mechanical transport coefficients. We *invent* a *fictitious* external field which interacts with the system in such a way as to precisely mimic the linear thermal transport process. The general procedure is outlined in Table 6.2. These methods are called *synthetic*, because the invented mechanical perturbation does not exist in Nature. It is our invention and its purpose is to produce a precise mechanical analog of a thermal transport process.

Table 6.2 *Synthetic NEMD*

1. For the transport coefficient of interest $L_{ij}, J_i \equiv L_{ij}X_j$. Identify the Green–Kubo relation for the transport coefficient	$L_{ij} = \int_0^\infty dt \langle J_i(t) \cdot J_j(0) \rangle$
2. Invent a fictitious field F_e and its coupling to the system (equations of motion) such that the dissipative flux	$J \equiv \dot{H}_0^{ad} = J_j$
3. Ensure AIΓ is satisfied that the equations of motion are homogeneous and that they are consistent with periodic boundary conditions.	
4. Apply a thermostat.	
5. Couple F_e to the system isothermally or isoenergetically and compute the steady state average, $\langle J_i(t) \rangle$, as a function of the external field, F_e. Linear response theory then proves,	$L_{ij} = \lim\limits_{F_e \to 0} \lim\limits_{t \to \infty} \dfrac{\langle J_i(t) \rangle}{F_e}$

With regard to step 3, it is not absolutely necessary to invent equations of motion which satisfy AIΓ (see Section 5.3). One can generalize response theory so that AIΓ is not required. However, it is simpler and more convenient to require AIΓ and thus far it has always proved possible to generate algorithms which satisfy AIΓ. Although AIΓ is satisfied, most sets of equations of motion used in synthetic NEMD are not derivable from a Hamiltonian. The preferred algorithms for thermal conductivity and shear viscosity are not derivable from Hamiltonians. In the case of thermal conductivity the Hamiltonian approach must be abandoned because of conflicts with the periodic boundary condition convention used in simulations. For shear viscosity the breakdown of the Hamiltonian approach occurs for deeper reasons.

Equations of motion generated by this procedure are not unique, and it is usually not possible a priori to predict which particular algorithm will be most efficient. It is important to realize that the algorithms generated by this procedure are only guaranteed to lead to the correct *linear* (limit $F_e \to 0$) transport coefficients. We have said nothing so far about generating the correct nonlinear response.

Many discussions of the relative advantages of NEMD and equilibrium molecular dynamics revolve around questions of numerical efficiency. For large fields, NEMD is *orders of magnitude* more efficient than equilibrium molecular dynamics. On the other hand, one can always make NEMD arbitrarily inefficient by choosing a sufficiently small field. At fields which are small enough for the response to be linear, there is no simple answer to the question of whether NEMD is more efficient than equilibrium MD. The number dependence of the errors for the two methods are very different – compared to equilibrium MD, the relative accuracy of NEMD can be made arbitrarily great by increasing the system size.

The discussions of efficiency ignore two major advantages of NEMD over equilibrium molecular dynamics. Firstly, by simulating a *nonequilibrium* system one can visualize microscopically the physical mechanisms that are important to the transport processes (this is true both for synthetic and *realistic* NEMD). One can readily study the distortions of the local molecular structure in nonequilibrium systems. For molecular systems under shear, one can watch the shear-induced processes of molecular alignment, rotation, and conformational change (Edberg *et al.*, 1987). Obtaining this sort of information from equilibrium time-correlation functions is possible, but it is so difficult that no one has yet attempted the task. Secondly, NEMD opens the door to studying the nonlinear response of systems far from equilibrium.

We will now give an extremely brief description of how one performs molecular dynamics simulations. We refer the reader to far more detailed treatments which can be found in the excellent monograph by Allen and Tildesley (1987) and in the review of NEMD by the present authors (Evans and Morriss, 1984b). Consider the potential energy Φ of a system of N interacting particles. The potential energy can always be expanded into a sum of pair, triplet, etc., interactions:

$$\Phi(r) = \frac{1}{2!} \sum \phi^{(2)}(\mathbf{r}_i, \mathbf{r}_j) + \frac{1}{3!} \sum \phi^{(3)}(\mathbf{r}_i, \mathbf{r}_j, \mathbf{r}_k) + \cdots \qquad (6.3)$$

For the inert gas fluids it is known that the total potential energy can be reasonably accurately written as a sum of effective pair interactions. The Lennard–Jones potential ϕ^{LJ} is frequently used as an effective pair potential:

$$\phi^{(2)}(\mathbf{r}_i, \mathbf{r}_j) \approx \phi^{LJ}(r_{ij}) = 4\varepsilon \left(\left(\frac{\sigma}{r_{ij}} \right)^{12} - \left(\frac{\sigma}{r_{ij}} \right)^{6} \right). \qquad (6.4)$$

The potential energy of the two particles i and j is solely a function of their separation distance r_{ij} and is independent of the relative orientation of their separation vector $\mathbf{r}_{ij}$. The Lennard–Jones potential is characterized by a well depth ε, which controls the energy of the interaction, and a length scale σ, which is the distance at which the potential energy is zero in Equation (6.4). If $\varepsilon/k_B = 119.8\,\text{K}$ and $\sigma = 3.405\text{Å}$, the Lennard–Jones potential forms a surprisingly accurate representation of liquid argon (Hansen and Verlet, 1969). For proper scaling during simulations, all calculations are performed in reduced units where $\varepsilon/k_B = \sigma = m = 1$. This amounts to measuring all distances in units of σ, all temperatures in units of ε/k_B and all masses in units of m. The Lennard–Jones potential is often truncated at a distance, $r_c = 2.5\,\sigma$. A typical state point for the dense fluid regime is the Lennard–Jones triple point where $\rho\sigma^3 = 0.8442$ and $kT/\varepsilon = 0.722$. Other potentials that are commonly used include the Weeks–Chandler–Andersen (WCA) potential which is the Lennard-Jones potential truncated at the position of the minimum potential energy $(2^{1/6}\sigma)$ and then shifted up so that the potential is zero at the cutoff:

$$\phi^{WCA}(r_{ij}) = \begin{cases} 4\varepsilon\left(\left(\sigma/r_{ij}\right)^{12} - \left(\sigma/r_{ij}\right)^{6}\right) + \varepsilon; & r < 2^{1/6}\sigma \\ 0; & r > 2^{1/6}\sigma \end{cases}. \qquad (6.5)$$

The main advantage of this potential is its extremely short range. This permits simulations to be carried out much more quickly than is possible with the longer-ranged Lennard–Jones potential. Another short-ranged potential that is often used is the soft sphere potential which omits the r^{-6} term from the Lennard–Jones potential. The soft sphere potential is often truncated at 1.5σ.

In molecular dynamics one simply solves the equations of motion for a system of N interacting particles. The force on particle i, due to particle j, $\mathbf{F}_{ij}$, is evaluated from the potential as:

$$\mathbf{F}_{ij} = -\frac{\partial \phi_{ij}}{\partial \mathbf{r}_i} = -\frac{\partial r_{ij}}{\partial \mathbf{r}_i}\frac{\partial \phi_{ij}}{\partial r_{ij}} = \hat{\mathbf{r}}_{ij}\phi'(r_{ij}), \qquad (6.6)$$

where $\mathbf{r}_{ij} = \mathbf{r}_j - \mathbf{r}_i$ and $\hat{\mathbf{r}}_{ij}$ is the unit vector in the direction of $\mathbf{r}_{ij}$. The N interacting particles are placed in a cubic cell which is surrounded by an infinite array of identical cells – so-called periodic boundary conditions. To compute the force on a given particle in the primitive cell one locates the closest (or minimum) image positions of each of the other $N - 1$ particles. The minimum image of particle i may be within the primitive cell, or in one of the surrounding image cells (see Figure 6.1). One then finds all the minimum images particles for i, that lie within the potential cutoff distance r_c and uses (6.6) to compute the contributions to the force on i, $\mathbf{F}_i = \sum \mathbf{F}_{ij}$.

Finally one solves Newton's or Hamilton's equations of motion for the system:

$$\dot{\mathbf{q}}_i = \frac{\mathbf{p}_i}{m}.$$
$$\dot{\mathbf{p}}_i = \mathbf{F}_i. \qquad (6.7)$$

If, during the course of the motion, particle i leaves the primitive cell, it will be replaced under the periodic boundary condition convention by an image of itself, travelling with exactly the same momentum, one lattice vector distant. We use Hamilton's form for the equations of motion because this form is more convenient than the Newtonian form both for NEMD and for equilibrium molecular dynamics with velocity-dependent forces (such as thermostats). Typically we solve the equations of motion using a fifth-order Gear predictor-corrector method (Gear, 1971). In this way the errors inherent in our model equations of motion are completely separate from the errors inherent in their numerical solution. In studies of

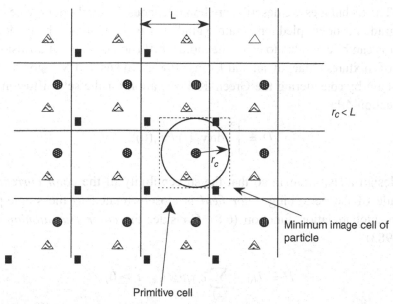

Figure 6.1 Orthogonal periodic boundary conditions

the stability or transient response of systems to external fields we use the less efficient Runge–Kutta methods. Unlike the Gear algorithms, Runge–Kutta methods are self-starting, achieving full accuracy in the first timestep.

We will now give a summary of some of the synthetic NEMD algorithms that have been used to calculate Navier–Stokes transport coefficients.

6.2 Self diffusion

The first NEMD algorithm for self diffusion was devised by Holian (Erpenbeck, 1977). In this elegant scheme the self-diffusion coefficient was evaluated as the limiting value of the mutual diffusion coefficient as the two species become identical. The two species differ only by a color label which plays no role in their subsequent dynamics, but which is reset in a probabilistic fashion as particles cross a labeling plane. A concentration gradient in colored species is set up and the mutual diffusion coefficient is calculated from the constitutive relation (color current/color gradient). If the labels or colors of the atoms are ignored, the simulation is an ordinary equilibrium molecular dynamics simulation. If one calculates the species density as a function of position, the periodic boundary conditions imply that it is discontinuous at the labeling plane. The technique is inhomogeneous and is not applicable to mutual diffusion of species which are really different molecules, as the relabeling process will obviously generate discontinuities in pressure and energy. The techniques we describe are homogeneous. They do not create concentration gradients or coupled temperature gradients as the Holian scheme does. The algorithms can be extended to calculate mutual diffusion or thermal diffusion coefficients of mixtures (Macgowan and Evans, 1986b; Evans and Macgowan, 1987).

We begin by considering the Green–Kubo relation for the self diffusion coefficient (Section 4.1):

$$D = \int_0^\infty dt \langle \mathbf{v}_{xi}(t) \cdot \mathbf{v}_{xi}(0) \rangle. \tag{6.8}$$

We design a Hamiltonian so that the susceptibility of the *color current* to the magnitude of the perturbing *color field* is closely related to the single-particle velocity autocorrelation function (6.8). Consider the *color Hamiltonian* (Evans *et al.*, 1983)

$$H = H_0 - \sum_{i=1}^{N} c_i x_i F(t), \quad t > 0, \tag{6.9}$$

where H_0 is the unperturbed Hamiltonian. The c_i are called color charges. We use color rather than charge to emphasize that H_0 is independent of the color charges $\{c_i\}$. At equilibrium, in the absence of the color field, the dynamics are color blind. For simplicity we consider an even number of particles N, with colours $c_i = (-1)^i$. The response function we consider is the color current density J_x:

$$J_x = \frac{1}{V} \sum_{i=1}^{N} c_i \dot{x}_i. \tag{6.10}$$

Since we are dealing with a Hamiltonian system, AIΓ (Section 5.3), is automatically satisfied. The dissipation function is:

$$\dot{H}_0^{ad} = F(t) \sum_{i=1}^{N} c_i v_{xi} = F(t) J_x V. \tag{6.11}$$

Linear response theory therefore predicts that (Sections 5.1 and 5.3):

$$\langle J_x(t) \rangle = \beta V \int_0^t dt \langle J_x(t-s) \cdot J_x(0) \rangle_0 F(s), \tag{6.12}$$

where the propagator implicit in $J_x(t-s)$ is the field-*free* equilibrium propagator. (Were we considering electrical rather than color conductivity, Equation (6.12) would give the Kubo expression for the electrical conductivity.) To obtain the diffusion coefficient we need to relate the color current autocorrelation function to the single particle velocity autocorrelation function. This relation, as we shall see, depends slightly on the choice of the equilibrium ensemble. If we choose the canonical ensemble then:

$$\langle J_x(t)J_x(0)\rangle_c = \frac{1}{V^2}\sum_{i,j}^{N} c_i c_j \langle v_{xi}(t)v_{xj}(0)\rangle_c.$$ (6.13)

In the thermodynamic limit, for the canonical ensemble, if $j \neq i$, then $\langle v_{xi}(t)v_{xj}(0)\rangle = 0$, $\forall t$. This is clear since if c is the sound speed, $v_{xj}(0)$ can only be correlated with other particles within its *sound cone* (i.e. within a volume with radius ct). In the thermodynamic limit there will always be infinitely more particles outside the sound cone than within it. Since the particles outside this cone cannot possibly be correlated with particle i, we find that:

$$\langle J_x(t)J_x(0)\rangle_c = \frac{1}{V^2}\sum_{i=1}^{N} c_i^2 \langle v_{xi}(t)v_{xi}(0)\rangle_c = \frac{N}{V^2}\langle v_x(t)v_x(0)\rangle_c.$$ (6.14)

Combining this equation with the Green–Kubo relation for self diffusion gives:

$$D = \frac{1}{\beta\rho}\lim_{t\to\infty}\lim_{F\to 0}\frac{\langle J_x(t)\rangle}{F}.$$ (6.15)

If we are working within the *molecular dynamics* ensemble where the total linear momentum of the system is zero, then v_{xi} are not independent. In this case there is an order N^{-1} correction to this equation and the self-diffusion coefficient becomes (Evans, 1983b):

$$D = \frac{N-1}{N}\frac{1}{\beta\rho}\lim_{t\to\infty}\lim_{F\to 0}\frac{\langle J_x(t)\rangle}{F}.$$ (6.16)

In the absence of a thermostat, the order of the limits in Equations (6.15) and (6.16) is important. They cannot be reversed. If a thermostat is applied to the system, a trivial application of the results of Section 5.3 allows the limits to be taken in either order.

As an example of the use of thermostats we will now derive the Gaussian isokinetic version of the color diffusion algorithm. Intuitively it is easy to see that as the heating effect is nonlinear (that is $O(F^2)$), it does not affect the linear response. The equations of motion we employ are:

$$\dot{\mathbf{q}}_i = \frac{\mathbf{p}_i}{m}$$

$$\dot{\mathbf{p}}_i = \mathbf{F}_i + \hat{\mathbf{x}}c_i F - \alpha\big(\mathbf{p}_i - \hat{\mathbf{x}}c_i m\bar{\dot{x}}_i\big),$$ (6.17)

where $\hat{\mathbf{x}}$ is the unit vector in the x-direction and $m\bar{\dot{x}}_i$ is the average momentum of particle i. If i is even, $m\bar{\dot{x}}_i > 0$ and if i is, odd $m\bar{\dot{x}}_i < 0$. The Gaussian multiplier required to thermostat the system is obtained from the constraint equation:

$$\frac{1}{m}\sum_{i=1}^{N}\left(\mathbf{p}_i - \hat{\mathbf{x}}c_i m\bar{\dot{x}}_i\right)^2 = 3Nk_BT. \qquad (6.18)$$

In this definition of the temperature we calculate the peculiar velocities of each particle relative to the streaming velocity of *the species*. If one imagined that the two species are physically separated, then this definition of the temperature is independent of the bulk velocity of the two species. Without this definition of the peculiar kinetic energy, the thermostat and the color field would work against each other and the temperature would have an explicit quadratic dependence on the color current. Combining Equation (6.17) and the time derivative of Equation (6.18) we identify the thermostatting multiplier as:

$$\alpha = \frac{\sum\left(\mathbf{F}_i + \hat{\mathbf{x}}c_i F\right)\cdot\left(\mathbf{p}_i - \hat{\mathbf{x}}c_i m\bar{\dot{x}}_i\right)}{\sum\left(\mathbf{p}_i - \hat{\mathbf{x}}c_i m\bar{\dot{x}}_i\right)^2}. \qquad (6.19)$$

In the original paper, (Evans *et al.*, 1983), the thermostat was only applied to the components of the velocity which were orthogonal to the color field. It can be shown that the *linear* response of these two systems is identical, provided the systems are at the same state point (in particular, if the systems have the same temperature).

The algorithm is homogeneous since, if we translate particle i and its interacting neighbors, the total force on i remains unchanged. The algorithm is also consistent with ordinary periodic boundary conditions (Figure 6.1). There is no change in the color charge of particles if they enter or leave the primitive cell. It may seem paradoxical that we can measure diffusion coefficients without the presence of concentration gradients, however, we have replaced the chemical potential gradient which drives real diffusion processes with a fictitious color field. A gradient in chemical potential implies a composition gradient and a coupled temperature gradient. Our color field acts homogeneously and leads to no temperature or density gradients. Linear response theory, when applied to our fictitious color field, tells us how the transport properties of our fictitious mechanical system relate to the thermal transport process of diffusion.

By applying a sinusoidal color field $F(t) = F_0 e^{i\omega t}$, we can calculate the entire equilibrium velocity autocorrelation function. Noting the amplitude and the relative phase of the color current, we can calculate the complex frequency dependent susceptibility:

$$\chi(\omega) = \int_0^{\infty} dt\, e^{-i\omega t}\chi(t) = \lim_{F\to 0}\frac{J(\omega)}{F(\omega)}. \qquad (6.20)$$

An inverse Fourier–Laplace transform gives of $\chi(t)$ gives the velocity autocorrelation function.

Figure 6.2 shows the results of computer simulations of the diffusion coefficient for the 108 particle Lennard–Jones fluid at $kT/\varepsilon = 1.08$ and $\rho\sigma^3 = 0.85$. The open circles were obtained using Equation (6.17). We see the color conductivity (left y-axis) and the diffusion coefficient (right y-axis), plotted as a function of the color current. The self-diffusion coefficient is obtained by extrapolating the current to zero. The arrow "EMD" shows the results of equilibrium molecular dynamics where the diffusion coefficient was obtained by integrating the velocity autocorrelation function (Section 4.1, Levesque and Verlet, 1970). The equilibrium and nonequilibrium simulations are in statistical agreement. Also shown in Figure 6.2, are the results of simulations performed at constant color current, rather than constant color field. We will return to this matter when we describe Norton ensemble methods in Section 6.7.

In terms of computational efficiency, the self-diffusion coefficient, being a single-particle property, is far more efficiently computed from equilibrium simulations rather than from the algorithm given above. The algorithm we have outlined above is useful for pedagogical reasons. It is the simplest NEMD algorithm. It is also the basis for developing algorithms for the mutual diffusion coefficients of mixtures (Evans and Macgowan, 1987). The mutual diffusion coefficient, being a collective transport property, is difficult to calculate using equilibrium molecular dynamics (Erpenbeck, 1989). If the two coloured species are distinct electrically charged species, the color conductivity is actually

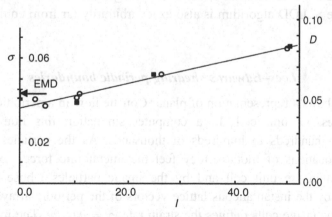

Figure 6.2 The filled squares are constant color simulations, while the open circles are constant color field simulations. On the left-hand vertical axis is the color conductivity and the right-hand vertical axis is the diffusion coefficient. The horizontal axis is the color field. Constant current results are plotted at their average color field

the electrical conductivity and the algorithm given above provides a simple means for its calculation, although H_0 is not independent of the electrical charge.

6.3 Couette flow and shear viscosity

We now describe a homogeneous algorithm for calculating the shear viscosity. Among the Navier–Stokes transport processes, shear viscosity is unique in that a steady, homogeneous algorithm is possible using only periodic boundary conditions to drive the system to a nonequilibrium state. Apart from the possible presence of a thermostat, the equations of motion can be simple Newtonian equations of motion. We will begin by describing how to adapt periodic boundary conditions for planar Couette flow. We will assume that the reader is familiar with the use of fixed orthogonal periodic boundary conditions in equilibrium molecular dynamics simulations (Allen and Tildesley, 1987). The *Lees–Edwards boundary conditions* are sufficient to define an algorithm for planar Couette flow – *boundary driven* shear flow. As this algorithm is the adaption of periodic boundary conditions to simulations of shear flow, the algorithm is exact arbitrarily far from equilibrium.

From a theoretical point of view the boundary driven algorithm is difficult to analyze, there is no external field in the equations of motion, and we cannot employ response theory to link the results obtained from these simulations to the Green–Kubo relations for shear viscosity. From a numerical point of view this algorithm also has some disadvantages. This will lead us to a discussion of the so-called SLLOD algorithm. This algorithm still employs Lees–Edwards boundary conditions, but it eliminates all of the disadvantages of the simple boundary-driven method. The SLLOD algorithm is also exact arbitrarily far from equilibrium.

Lees–Edwards shearing periodic boundaries

Figure 6.3 shows a representation of planar Couette flow in a periodic system with two particles per unit cell. In a computer simulation this number typically ranges from hundreds to hundreds of thousands. As the particles move under Newton's equations of motion, they feel the interatomic forces exerted by the particles within the unit cell and by the image particles whose positions are determined by the instantaneous lattice vectors of the periodic array of cells. The motion of the image cells defines the strain rate, $\gamma \equiv \partial u_x / \partial y$. The individual cells move so that their origins are at the local streaming velocity of the fluid, given by:

$$\mathbf{u}(\mathbf{r}, t) = \hat{\mathbf{x}} \gamma y. \tag{6.21}$$

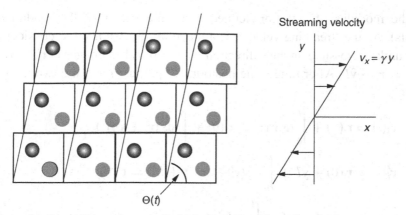

Figure 6.3 Lees–Edwards periodic boundary conditions for planar Couette flow

If the Reynolds number is sufficiently small and turbulence does not occur, we expect that the motion of image particles above and below any given cell will induce a linear velocity profile $\mathbf{u}(\mathbf{r})$ across the system.

If a particle moves out of the simulation cell it will be replaced by one of its periodic images. If the particle moves through a y-face of a cell (that is, through the planes $y = 0$ or $y = L$) the replacing image particle will not have the same laboratory velocity, nor necessarily the same x-coordinate. This movement of particles into and out of the primitive cell promotes the generation of a stable linear streaming-velocity profile.

Although there are jump discontinuities in both the laboratory coordinates and the laboratory velocities of particles between cells, there is no way in which the particles can actually sense the boundaries of any given cell. They are merely book-keeping devices. The system is spatially homogeneous. As we shall see, those components of particle velocity and position which are discontinuous have *NO* thermodynamic meaning. We have depicted the Lees–Edwards boundary conditions in the so-called *sliding brick* representation.

We now consider the motion of particles under Lees–Edwards boundary conditions in more detail. Consider a simulation cube of side L, located so that the streaming velocity at the cube origin is zero (that is, the cube $0 < \{x, y, z\} < L$). The laboratory velocity of a particle i is then the sum of two parts: a peculiar or thermal velocity $\mathbf{c}_i$, and a streaming velocity $\mathbf{u}(\mathbf{r}_i)$, so:

$$\dot{\mathbf{r}}_i = \mathbf{c}_i + \mathbf{u}(\mathbf{r}_i). \tag{6.22}$$

At $t = 0$ we have the usual periodic replication of the simulation cube where the boundary condition is:

$$\mathbf{r}_i = (\mathbf{r}_i)_{\mathrm{mod}\,L}, \tag{6.23}$$

(with the modulus of a vector defined to be the vector of the moduli of the elements). As the streaming velocity is a function of y, we need to consider explicitly boundary crossings in the y direction. At $t = 0$, $\mathbf{r}_i$ has images at $\mathbf{r}'_i = \mathbf{r}_i + \hat{\mathbf{y}}L$, and $\mathbf{r}''_i = \mathbf{r}_i - \hat{\mathbf{y}}L$. After time t, the positions of particle i and the two images are given by:

$$\mathbf{r}_i(t) = \mathbf{r}_i(0) + \int_0^t ds\, \dot{\mathbf{r}}_i(s) = \mathbf{r}_i(0) + \int_0^t ds\, (\mathbf{c}_i + \hat{\mathbf{x}}\gamma y_i),$$

$$\mathbf{r}'_i(t) = \mathbf{r}_i(0) + \hat{\mathbf{y}}L + \int_0^t ds\, (\mathbf{c}'_i + \hat{\mathbf{x}}\gamma(y_i + L)) = \mathbf{r}_i(0)$$

$$+ \hat{\mathbf{x}}\gamma t L + \hat{\mathbf{y}}L + \int_0^t ds\, (\mathbf{c}'_i + \hat{\mathbf{x}}\gamma y_i), \tag{6.24}$$

$$\mathbf{r}''_i(t) = \mathbf{r}_i(0) - \hat{\mathbf{y}}L + \int_0^t ds\, (\mathbf{c}''_i + \hat{\mathbf{x}}\gamma(y_i - L)) = \mathbf{r}_i(0) - \hat{\mathbf{x}}\gamma t L$$

$$- \hat{\mathbf{y}}L + \int_0^t ds\, (\mathbf{c}''_i + \hat{\mathbf{x}}\gamma y_i),$$

where $\mathbf{c}_i$ and y_i are functions of time. By definition, the peculiar velocities of a particle and its periodic images are equal, $\mathbf{c}_i = \mathbf{c}'_i = \mathbf{c}''_i$, so that:

$$\mathbf{r}_i(t) = \mathbf{r}_i(0) + \int_0^t ds\, (\mathbf{c}_i + \hat{\mathbf{x}}\gamma y_i),$$

$$\mathbf{r}'_i(t) = \mathbf{r}_i(t) + \hat{\mathbf{x}}\gamma t L + \hat{\mathbf{y}}L, \tag{6.25}$$

$$\mathbf{r}''_i(t) = \mathbf{r}_i(t) - \hat{\mathbf{x}}\gamma t L - \hat{\mathbf{y}}L.$$

If $\mathbf{r}_i(t)$ moves out the bottom of the simulation cube, it is replaced by the image particle at $\mathbf{r}'_i(t)$ and $\mathbf{r}_i^{new} = (\mathbf{r}'_i)_{\mathrm{mod}\,L} = (\mathbf{r}_i + \hat{\mathbf{x}}\gamma Lt)_{\mathrm{mod}\,L}$. Else, if $\mathbf{r}_i(t)$ moves out of the top of the simulation cube, it is replaced by the image particle at $\mathbf{r}''_i(t)$, $\mathbf{r}_i^{new} = (\mathbf{r}''_i)_{\mathrm{mod}\,L} = (\mathbf{r}_i - \hat{\mathbf{x}}\gamma Lt)_{\mathrm{mod}\,L}$. The change in the laboratory velocity of a particle is given by the time derivative of these two equations. These rules for imaging particles and their velocities are shown schematically in Figure 6.4.

There is a major difficulty with the boundary-driven algorithm. The way in which the boundaries induce a shearing motion to the particles takes time to occur, approximately given by the sound traversal time for the primitive cell. This is the minimum time taken for the particles to realize that the shear is taking place. The boundary-driven method above, therefore *cannot* be used to study time-dependent flows. The most elegant solution to this problem introduces the SLLOD algorithm. We will defer a discussion of thermostats and the evaluation of thermodynamic properties until after we have discussed the SLLOD algorithm.

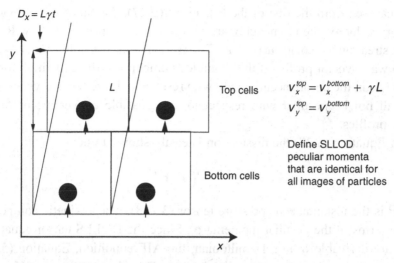

Top cells $v_x^{top} = v_x^{bottom} + \gamma L$
             $v_y^{top} = v_y^{bottom}$

Bottom cells

Define SLLOD
peculiar momenta
that are identical for
all images of particles

Figure 6.4 A particle moving out of the top of a cell is replaced by its image from the cell below

The SLLOD algorithm

The boundary-driven shear-flow algorithm has a number of disadvantages, the principle one being its lack of contact with response theory. We will now describe two synthetic field algorithms for simulating any form of flow deformation. Historically the first fictitious-force method proposed for viscous flow calculations was the DOLLS tensor method (Hoover *et al.*, 1980). This method can be derived from the DOLLS tensor Hamiltonian:

$$H = H_0 + \sum_{i=1}^{N} \mathbf{q}_i \mathbf{p}_i : (\nabla \mathbf{u}(t))^{\mathrm{T}}. \tag{6.26}$$

It generates the following equations of motion:

$$\dot{\mathbf{q}}_i = \frac{\mathbf{p}_i}{m} + \mathbf{q}_i \cdot \nabla \mathbf{u},$$
$$\dot{\mathbf{p}}_i = \mathbf{F}_i - \nabla \mathbf{u} \cdot \mathbf{p}_i. \tag{6.27}$$

These equations of motion *must* be implemented with compatible periodic boundary conditions. If the shear-rate tensor has only one nonzero element and it is off-diagonal, the deformation is planar Couette flow and Lees–Edwards boundary conditions must be used. If the shear-rate tensor is isotropic then the flow is dilational and the appropriate variation of Lees–Edwards boundaries must be used. Other flow geometries can also be simulated using these equations.

One can see from the first of the Equations (6.27), that since $\dot{\mathbf{q}}_i$ is obviously a laboratory velocity, the momenta $\mathbf{p}_i$ are *peculiar* with respect to the low Reynolds number streaming velocity $\mathbf{u}(\mathbf{r}) = \mathbf{r} \cdot \nabla\mathbf{u}$. We call this streaming velocity profile the zero wave-vector profile. If the Reynolds number is sufficiently high for turbulence to occur, the $\mathbf{p}_i$s are peculiar only with respect to the zero wave-vector profile. They will not be peculiar with respect to any possible convective or turbulent velocity profiles.

From Equation (6.27) the dissipation is easily shown to be:

$$\dot{H}_0^{ad} = -\nabla\mathbf{u} : \mathsf{P}V, \tag{6.28}$$

where P is the instantaneous pressure tensor (3.138), whose kinetic component is given in terms of the peculiar momenta $\mathbf{p}_i$. Since the DOLLS tensor equations of motion are derivable from a Hamiltonian, the AIΓ condition, Equation (5.73), is clearly satisfied and we see immediately from Equations (6.28) and (5.73), that in the linear regime, close to equilibrium, the shear and bulk viscosities will be related to equilibrium fluctuations via the Green–Kubo formula (T.6.3). This *proves* that the DOLLS tensor algorithm is correct for the limiting linear regime. The linear response of the pressure tensor is therefore:

$$\langle \mathsf{P}(t) \rangle = -\beta V \int_0^t ds \langle \mathsf{P}(t-s)\mathsf{P} \rangle : \nabla\mathbf{u}(\mathbf{s}). \tag{6.29}$$

The DOLLS tensor method has now been replaced by the SLLOD algorithm (Evans and Morriss, 1984c). The only difference between the SLLOD algorithm and the DOLLS tensor equations of motion involves the equation of motion for the momenta. The Cartesian components that couple to the shear rate tensor are transposed. Unlike the DOLLS tensor equations, the SLLOD equations of motion *cannot* be derived from a Hamiltonian:

$$\dot{\mathbf{q}}_i = \frac{\mathbf{p}_i}{m} + \mathbf{q}_i \cdot \nabla\mathbf{u},$$
$$\dot{\mathbf{p}}_i = \mathbf{F}_i - \mathbf{p}_i \cdot \nabla\mathbf{u}. \tag{6.30}$$

It is easy to see that the dissipation function for the SLLOD algorithm is precisely the same as for the DOLLS tensor equations of motion. In spite of the absence of a generating Hamiltonian, the SLLOD equations also satisfy AIΓ. This means that the linear response for both systems is identical and is given by Equation (6.29). By taking the limit $\gamma \to 0$, followed by the limit $t \to \infty$, we see that the linear shear viscosity can be calculated from a nonequilibrium simulation, evolving under either the SLLOD or the DOLLS tensor equations of motion. With $\nabla\mathbf{u} = \hat{\mathbf{y}}\hat{\mathbf{x}}(\partial u_x/\partial y)$, and calculating the ratio of stress to shear rate we calculate:

$$\eta = \lim_{t\to\infty} \lim_{\gamma\to 0} \frac{-\langle P_{xy}(t)\rangle}{\gamma}. \qquad (6.31)$$

From Equation (6.29) we see that the susceptibility is precisely the Green–Kubo expression for the shear viscosity (Table 6.1). Because the linear response of the SLLOD and DOLLS tensor algorithms are related to equilibrium fluctuations by the Green–Kubo relations, these algorithms can be used to calculate the reaction of systems to time-varying shear rates. If the shear rate is a sinusoidal function of time, then the Fourier transform of the susceptibility gives the complex, frequency-dependent shear viscosity measured in viscoelasticity (Sections 2.4 and 4.3).

If the shear rate tensor is isotropic then the equations of motion describe adiabatic dilation of the system. If this dilation rate is sinusoidal then the limiting small-field bulk viscosity can be calculated by monitoring the amplitude and phase of the pressure response and extrapolating both the amplitude and frequency to zero (Hoover et al., 1980). It is again easy to see from Equation (6.32) that the susceptibility for the dilation-induced pressure change is precisely the Green–Kubo transform of the time-dependent equilibrium fluctuations in the hydrostatic pressure (Table 6.1).

Although the DOLLS tensor and SLLOD algorithms have the same dissipation and give the correct *linear* behavior, the DOLLS tensor algorithm begins to yield incorrect results at quadratic order in the shear rate. These errors show up first as errors in the normal stress differences. For irrotational flows ($\nabla\mathbf{u} = (\nabla\mathbf{u})^{\mathrm{T}}$) so the SLLOD and DOLLS tensor methods are identical, as can easily be seen from their equations of motion.

We will now show that the SLLOD algorithm gives an exact description of shear flow *arbitrarily far from equilibrium*. This method is also correct in the high Reynolds-number regime in which laminar flow is unstable. Consider superimposing a linear velocity profile on a canonical ensemble of N-particle systems. This will generate the *local equilibrium* distribution function for Couette flow, f_1:

$$f_1 = \frac{\exp\left[-\beta\left(\frac{1}{2}m(\mathbf{v}_i - \hat{\mathbf{x}}\gamma y_i)^2 + \Phi\right)\right]}{\int \mathrm{d}\Gamma \exp\left[-\beta\left(\frac{1}{2}m(\mathbf{v}_i - \hat{\mathbf{x}}\gamma y_i)^2 + \Phi\right)\right]}. \qquad (6.32)$$

Macroscopically such an ensemble is described by a linear streaming-velocity profile:

$$\mathbf{u}(\mathbf{r}, t) = \hat{\mathbf{x}}\gamma y, \qquad (6.33)$$

so that the second-rank shear-rate tensor, $\nabla \mathbf{u}$, has only one nonzero element, $(\nabla \mathbf{u})_{yx} = \gamma$. The local equilibrium distribution function is *not* the same as the steady-state distribution. This is easily seen when we realize that the shear stress evaluated for f_1, is zero. The local distribution function is no more than a canonical distribution with a superimposed linear velocity profile. No molecular relaxation has yet taken place.

If we allow this relaxation to take place by advancing time using Newton's equations (possibly supplemented with a thermostat), the system will go on shearing forever. This is because the linear velocity profile of the local distribution generates a *zero* wave-vector transverse-momentum current. As we saw in Section 3.8, the zero wave-vector momentum densities are conserved. The transverse-momentum current will persist forever, at least for an infinite system.

Now let us see what happens under the SLLOD equations of motion (6.30), when the shear-rate tensor is given by (6.33). Differentiating the first equation, then substituting for $\dot{\mathbf{p}}_i$ using the second equation gives:

$$m\ddot{\mathbf{q}}_i = \mathbf{F}_i - \hat{\mathbf{x}}\gamma p_{yi} + \hat{\mathbf{x}}(\gamma p_{yi} + m\dot{\gamma}y_i) = \mathbf{F}_i + \hat{\mathbf{x}}m\dot{\gamma}y_i. \qquad (6.34)$$

If the shear rate γ is switched on at time zero, and remains steady thereafter:

$$\gamma(t) = \gamma\Theta(t) \implies \dot{\gamma} = \gamma\delta(t). \qquad (6.35)$$

Thus $\dot{\gamma}$ is a delta function at $t = 0$. Now consider subjecting a canonical ensemble to these transformed SLLOD equations of motion, Equation (6.34). If we integrate the velocity of particle i, over an infinitesimal time interval about zero. We see that:

$$\mathbf{v}_i(0^+) - \mathbf{v}_i(0) = \int_0^{0^+} ds\, \dot{\mathbf{v}}(s) = \hat{\mathbf{x}}\gamma y_i. \qquad (6.36)$$

So at time 0^+ the x-velocity of every particle is incremented by an amount proportional to the product of the shear rate times its y-coordinate. At time 0^+, the other components of the velocity and positions of the particles are unaltered because there are no delta-function singularities in their equations of motion. Applying Equation (6.36) to a canonical ensemble of systems will clearly generate the local equilibrium distribution for planar Couette flow.

The application of SLLOD dynamics to the canonical ensemble is thus seen to be equivalent to applying Newton's equations to the local distribution function. The SLLOD equations of motion have therefore succeeded in transforming the boundary condition expressed in the form of the local distribution function into the form of a smooth mechanical force which appears as a *mechanical* perturbation in the equations of motion. This property is unique to SLLOD dynamics. It is not satisfied

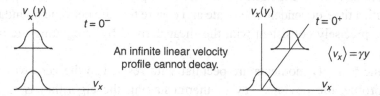

$$\frac{dq_i}{dt} = \frac{p_i}{m} + \hat{x}\gamma y_i$$

$$\frac{dp_i}{dt} = F_i - \hat{x}\gamma p_{yi}$$

are equivalent to:

$$m\frac{d^2q_i}{dt^2} = F_i + \hat{x}m\frac{d\gamma}{dt}y_i$$

If the shear rate is a step function, then $\frac{d\gamma}{dt} = \gamma\delta(t)$ and the $t = 0$ velocities are incremented.

$v_x(y)$ $t = 0^-$

An infinite linear velocity profile cannot decay.

$v_x(y)$ $t = 0^+$

$\langle v_x\rangle = \gamma y$

Figure 6.5 SLLOD equations of motion give an exact representation of planar Couette flow

by the DOLLS tensor equations of motion for example. Since one cannot really call into question, the validity of the application of Newtonian dynamics to the local distribution as a correct description of Couette flow, we are led to the conclusion that the adiabatic application of SLLOD dynamics to the canonical ensemble gives an exact description of Couette flow (Figure 6.5).

Knowing that the SLLOD equations are exact, and that they generate Green–Kubo expressions for the shear and bulk viscosities, provides a *proof* of the validity of the Green–Kubo expressions themselves. The SLLOD transformation of a thermal transport process into a mechanical one provides us with a direct route to the Green–Kubo relations for the viscosity coefficients. From Equation (6.31) we see that we already have these relations for both the shear and bulk viscosity coefficients. We also see that these expressions are identical to those we derived in Chapter 4, using the generalized Langevin equation. It is clear that the present derivation is simpler and gives greater physical insight into the processes involved.

Compared to the boundary-driven methods, the advantages of using the SLLOD algorithm in computer simulations are many. Under periodic boundaries, the SLLOD momenta are peculiar with respect to the zero wave-vector velocity field, and are continuous functions of time and space. This is not so for the laboratory velocities v_i. The internal energy and the pressure tensor of the system are more simply expressed in terms of SLLOD momenta rather than laboratory momenta. The internal energy E is given as:

$$E(T, \rho, N, \gamma) = \langle H_0\rangle = \left\langle \sum_{i=1}^{N} \frac{p_i^2}{2m} + \frac{1}{2}\sum_{i,j}^{N} \phi_{ij}\right\rangle, \tag{6.37}$$

while the ensemble averaged pressure tensor is:

$$P(T, \rho, N, \gamma)V = \left\langle \sum_{i=1}^{N} \frac{\mathbf{p}_i \mathbf{p}_i}{m} - \frac{1}{2} \sum_{i,j}^{N} \mathbf{r}_{ij} \mathbf{F}_{ij} \right\rangle. \tag{6.38}$$

For simulations of viscoelasticity special measures have to be taken in the boundary-driven algorithm to ensure that the time-varying shear rate is actually what you expect it to be. In the SLLOD method, no special techniques are required for simulations of time-dependent flows. One simply has to solve the equations of motion with a time-dependent shear rate and ensure that the periodic boundary conditions are precisely consistent with the shear derived by integrating the imposed shear rate $\gamma(t)$.

Since the SLLOD momenta are peculiar with respect to the zero wave-vector velocity profile, the obvious way of thermostatting the algorithm is to use the equations:

$$\dot{\mathbf{q}}_i = \frac{\mathbf{p}_i}{m} + \hat{\mathbf{x}} \gamma y_i.$$

$$\dot{\mathbf{p}}_i = \mathbf{F}_i - \hat{\mathbf{x}} \gamma p_{yi} - \alpha \mathbf{p}_i. \tag{6.39}$$

The thermostatting multiplier α is calculated in the usual way by ensuring that $\frac{d}{dt}\left(\sum p_i^2\right) = 0$:

$$\alpha = \frac{\sum \left(\mathbf{F}_i \cdot \mathbf{p}_i - \gamma p_{xi} p_{yi}\right)}{\sum p_i^2}. \tag{6.40}$$

The temperature is assumed to be related to the peculiar kinetic energy. These equations *assume* that a linear velocity profile is stable. However, as we have mentioned a number of times, the linear velocity profile is only stable at low Reynolds number ($Re = \rho m \gamma L^2 / \eta$).

In Figure 6.6 we show the shear viscosity of 2048 WCA particles as a function of shear rate close to the Lennard–Jones triple point. The simulations use the Gaussian isokinetic SLLOD algorithm. There is a substantial change in the viscosity with shear rate. Evidently WCA fluids are shear thinning as the viscosity decreases with increasing shear rate. This is common for all simple fluids at all thermodynamic state points. Shear thinning is also a widely observed phenomenon in the rheology of complex molecular fluids.

The imposed shear causes a major change in the microscopic fluid structure with *all* the thermodynamic properties of the system changing with shear rate, for example the internal energy changes with shear rate (see Figure 6.7). For reduced shear rates in the range 0–1.5, the shear viscosity and the internal energy change by approximately 50% compared to their equilibrium values. Furthermore the viscosity coefficient appears to vary as the square root of the

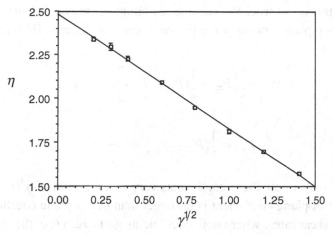

Figure 6.6 Viscosity of 2048 WCA particles at the triple point

shear rate while the energy appears to change with the $\frac{3}{2}$ power of the shear rate. Over the range of shear rates studied, the maximum deviation from the functional forms is 2.5 % for the viscosity, and 0.1 % for the internal energy. There has been much discussion of these nonanalytic dependences in relation to mode-coupling theory (see Kawasaki and Gunton, 1973; Yamada and Kawasaki, 1975a; Yamada and Kawasaki, 1975b; Ernst *et al.*, 1978; Evans, 1983b; Kirkpatrick, 1984; van Beijeren, 1984; de Schepper *et al.*, 1986). It is clear that the final resolution of this matter is still a long way off.

One of the most interesting and subtle rheological effects concerns the diagonal elements of the pressure tensor. For Newtonian fluids (i.e. fluids characterized by a shear rate-independent and frequency-independent viscosity), the diagonal elements are equal to each other and to their equilibrium values. Far from

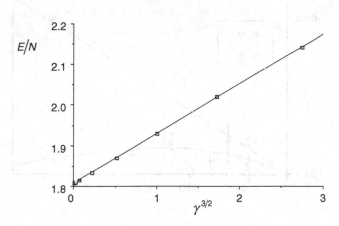

Figure 6.7 The internal energy as a function of shear rate

equilibrium, this is not true. We can define normal stress coefficients, η_0, η_-, (the so-called out-of-plane and in-plane normal stress coefficients [Hess *et al.*, 1984; Hess, 1987] as:

$$P_0 \equiv \frac{1}{2}\left[P_{zz} - \frac{1}{2}(P_{xx} + P_{yy})\right] = -\eta_0 \gamma, \qquad (6.41)$$

$$P_- \equiv \frac{1}{2}(P_{xx} - P_{yy}) = -\eta_- \gamma. \qquad (6.42)$$

Figure 6.8 shows how these coefficients vary as a function of $\gamma^{1/2}$ for the WCA fluid. The out-of-plane coefficient is far larger than the in-plane coefficient, except at very small shear rates, where both coefficients go to zero (i.e. the fluid becomes Newtonian). These coefficients are very difficult to compute accurately. They require both larger and longer simulations to achieve an accuracy that is comparable to that for the shear viscosity. In terms of the macroscopic hydrodynamics of nonNewtonian fluids, these normal stress differences are responsible for a wide variety of interesting phenomena (e.g. the Weissenberg effect, see Rainwater and Hanley, 1985; Rainwater *et al.* 1985).

If one allows the shear rate to be a sinusoidal function of time and one extrapolates the system response to zero amplitude, one can calculate the linear viscoelastic response of a fluid. Figure 6.9 shows complex frequency dependent shear viscosity for the Lennard–Jones fluid (Evans, 1980), at its triple point.

If one compares Figure 6.9 with the Maxwell model for viscoelasticity, Figure 2.4, one sees a qualitative similarity with the low-frequency response

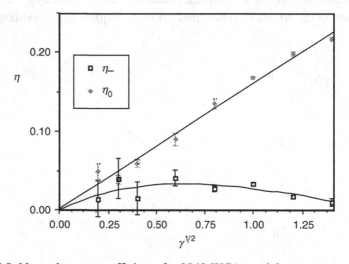

Figure 6.8 Normal stress coefficients for 2048 WCA particles

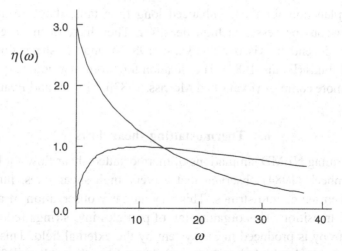

Figure 6.9 Frequency-dependent shear viscosity at the Lennard–Jones triple point

being viscous and the high-frequency response being elastic. The shape of the two sets of curves is, however, quite different. This is particularly so at low frequencies. An analysis of the low-frequency data shows that it is consistent with a nonanalytic square-root dependence on frequency:

$$\tilde{\eta}_R(\omega) = \eta(0) - \eta_{\omega 1}\omega^{1/2} + O(\omega),$$
$$\tilde{\eta}_I(\omega) = \eta_{\omega 1}\omega^{1/2} + O(\omega),$$

(6.43)

where $\tilde{\eta}_R$, $\tilde{\eta}_I$ are the real and imaginary parts of the viscosity coefficient. Since the frequency-dependent viscosity is the Fourier–Laplace transform of the memory function Equation (2.76), we can use the Tauberian theorems (Doetsch, 1961), to show that if Equation (6.46) represents the asymptotic low-frequency behavior of the frequency-dependent viscosity, then the memory function must have the form:

$$\lim_{t \to \infty} \eta(t) = \frac{\eta_{\omega 1} t^{-3/2}}{\sqrt{2\pi}}.$$

(6.44)

This time dependence is again consistent with the time dependence predicted by mode-coupling theory (Pomeau and Resibois, 1975). However, as was the case for the shear-rate dependence, the amplitude of the effect shown in Figure 6.9 is orders of magnitude larger than theoretical predictions. This matter is also the subject of much current research and investigation.

Similar *enhanced* long time tails have been observed subsequently in Green–Kubo calculations for the shear viscosity (Erpenbeck and Wood, 1981). Whatever the final explanation for these enhanced long time tails, they are a ubiquitous feature of viscous processes at high densities. They have been observed in the wave-vector dependent viscosity (Evans, 1982b) and in shear flow of four-dimensional fluids (Evans, 1984). The situation for two-dimensional liquids is apparently even more complex (Evans and Morriss, 1983b; Morriss and Evans, 1989).

6.4 Thermostatting shear flows

While performing NEMD simulations of thermostatted shear flow for hard-sphere fluids, Erpenbeck (1984) observed that at very high shear rates, fluid particles organized themselves into strings. This was an early observation of a nonequilibrium phase transition. This organization of particles into strings reduces the rate at which entropy is produced in the system by the external field. This effect is in competition with the kink instability of the strings themselves. If the strings move too slowly across the simulation cell, thermal fluctuations in the curvature of the strings lead to their destruction. A snapshot of a string phase is shown in Figure 6.10. The velocity gradient is vertical and the streaming velocity is horizontal. The system is 896 soft discs at a state point close to freezing and a reduced shear rate of 17.

The string phase is, in fact, stabilized by the use of a thermostat which *assumes* that a linear velocity profile, (implicit in Equation 6.33), is stable. Thermostats which make some assumption about the form of the streaming velocity profile are called profile-biased thermostats (PBT). All the thermostats we have met so far are profile-biased. At equilibrium there can be little cause for worry, the streaming velocity must be zero. Away from equilibrium we must be more careful.

Any kink instability that might develop in Erpenbeck's strings, leading to their breakup, would necessarily lead to the formation of large-scale *eddies* in the streaming velocity of the fluid. The profile-biased thermostat would interpret any incipient eddy motion as *heat*, and then thermostat would try to *cool* the system by suppressing the eddy formation. This, in effect, stabilizes the string phase (Evans and Morriss, 1986).

Thermostats for secondary or convecting flows

Profile-biased thermostats (PBT) for shear flow assume that the kinetic temperature T_B, for a system undergoing planar Couette flow, can be defined from the equation:

$$dNk_BT_B = \left\langle \sum_{i=1}^{N} m(\mathbf{v}_i - \hat{\mathbf{x}}\gamma y_i)^2 \right\rangle. \tag{6.45}$$

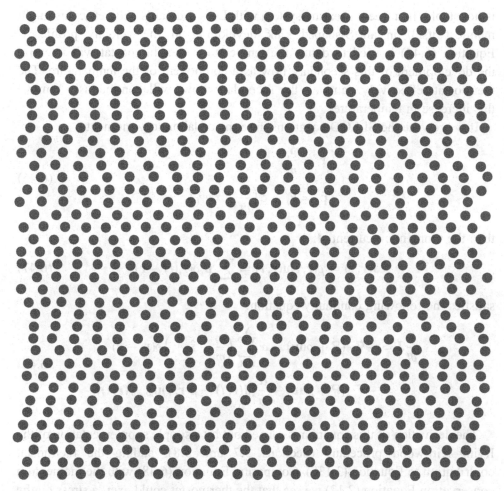

Figure 6.10 The high shear-rate string phase for soft disks

In this equation d is the number of spatial dimensions and N is the number of particles. The term $\hat{x}\gamma y_i$ is the *presumed* streaming velocity at the location of particle i. Once the form of the streaming-velocity profile is established it is a simple matter to use peculiar velocity scaling, Gaussian isokinetic, or Nosé methods to thermostat the shearing system.

At small shear rates and low Reynolds number, the Lees–Edwards shearing periodic boundary conditions do indeed lead to a planar velocity profile of the form assumed in Equation (6.39). In Erpenbeck's (1984) simulations the Reynolds numbers ($\text{Re} = \rho m \gamma L^2 / \eta$) were very large ($10^3$–$10^5$). The assumption of a linear streaming-velocity profile under these conditions is extremely dubious. Suppose that at high Reynolds number the linear velocity profile assumed in Equation (6.39) is not stable. In a freely shearing system with Lees–Edwards geometry, this might manifest itself in an S-shaped kink developing in the velocity profile. If Equation (6.45) is used to define the temperature, the thermostat will interpret the development of this secondary flow as a component of the temperature. The thermostat, effectively damping the secondary flow, will continuously remove this increase in temperature.

If we rewrite the SLLOD equations in terms of laboratory momenta:

$$\dot{\mathbf{r}}_i = \frac{\mathbf{p}_i}{m},$$

$$\dot{\mathbf{p}}_i = \mathbf{F}_i - \alpha\left(\frac{\mathbf{p}_i}{m} - \hat{\mathbf{x}}\gamma y_i\right),$$
(6.46)

then the momentum current, $\mathbf{J}$:

$$\mathbf{J}(\mathbf{r}, t) \equiv \rho(\mathbf{r}, t)\mathbf{u}(\mathbf{r}, t) = \sum \mathbf{p}_i \delta(\mathbf{r}_i(t) - \mathbf{r}),$$
(6.47)

satisfies the following continuity equation:

$$\frac{\partial}{\partial t}\mathbf{J} = -\nabla \cdot (\mathsf{P} + \rho\mathbf{u}\mathbf{u}) - \alpha \sum_{i=1}^{N} \left(\frac{\mathbf{p}_i}{m} - \hat{\mathbf{x}}\gamma y_i\right)\delta(\mathbf{r}_i - \mathbf{r})$$

$$= -\nabla \cdot (\mathsf{P} + \rho\mathbf{u}\mathbf{u}) - \frac{\alpha}{m}(\mathbf{J}(\mathbf{r}, t) - \rho(\mathbf{r}, t)\mathbf{u}_{linear}(\mathbf{r}, t)).$$
(6.48)

The derivation of this equation is carried out by a simple supplementation of the Irving–Kirkwood procedure (Sections 3.7 and 3.8), adding the thermostat contribution to Equation (3.130). Comparing Equation (6.48) with the momentum conservation Equation (2.12) we see that the thermostat could exert a stress on the system. The expected divergence terms ($\rho\mathbf{u}\mathbf{u} + \mathsf{P}$) are present on the right-hand side of Equation (6.48). However, the term involving α, the thermostatting term, is new and represents the force exerted on the fluid by the thermostat. It will only vanish if a linear velocity profile is *stable* and:

$$\mathbf{J}(\mathbf{r}, t) = \rho(\mathbf{r}, t)\mathbf{u}_{linear}(\mathbf{r}, t) = \hat{\mathbf{x}}m\gamma y, \quad \forall \mathbf{r}.$$
(6.49)

At high Reynolds number this condition may not be true. For simulations at high Reynolds numbers, a thermostat that makes no assumptions about the form of the streaming-velocity profile is needed. The thermostat should not even assume that a stable profile exists. These ideas led to development (Evans and Morriss, 1986) of profile-unbiased thermostats (PUT).

The PUT thermostat begins by letting the simulation itself define the local streaming velocity $\mathbf{u}(\mathbf{r}, t)$. This is done by considering small cells and defining the temperature of a particular cell at $\mathbf{r}$, $T(\mathbf{r}, t)$, to be:

$$\frac{dn(\mathbf{r}, t) - d}{2} k_B T(\mathbf{r}, t) \equiv \sum_{i \in cell} \frac{m}{2} (\mathbf{v}_i - \mathbf{u}(\mathbf{r}, t))^2, \tag{6.50}$$

where $n(\mathbf{r}, t)$ is the number density at $\mathbf{r}, t$. The number of degrees of freedom in the cell is $dn(\mathbf{r}, t) - d$, because d degrees of freedom are used to determine the streaming velocity of the cell.

The PUT thermostatted SLLOD equations of motion can be written as:

$$\frac{d\mathbf{r}_i}{dt} = \frac{\mathbf{p}_i}{m},$$

$$\frac{d\mathbf{p}_i}{dt} = \mathbf{F}_i - \alpha \left(\frac{\mathbf{p}_i}{m} - \mathbf{u}(\mathbf{r}, t) \right) \delta(\mathbf{r}_i - \mathbf{r}). \tag{6.51}$$

The streaming velocity, $\mathbf{u}(\mathbf{r}, t)$, is not known in advance, but is computed as time progresses from its definition, Equation (6.47). The thermostat multiplier α could be a Gaussian multiplier chosen to fix the peculiar kinetic energy (Equation 6.50). Equally well the multiplier could be a Nosé–Hoover multiplier. The equation of motion for the momentum density is then:

$$\frac{\partial}{\partial t} \mathbf{J} = -\nabla \cdot (\mathbf{P} + \rho \mathbf{u}\mathbf{u}) - \alpha \sum_{i=1}^{N} \left(\frac{\mathbf{p}_i}{m} - \hat{\mathbf{x}} \gamma y_i \right) \delta(\mathbf{r}_i - \mathbf{r})$$

$$= -\nabla \cdot (\mathbf{P} + \rho \mathbf{u}\mathbf{u}) - \frac{\alpha}{m} (\mathbf{J}(\mathbf{r}, t) - \rho(\mathbf{r}, t)\mathbf{u}(\mathbf{r}, t)). \tag{6.52}$$

From the definition of the streaming velocity of a cell we know that:

$$\sum_{i=1}^{N} (\mathbf{p}_i - \hat{\mathbf{x}} m \gamma y_i) \delta(\mathbf{r}_i - \mathbf{r}) = \sum_{i=1}^{N} \mathbf{p}_i \delta(\mathbf{r}_i - \mathbf{r}) - \sum_{i=1}^{N} \mathbf{u}(\mathbf{r}, t) \delta(\mathbf{r}_i - \mathbf{r})$$

$$= \mathbf{J}(\mathbf{r}, t) - \rho(\mathbf{r}, t)\mathbf{u}(\mathbf{r}, t) = 0.$$

Thus the thermostatting term in Equation (6.52) vanishes for all values of $\mathbf{r}$. In terms of practical implementation in computer programs, PUT thermostats can only be used in simulations with sufficient particles to allow a break up into cells, and sufficient particles per cell to allow the definition of the local streaming velocity and the temperature. Thus far their use has been restricted to simulations of two-dimensional systems. At low Reynolds numbers where no strings are observed in profile-biased simulations, it is found that profile-unbiased simulations yield the same results as PBT methods. However at high shear rates the results obtained using the two different thermostatting methods are quite different. No one has observed a string phase while using a PUT thermostat.

6.5 Elongational flows

The difficulty in simulating elongational flows is matching the flow to a set of boundary conditions which allows a continuous deformation, particularly in the direction of the compression. This problem was solved by Todd and Daivis (1998; 1999) and also by Baranyai and Cummings (1999). Their method is based upon the Kraynik–Reinelt boundary conditions (1992) that allow a square lattice to be deformed continuously, and then at periodic intervals, mapped back onto the initial square lattice. For a review of the method see (Todd and Daivis, 2007). The derivation we give here is not lattice based, but rather treats the deformed initial simulation cell and maps it back to the initial square (for a two-dimensional system). We consider a cold, noninteracting system to see how the simulation cell is deformed by the flow. After that we return to the warm interacting system to obtain the full equations of motion.

The equations of motion for planar elongational flow can be obtained from the appropriate shear-rate tensor:

$$\nabla \mathbf{u} = \begin{pmatrix} \varepsilon & 0 & 0 \\ 0 & -\varepsilon & 0 \\ 0 & 0 & 0 \end{pmatrix}, \tag{6.53}$$

and the SLLOD equations of motion (6.30). For this shear-rate tensor, the expansion is directed along the x-axis and the contraction directed along the y-axis. To make the simulation process continuous in time and allow unlimited simulation times, we consider a geometry where the expanding direction is at an angle θ to the x-axis. The expanding and contracting directions define a new reference frame (x', y'), and these primed coordinates are related to the space-fixed coordinates by a rotation matrix R_θ, so:

$$\begin{pmatrix} x' \\ y' \end{pmatrix} = R_\theta \begin{pmatrix} x \\ y \end{pmatrix} = \begin{pmatrix} c & -s \\ s & c \end{pmatrix} \begin{pmatrix} x \\ y \end{pmatrix}, \tag{6.54}$$

where $s = \sin(\theta)$ and $c = \cos(\theta)$. The unprimed coordinates can be obtained from the primed coordinates by the inverse rotation R_θ^{-1}. The equations of motion in the primed coordinates are the SLLOD equations with the shear-rate tensor given in Equation (6.53). If we assume that the system is cold ($\mathbf{p}_i = 0$ for all i) and noninteracting ($\mathbf{F}_i = 0$ for all i), then the time evolution of the primed coordinates is given by the solution of $\dot{\mathbf{q}}'_i = \nabla \mathbf{u} \cdot \mathbf{q}'_i$ where $\nabla \mathbf{u}$ is the shear rate tensor in Equation (6.53). The time evolution is thus the solution of $\mathbf{q}'(t) = \exp[\nabla \mathbf{u} t]\mathbf{q}'(0)$, where it can be shown that

$$\exp[\nabla \mathbf{u} t] = \begin{pmatrix} \exp[\varepsilon t] & 0 \\ 0 & \exp[-\varepsilon t] \end{pmatrix}. \tag{6.55}$$

The time evolution in the unprimed frame is given by rotating the primed evolution, thus:

$$\begin{pmatrix} x(t) \\ y(t) \end{pmatrix} = R_\theta^{-1} \exp[\nabla \mathbf{u}t] R_\theta \begin{pmatrix} x \\ y \end{pmatrix}$$

$$= \begin{pmatrix} c^2\exp[\varepsilon t] + s^2\exp[-\varepsilon t] & -sc(\exp[\varepsilon t] - \exp[-\varepsilon t]) \\ -sc(\exp[\varepsilon t] - \exp[-\varepsilon t]) & s^2\exp[\varepsilon t] + c^2\exp[-\varepsilon t] \end{pmatrix} \begin{pmatrix} x \\ y \end{pmatrix}. \quad (6.56)$$

The key to making the simulation a continuous time process is to be able to map the system back into square boundaries. It is sufficient that for some value of θ, and some value of t_{MOD}, the time-evolution matrix in Equation (6.56) is equal to an integer matrix M:

$$\begin{pmatrix} c^2\exp[\varepsilon t] + s^2\exp[-\varepsilon t] & -sc(\exp[\varepsilon t] - \exp[-\varepsilon t]) \\ -sc(\exp[\varepsilon t] - \exp[-\varepsilon t]) & s^2\exp[-\varepsilon t] + c^2\exp[-\varepsilon t] \end{pmatrix}$$

$$= \begin{pmatrix} m_1 & m_2 \\ m_2 & m_3 \end{pmatrix} = M. \quad (6.57)$$

The matrix M has integer entries m_1, m_2, m_3. Notice that both matrices are symmetric, area preserving, and both have determinant one. Further, the eigenvalues and eigenvectors of both matrices must be equal and this condition determines the values of the angle θ and the time t_{MOD}. The eigenvalues of the two matrices in Equation (6.57) are:

$$\lambda = \exp[\pm\varepsilon t] = \frac{(m_1 + m_3) \pm \sqrt{(m_1 + m_3)^2 - 4}}{2}, \quad (6.58)$$

and if the eigenvector corresponding to the largest eigenvalue is $\begin{pmatrix} 1 \\ a \end{pmatrix}$, then:

$$a = -\tan\theta = \frac{(m_3 - m_1) + \sqrt{(m_1 + m_3)^2 - 4}}{2\,m_2}. \quad (6.59)$$

Choosing values for m_1, m_2, m_3, the time at which the boundary conditions are applied t_{MOD} and angle of the rotation between the primed and unprimed coordinates θ are given by Equations (6.58) and (6.59). Positive integer values of m_1, m_2, m_3 ensure that the torus can be mapped back onto the original simulation square (Figure 6.11), although noninteger values are also possible. Many choices of m_1, m_2, m_3 are possible and each will lead to a different value for t_{MOD} and θ. However, there are constraints on the values of m_1, m_2, m_3 and hence these cannot be chosen arbitrarily.

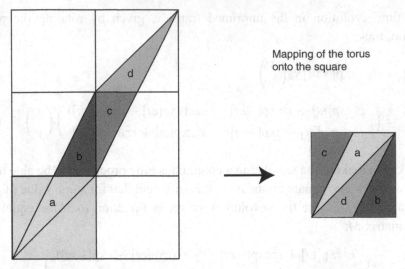

Mapping of the torus
onto the square

Figure 6.11 The initial coloured square is mapped to the region a, b, c, d under the action of the integer matrix for the cat map. The final square shows how each of the pieces a, b, c, d map are backed to the original square using the mod function

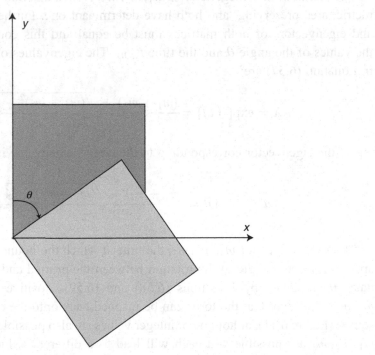

Figure 6.12 The original square in the Figure 6.11 is rotated before the simulation begins

As an example of this approach, the Arnold cat map (Hunt and Todd, 2003) is obtained when $m_1 = m_2 = 1$ and $m_3 = 2$. The value of the time and the angle of rotation (Figure 6.12) are then:

$$t_{MOD} = \frac{1}{\varepsilon} \ln\left(\frac{3 + \sqrt{5}}{2}\right) \text{ and } \tan\theta = \frac{1 + \sqrt{5}}{2}. \tag{6.60}$$

In a computer simulation it is simpler to rotate the elongation direction so that it coincides with the x-axis. This requires that the simulation square be also rotated through the same angle. The observables of interest are then simply $\langle P_{xx} \rangle$ and $\langle P_{yy} \rangle$. The equations of motion are:

$$\dot{\mathbf{q}}_i = \frac{1}{m_i}\mathbf{p}_i + \varepsilon(\hat{\mathbf{x}}x_i - \hat{\mathbf{y}}y_i),$$

$$\dot{\mathbf{p}}_i = \mathbf{F}_i - \varepsilon(\hat{\mathbf{x}}p_{xi} - \hat{\mathbf{y}}p_{yi}) - \alpha\mathbf{p}_i, \tag{6.61}$$

where:

$$\alpha = \frac{\sum_{i=1}^{N}\left(\mathbf{p}_i \cdot \mathbf{F}_i - \varepsilon\left(p_{xi}^2 - p_{yi}^2\right)\right)}{\sum_{i=1}^{N}\mathbf{p}_i^2}. \tag{6.62}$$

The usual minimum image convention is used to determine the forces, and periodic boundary conditions appropriate for the map M are only used at multiples of $t = t_{MOD}$. The periodic boundary conditions require the system to be rotated back to their original position, then the appropriate mod function is used to map the system back to the square, and finally the cell is rotated again and the simulation continues.

A trivial consequence of the SLLOD equations of motion for the shear tensor (Equation 6.53) is that they imply equations of motion for the total momentum (Todd and Daivis, 2000). Summing over all particles:

$$\dot{p}_x^{TOT} = F_x^{TOT} - \varepsilon p_x^{TOT},$$

$$\dot{p}_y^{TOT} = F_y^{TOT} + \varepsilon p_y^{TOT}. \tag{6.63}$$

If the total force is zero in each coordinate direction, that is $F_x^{TOT} = F_y^{TOT} = 0$, then the total momentum has exponential solutions. In particular, the y-component of the total momentum grows exponentially. To prevent this, the initial momentum in the y-direction needs to be zero, and some care must be taken to see that it remains zero during the simulation.

6.6 Thermal conductivity

Thermal conductivity has proven to be one of the most difficult transport coefficients to calculate. Green–Kubo calculations are notoriously difficult to perform. Natural NEMD, where one might simulate heat flow between walls maintained at different temperatures (Tenenbaum *et al.*, 1982), is also fraught with major difficulties. Molecules stack against the walls leading to a major change in the microscopic fluid structure. This means that the results can be quite different from those characteristic of the bulk fluid. In order to measure a statistically significant heat flux, one must use enormously large temperature gradients. These gradients are so large that the absolute temperature of the system may change by 50 % in a few nanometers. The thermal conductivity that one obtains from such simulations is an average over the wide range of temperatures and densities present in the simulation cell.

We will now describe the most efficient presently known algorithm for calculating the thermal conductivity (Evans, 1982a). This technique is synthetic, in that a fictitious field replaces the temperature gradient as the force driving the heat flux. Unlike real heat flow, this technique is homogeneous with no temperature or density gradients. We start with the Green–Kubo expression for the thermal conductivity (Section 4.4):

$$\lambda = \frac{V}{k_B T^2} \int_0^\infty dt \langle J_{Q_z}(t) J_{Q_z}(0) \rangle, \tag{6.64}$$

where J_{Q_z}, is the z-component of the heat-flux vector. It appears to be impossible to construct a Hamiltonian algorithm for the calculation of thermal conductivity. This is because the equations of motion are discontinuous when used in conjunction with periodic boundary conditions. We shall instead invent an external field and its coupling to the N-particle system, so that the heat flux generated by this external field is trivially related to the magnitude of the heat flux induced by a real temperature gradient.

Aided by the realization that the heat-flux vector is the diffusive energy flux, computed in a comoving coordinate frame (see Equation 3.145), we proposed the following equations of motion:

$$\dot{\mathbf{q}}_i = \frac{\mathbf{p}_i}{m},$$

$$\dot{\mathbf{p}}_i = \mathbf{F}_i + (E_i - \overline{E})\mathbf{F}(t) - \frac{1}{2}\sum_{j=1}^N \mathbf{F}_{ij}(\mathbf{q}_{ij} \cdot \mathbf{F}(t)) + \frac{1}{2N}\sum_{j,k}^N \mathbf{F}_{jk}(\mathbf{q}_{jk} \cdot \mathbf{F}(t)), \tag{6.65}$$

where E_i is the energy of particle i and:

$$\overline{E} = \frac{1}{N}\left\{ \sum_{i=1}^N \frac{\mathbf{p}_i^2}{2m} + \frac{1}{2}\sum_{i \neq j} \phi_{ij} \right\}, \tag{6.66}$$

the *instantaneous* average energy per particle.

There is no known Hamiltonian which generates these equations, but they do satisfy AIΓ. This means that linear response theory can be applied in a straightforward fashion. The equations of motion conserve momentum, are homogeneous and compatible with the usual periodic boundary conditions. It is clear from the term $(E_i - \overline{E})\mathbf{F}(t)$ that these equations of motion will drive a heat current. A particle whose energy is greater than the average energy will experience a force in the direction of $\mathbf{F}$, while a particle whose energy is lower than the average will experience a force in the $-\mathbf{F}$ direction. *Hotter* particles are driven with the field; *colder* particles are driven against the field.

If the total momentum is zero it will be conserved and the dissipation is:

$$\dot{H}_0^{ad} = \mathbf{F}(t) \cdot \left\{ \sum_{i=1}^N \frac{\mathbf{p}_i E_i}{m} - \frac{1}{2} \sum_{i,j} \mathbf{q}_{ij} \left(\frac{\mathbf{p}_i \cdot \mathbf{F}_{ij}}{m} \right) \right\} = \mathbf{F}(t) \cdot \mathbf{J}_Q V. \tag{6.67}$$

Using linear response theory we have:

$$\langle \mathbf{J}_Q(t) \rangle = -\beta V \int_0^t ds \langle \mathbf{J}_Q(t-s) \mathbf{J}_Q(0) \rangle \cdot \mathbf{F}(s). \tag{6.68}$$

If the field is $\mathbf{F} = (0, 0, F_z)$, then in the limit $t \to \infty$, the ratio of the induced heat flux to the product of the temperature and the external field F_z is the thermal conductivity:

$$\lambda = \frac{V}{k_B T^2} \int_0^\infty dt \langle J_{Q_z}(t) J_{Q_z}(0) \rangle = \lim_{F \to 0} \frac{-\langle J_{Q_z}(\infty) \rangle}{TF}. \tag{6.69}$$

In the linear limit, the effect the heat field has on the system is identical to that of a logarithmic temperature gradient ($F = \partial \ln T / \partial z$). The theoretical justification for this algorithm is tied to *linear* response theory. No meaning is known for the finite field susceptibility.

Gillan and Dixon (1983) have introduced a slightly different synthetic method for computing the thermal conductivity. Although their algorithm is considerably more complex to apply in computer simulations, their equations of motion look quite similar to those given above. Gillan's synthetic algorithm is of some theoretical interest since it is the only known algorithm which violates momentum conservation and AIΓ, (Macgowan and Evans, 1986a).

In Figure 6.13 the thermal conductivity of the Lennard–Jones fluid at the triple point is shown. Assuming that argon can be modeled by the Lennard–Jones fluid with $\varepsilon/k_B = 119.8$ K and $\sigma = 3.405$ Å, this method gives a more accurate estimate of the thermal conductivity than is presently possible by experiment. There is a hydrodynamic instability in the Evans heat-flow algorithm that occurs at some threshold in field strength (Evans and Hanley, 1989). A method of combating this problem can be found in Hansen and Evans (1994).

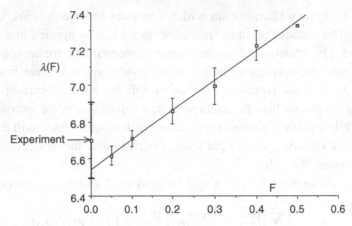

Figure 6.13 Thermal conductivity: Lennard–Jones triple point

6.7 Norton ensemble methods

Norton and Thévenin's theorems are of fundamental importance in electrical circuit theory (Brophy, 1966). They prove that any network of resistors and power supplies can be analyzed in terms of equivalent circuits which include either ideal-current or ideal-voltage sources. These two theorems are an example of the macroscopic duality that exists between what are generally recognized as thermo-dynamic fluxes, and thermodynamic forces – in the electrical circuit case, electrical currents, and the electromotive force. Indeed in our earlier introduction to linear irreversible thermodynamics (Chapter 2), there was an apparent arbitrariness with respect to our definition of forces and fluxes. At no stage did we give a con-vincing macroscopic distinction between the two.

Microscopically one might think that there is a clear and unambiguous distinc-tion that can be drawn. For an arbitrary mechanical system subject to a perturbing external field, the dissipation can be written as: $\dot{H}_0^{ad} = -J(\Gamma)F_e(t)$. The dissipative flux is the phase variable $J(\Gamma)$ and the force is the time-dependent variable, $F_e(t)$. This might seem to remove the arbitrariness. However, suppose that we complicate matters a little and regard the external field $F_e(t)$ as a Gaussian multiplier in a feed-back scheme designed to stop the flux $J(\Gamma)$ from changing. We might wish to perform a constant-current simulation where the imposed *external* field $F_e(t)$, is in fact a phase variable, $F_e(\Gamma)$. Even microscopically, the distinction between forces and fluxes is more complex than is often thought.

In this section we will explore the statistical mechanical consequences of this duality. Until recently the Green–Kubo relations were only known for the conven-tional Thévenin ensemble in which the forces are the independent state-defining variables. We will derive their Norton ensemble equivalents. We will then show

how these ideas have been applied to algorithms for isobaric molecular-dynamics simulations. We have given a statistical mechanical proof of the Norton–Thévenin equivalence (Evans, 1993).

Gaussian constant color current algorithm

From the color Hamiltonian (6.9) we see that the equations of motion for color conductivity in the Thévenin ensemble are:

$$\dot{\mathbf{q}}_i = \frac{\mathbf{p}_i}{m}$$
$$\dot{\mathbf{p}}_i = \mathbf{F}_i + c_i \mathbf{F}(t). \tag{6.70}$$

These equations are the adiabatic version of Equation (6.17). We will now treat the color field as a Gaussian multiplier chosen to fix the color current and introduce a thermostat. Our first step is to redefine the momenta (Evans and Morriss, 1985) so that they are measured with respect to the species current of the particles. Consider the following set of equations of motion:

$$\dot{\mathbf{q}}_i = \frac{\mathbf{p}_i}{m} + \frac{c_i \mathbf{I}(t)}{\sum_{i=1}^{N} c_i^2},$$
$$\dot{\mathbf{p}}_i = \mathbf{F}_i - c_i \boldsymbol{\lambda} - \alpha \mathbf{p}_i, \tag{6.71}$$

where α is the thermostatting multiplier and $\boldsymbol{\lambda}$ is the current multiplier. These equations are easily seen to be equivalent to Equation (6.17). We distinguish two types of current, a canonical current $\mathbf{J}$ defined in terms of the canonical momenta:

$$\mathbf{J} \equiv \sum_i \frac{c_i \mathbf{p}_i}{m}, \tag{6.72}$$

and a kinetic current $\mathbf{I}$, where:

$$\mathbf{I} \equiv \sum_i c_i \dot{\mathbf{q}}_i. \tag{6.73}$$

We choose $\boldsymbol{\lambda}$ so that the canonical current is always zero, and α so that the canonical (i.e. peculiar) kinetic energy is fixed. Our constraint equations are therefore:

$$g_d = \sum_i \frac{c_i \mathbf{p}_i}{m} - \mathbf{J} = 0, \tag{6.74}$$

and

$$g_T = \frac{1}{m} \sum_i \mathbf{p}_i^2 - 3Nk_B T = 0. \tag{6.75}$$

The Gaussian multipliers may be evaluated in the usual way by summing moments of the equations of motion and eliminating the accelerations using the differential forms of the constraints. We find that:

$$\lambda = \frac{\sum_{i=1}^{N} c_i \mathbf{F}_i}{\sum_{i=1}^{N} c_i^2},$$

(6.76)

and

$$\alpha = \frac{\sum_{i=1}^{N} \mathbf{F}_i \cdot \mathbf{p}_i}{\sum_{i=1}^{N} \mathbf{p}_i^2}.$$

(6.77)

If we compare the Gaussian equations of motion with the corresponding Hamiltonian equations we see that the Gaussian multiplier λ can be identified as a fluctuating external color field which maintains a constant color current. It is, however, a phase variable. Gauss' principle has enabled us to go from a constant-field nonequilibrium ensemble to the conjugate ensemble where the current is fixed. The Gaussian multiplier fluctuates in the precise manner required to fix the current. The distinction drawn between canonical and kinetic currents has allowed us to decouple the Lagrange multipliers appearing in the equations of motion. Furthermore, setting the canonical current to zero is equivalent to setting the kinetic current to the required value $\mathbf{I}$. This can be seen by taking the charge moment of Equation (6.71). If the canonical current is zero, then:

$$\sum c_i \dot{\mathbf{q}}_i = \frac{\sum c_i^2 \mathbf{I}(t)}{\sum c_i^2} = \mathbf{I}(t).$$

(6.78)

In this equation the current, which was formerly a phase variable, has now become a possibly time-dependent external force.

In order to be able to interpret the response of this system to the external current field, we need to compare the system's equations of motion with a macroscopic constitutive relation. Under adiabatic conditions, the second-order form of the equations of motion is:

$$m\ddot{\mathbf{q}}_i = \mathbf{F}_i + \frac{c_i m \dot{\mathbf{I}}(t)}{\sum c_i^2} - \lambda e_i.$$

(6.79)

We see that to maintain a constant current $\mathbf{I}(t)$ we must apply a fluctuating colour field $\mathbf{E}_{eff}$:

$$\mathbf{E}_{eff}(t) = \frac{m \dot{\mathbf{I}}(t)}{\sum c_i^2} - \lambda.$$

(6.80)

The adiabatic rate of change of internal energy H_0 is given by:

$$\dot{H}_0^{ad} = -\sum_i \left\{ \frac{c_i \mathbf{p}_i}{m} \cdot \boldsymbol{\lambda} + \frac{c_i \mathbf{F}_i \cdot \mathbf{I}}{\sum c_i^2} \right\} = -\mathbf{J} \cdot \boldsymbol{\lambda} - \mathbf{I} \cdot \boldsymbol{\lambda}. \tag{6.81}$$

As the current, $\mathbf{J} = \mathbf{J}(\Gamma)$ is fixed at the value zero, the dissipation is $-\mathbf{I}(t) \cdot \boldsymbol{\lambda}(\Gamma)$. As expected, the current is now an external time-dependent field while the color field is a phase variable. Using linear response theory we have:

$$\langle \boldsymbol{\lambda}(t) \rangle = \beta \int_0^t ds \langle \boldsymbol{\lambda}(t-s) \boldsymbol{\lambda} \rangle \cdot \mathbf{I}(s), \tag{6.82}$$

which gives the linear response result for the phase variable component of the effective field. Combining Equation (6.82) with Equation (6.80), the effective field is therefore:

$$\mathbf{E}_{eff}(t) = \int_0^t ds \chi(t-s) \mathbf{I}(s) + \frac{m \dot{\mathbf{I}}(t)}{\sum c_i^2}, \tag{6.83}$$

where the susceptibility χ is the equilibrium $\boldsymbol{\lambda}$ autocorrelation function:

$$\chi(t) = \beta \langle \boldsymbol{\lambda}(t) \boldsymbol{\lambda}(0) \rangle. \tag{6.84}$$

By doing a Fourier–Laplace transform on Equation (6.83), we obtain the frequency-dependent color resistance, $\mathbf{E} \equiv R\mathbf{I}$:

$$\tilde{R}(\omega) = \tilde{\chi}(\omega) + \frac{i\omega m}{\sum_{i=1}^N c_i^2}. \tag{6.85}$$

To compare with the usual Green–Kubo relations, which have always been derived for conductivities rather than resistances, we find:

$$\tilde{\sigma}(\omega) = \frac{1}{V\left(\tilde{\chi}(\omega) + \frac{i\omega m}{\sum_{i=1}^N c_i^2} \right)}. \tag{6.86}$$

This equation shows that the Fourier–Laplace transform of $\chi(t)$ is the memory function of the complex frequency-dependent conductivity. In the conjugate constant-force ensemble the frequency-dependent conductivity is related to the current-autocorrelation function:

$$\tilde{\sigma}(\omega) = \frac{1}{3Vk_BT} \int_0^\infty dt \, \exp[-i\omega t] \langle \mathbf{J}(t) \cdot \mathbf{J} \rangle_{\mathbf{E}=0}. \tag{6.87}$$

From Equations (6.84)–(6.87) we see that at *zero* frequency, the color conductivity is given by the integral of the Thévenin ensemble current-correlation function

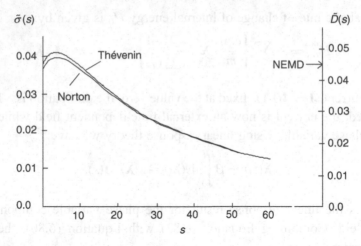

Figure 6.14 The color conductivity as a function of the Laplace transform variable s

while the resistance, which is the reciprocal of the conductivity, is given by the integral of the color-field autocorrelation function computed in the Norton ensemble. Thus at *zero frequency* the integral of the Thévenin ensemble current correlation function is the reciprocal of the integral of the Norton ensemble field correlation function. Figure 6.2 gave a comparison of Norton and Thévenin algorithms for computing the color conductivity. The results obtained for the conductivity are ensemble independent – even in the nonlinear regime far from equilibrium.

In Figure 6.14 we show the reduced color conductivity plotted as a function of frequency (Evans and Morriss, 1985). The system is identical to the Lennard–Jones system studied in Figure 6.2. The curves were calculated by taking the Laplace transforms of the appropriate equilibrium time-correlation functions computed in both the Thévenin and Norton ensembles. Within statistical uncertainties, the results are in agreement. The arrow shows the zero-frequency color conductivity computed using NEMD. The value is taken from Figure 6.2.

6.8 Constant-pressure ensembles

For its first 30 years, molecular dynamics was limited to the microcanonical ensemble with fixed total momentum. We have already seen how the development of thermostats has enabled simulations to be performed in the isochoric, canonical, and isokinetic ensembles. We will now describe molecular-dynamics algorithms for performing simulations at constant pressure or constant enthalpy. The technique used to make the pressure, rather than the volume, the independent state variable, uses essentially the same ideas as those employed in Section 6.6 to design Norton ensemble algorithms. The methods we describe now are of use for both equilibrium and nonequilibrium simulations.

It is often advantageous, particularly in studies of phase transitions, to work within the isobaric ensemble. It is possible to stabilize the pressure in a number of ways: we will describe the Gaussian method (Evans and Morriss, 1983a), since it was both the first deterministic isobaric technique to be developed and it is conceptually simpler than the corresponding Nosé–Hoover (Hoover, 1985) and Parrinello–Rahman (1980a; 1980b; 1981) schemes. Although it may be slightly more difficult to write the computer programs, once written they are certainly easier to use. The Gaussian method has the distinct advantage that the pressure is a rigorous constant of the motion, whereas the Nosé-based schemes (1984a) and those of Parrinello and Rahman allow fluctuations in both the pressure and the volume.

If one makes a poor initial guess for the density, Nosé–Hoover isobaric algorithms induce sharp density changes in an attempt to correct the density to that appropriate for the specified mean pressure. Because bulk oscillations damp quite slowly, Nosé–Hoover methods can easily result in the system exploding – a situation that cannot be reversed due to the finite range of the interaction potentials. Gaussian isobaric algorithms are free of these instabilities.

Isothermal-isobaric molecular dynamics

Consider the SLLOD equations of motion where the shear-rate tensor $\nabla \mathbf{u}$ is isotropic. The equations of motion become:

$$\dot{\mathbf{q}}_i = \frac{\mathbf{p}_i}{m} + \dot{\varepsilon}\mathbf{q}_i,$$

$$\dot{\mathbf{p}}_i = \mathbf{F}_i - \dot{\varepsilon}\mathbf{p}_i. \tag{6.88}$$

Now if the system was *cold* ($\mathbf{p}_i = 0$ for all i) and noninteracting ($\phi_{ij} = 0$), these equations would reduce to:

$$\dot{\mathbf{q}}_i = \dot{\varepsilon}\mathbf{q}_i. \tag{6.89}$$

Since this equation is true for all particles i, it describes a uniform dilation or contraction of the system. This dilation or contraction is the same in each coordinate direction, so if the system initially occupied a cube of volume V, then the volume would satisfy the following equation of motion:

$$\dot{V} = 3V\dot{\varepsilon}. \tag{6.90}$$

For *warm* interacting systems, the equation of motion for $\mathbf{q}_i$ shows that the canonical momentum $\mathbf{p}_i$ is in fact peculiar with respect to the streaming velocity $\dot{\varepsilon}\mathbf{q}_i$. The dissipation for the system (Equations 6.88 and 6.89) is:

$$\dot{H}_0 = -\dot{\varepsilon} \sum_{i=1}^{N} \left\{ \frac{1}{m}\mathbf{p}_i \cdot \mathbf{p}_i + \mathbf{F}_i \cdot \mathbf{q}_i \right\} = -3pV\dot{\varepsilon}. \tag{6.91}$$

Since H_0 is the internal energy of the system we can combine Equation (6.91) with the equation of motion for the volume to obtain the first law of thermodynamics for adiabatic compression:

$$\dot{H}_0 = -3pV\dot{\varepsilon} = -p\dot{V}. \tag{6.92}$$

It is worth noting that these equations are true *instantaneously*. One does not need to employ any ensemble averaging to obtain Equation (6.92). By choosing the dilation rate $\dot{\varepsilon}$ to be a sinusoidal function of time, these equations of motion can be used to calculate the bulk viscosity. Our purpose is, however, to use the dilation rate as a multiplier to maintain the system at a constant hydrostatic pressure. Before we do this, however, we will introduce a Gaussian thermostat into the equations of motion:

$$\dot{\mathbf{q}}_i = \frac{\mathbf{p}_i}{m} + \dot{\varepsilon}\mathbf{q}_i,$$
$$\dot{\mathbf{p}}_i = \mathbf{F}_i - \dot{\varepsilon}\mathbf{p}_i - \alpha\mathbf{p}_i. \tag{6.93}$$

The form for the thermostat multiplier is determined by the fact that the momenta in Equation (6.93) are peculiar with respect to the dilating coordinate frame. By taking the moment of Equation (6.93) with respect to $\mathbf{p}_i$, and setting the time derivative of the peculiar kinetic energy to zero, we observe that:

$$\alpha = -\dot{\varepsilon} + \frac{\sum\limits_{i=1}^{N} \mathbf{F}_i \cdot \mathbf{p}_i}{\sum\limits_{i=1}^{N} \mathbf{p}_i^2}. \tag{6.94}$$

Differentiating the product pV, Equation (6.91), with respect to time gives:

$$3\dot{p}V + 3p\dot{V} = \sum_{i=1}^{N} \left\{ \frac{2}{m}\dot{\mathbf{p}}_i \cdot \mathbf{p}_i + \dot{\mathbf{q}}_i \cdot \mathbf{F}_i + \mathbf{q}_i \cdot \frac{\partial \mathbf{F}_i}{\partial \mathbf{q}_i} \cdot \dot{\mathbf{q}}_i + \sum_{j \neq i} \mathbf{q}_i \cdot \frac{\partial \mathbf{F}_i}{\partial \mathbf{q}_j} \cdot \dot{\mathbf{q}}_j \right\}. \tag{6.95}$$

The first term on the left-hand side is zero because the pressure is constant, and the first term on the right-hand side is zero because the peculiar kinetic energy is constant. Substituting the equations of motion for $\dot{\mathbf{q}}_i$ and $\dot{V}$, we can solve for the dilation rate:

$$\dot{\varepsilon} = \frac{\frac{1}{2m}\sum_{i \neq j} \mathbf{q}_{ij} \cdot \mathbf{p}_{ij}\left(\phi''_{ij} + \frac{\phi'_{ij}}{q_{ij}}\right)}{\frac{1}{2}\sum_{i \neq j} \mathbf{q}^2_{ij}\left(\phi''_{ij} + \frac{\phi'_{ij}}{q_{ij}}\right) + 9pV}. \tag{6.96}$$

Combining this equation with Equation (6.94) gives a closed expression for the thermostat multiplier α.

In summary our isothermal–isobaric molecular-dynamics algorithm involves solving $6N + 1$ first-order equations of motion (Equations 6.90, 6.93 and 6.94). There are two subtleties to be aware of before implementing this method. Firstly the pressure is sensitive to the long-range tail of the interaction potential. In order to obtain good pressure stability the long-range truncation of the potential needs to be handled carefully. Secondly, if a Gear predictor-corrector scheme is used to integrate the equations of motion then some care must be taken in handling the higher-order derivatives of the coordinates and momenta under periodic boundary conditions. More details are given in Evans and Morriss (1983a; 1984b).

Isobaric–isoenthalpic molecular dynamics

For the adiabatic constant-pressure equations of motion we have already shown that the first law of thermodynamics for compression is satisfied:

$$\dot{H}_0 = -p\dot{V}. \tag{6.97}$$

It is now easy to construct equations of motion for which the enthalpy $I = H_0 + pV$ is a constant of the motion. The constraint we wish to impose is that:

$$\dot{I} = \dot{H}_0 + \dot{p}V + p\dot{V} = 0. \tag{6.98}$$

Combining these two equations we see that, for our adiabatic constant-pressure equations of motion, the rate of change of enthalpy is simply:

$$\dot{I} = \dot{p}V. \tag{6.99}$$

This equation says that if our adiabatic equations preserve the pressure, then the enthalpy is automatically constant. The isobaric–isoenthalpic equations of motion are simply obtained from the isothermal–isobaric equations by dropping the constant-temperature constraint. The isoenthalpic dilation rate can be shown to be (Evans and Morriss, 1984b):

$$\dot{\varepsilon} = \frac{\dfrac{2}{m}\displaystyle\sum_{i=1}^{N}\mathbf{p}_i \cdot \mathbf{F}_i - \dfrac{1}{2m}\displaystyle\sum_{i \neq j}\mathbf{q}_{ij} \cdot \mathbf{p}_{ij}\left(\phi_{ij}'' + \dfrac{\phi_{ij}'}{q_{ij}}\right)}{\dfrac{2}{m}\displaystyle\sum_{i=1}^{N}\mathbf{p}_i^2 + \dfrac{1}{2}\displaystyle\sum_{i \neq j}\mathbf{q}_{ij}^2\left(\phi_{ij}'' + \dfrac{\phi_{ij}'}{q_{ij}}\right) + 9pV}. \tag{6.100}$$

6.9 Constant stress ensembles

We will now give another example of the usefulness of the Norton ensemble. Suppose we wish to calculate the yield stress of a Bingham plastic – a solid with a yield stress. If we use the SLLOD method outlined above the Bingham plastic will always yield simply because the shear *rate* is an input into the simulation. It would not be easy to determine the yield stress from such a calculation. For simulating yield phenomena, one would prefer the shear stress as the input variable. If this were the case, simulations could be run for a series of incremented values of the shear stress. If the stress was less than the yield stress, the solid would strain elastically under the stress. Once the yield stress was exceeded, the material would shear.

Here we discuss a simple method for performing NEMD simulations in the stress ensemble. We will use this as an opportunity to illustrate the use the Nosé–Hoover feedback mechanism. We will also derive linear response expressions for the viscosity within the context of the Norton ensemble. The equations of motion for shear flow, thermostatted using the Nosé–Hoover thermostat are:

$$\dot{\mathbf{q}}_i = \frac{\mathbf{p}_i}{m} + \hat{\mathbf{x}}\gamma y_i,$$

$$\dot{\mathbf{p}}_i = \mathbf{F}_i - \hat{\mathbf{x}}\gamma p_{yi} - \xi\mathbf{p}_i, \tag{6.101}$$

$$\dot{\xi} = \frac{K(\Gamma) - K_0}{Q_\xi} = \frac{1}{\tau_\xi^2}\left(\frac{K(\Gamma)}{K_0} - 1\right).$$

Using the Nosé–Hoover feedback mechanism we relate the rate of change of the shear rate, γ, to the degree to which the instantaneous shear stress, $-P_{xy}(\Gamma)$ differs from a specified mean value, $-S_{xy}(t)$. We therefore determine the shear rate from the differential equation:

$$\dot{\gamma} = \frac{\left(P_{xy}(\Gamma) - S_{xy}(t)\right)V}{Q_\gamma}. \tag{6.102}$$

If the instantaneous stress is greater (i.e. more negative) than the specified value, the shear rate will decrease in an attempt to make the two stresses more nearly equal. The relaxation constant Q_γ should be chosen so that the timescale for feedback fluctuations is roughly equal to the natural relaxation time of the system.

From the equations of motion, the time derivative of the internal energy $H_0 = \sum_i p_i^2/2m + \Phi$, is easily seen to be:

$$\dot{H}_0 = -P_{xy}V\gamma - 2\xi K. \tag{6.103}$$

The Nosé constant-stress, constant-temperature dynamics satisfy a Liouville equation in which phase space behaves as a compressible $(6N + 2)$-dimensional fluid. The equilibrium distribution function is a function of the $3N$-particle coordinates, the $3N$-particle momenta, the thermostatting multiplier ξ, and shear rate γ, $f_0 = f_0(\Gamma, \xi, \gamma)$. The Liouville equation for this system is then:

$$\frac{df_0}{dt} = -f_0 \left(\frac{\partial}{\partial \Gamma} \cdot \dot{\Gamma} + \frac{\partial}{\partial \xi} \dot{\xi} + \frac{\partial}{\partial \gamma} \dot{\gamma} \right). \tag{6.104}$$

Since $\dot{\xi} = \dot{\xi}(\Gamma)$ and $\dot{\gamma} = \dot{\gamma}(\Gamma)$ then:

$$\frac{\partial}{\partial \xi} \dot{\xi} = \frac{\partial}{\partial \gamma} \dot{\gamma} = 0, \tag{6.105}$$

the phase-space compression factor $\Lambda(\Gamma)$ is easily seen to be $-3N\xi$. If we consider the time derivative of the extended internal energy $H_0 + \frac{1}{2}Q_\xi\xi^2 + \frac{1}{2}Q_\gamma\gamma^2$, we find that:

$$\begin{aligned}
\frac{d}{dt} \left(H_0 + \tfrac{1}{2}Q_\xi\xi^2 + \tfrac{1}{2}Q_\gamma\gamma^2 \right) &= \dot{H}_0 + Q_\xi\xi\dot{\xi} + Q_\gamma\gamma\dot{\gamma} \\
&= -P_{xy}V\gamma - \xi K + \xi(K - K_0) + (P_{xy} - S_{xy})V\gamma \\
&= -\xi K_0 - S_{xy}V\gamma. \tag{6.106}
\end{aligned}$$

If we consider the situation at equilibrium, when the set value of the shear stress $-S_{xy}(t)$ is zero and $K_0 = 3N/2\beta$, the Liouville equation becomes:

$$\frac{df_0}{dt} = \beta\xi K_0 f_0 = -\beta f_0 \frac{d}{dt} \left(H_0 + \tfrac{1}{2}Q_\xi\xi^2 + \tfrac{1}{2}Q_\gamma\gamma^2 \right). \tag{6.107}$$

Integrating both sides with respect to time gives the equilibrium distribution function for the constant stress Norton ensemble to be:

$$f_0 = \frac{\exp\left[-\beta\left(H_0 + \tfrac{1}{2}Q_\xi\xi^2 + \tfrac{1}{2}Q_\gamma\gamma^2\right)\right]}{\int d\Gamma \int d\gamma \int d\xi \exp\left[-\beta\left(H_0 + \tfrac{1}{2}Q_\xi\xi^2 + \tfrac{1}{2}Q_\gamma\gamma^2\right)\right]}. \tag{6.108}$$

The equilibrium distribution function is thus a generalized canonical distribution, permitting shear rate fluctuations. Indeed the mean-square shear rate is:

$$\langle \gamma^2 \rangle_{S_{xy}=0} = \frac{1}{\beta Q_\gamma}, \tag{6.109}$$

so the amplitude of the shear-rate fluctuations are controlled by the adjustable constant Q_γ.

We wish to calculate the linear response of an equilibrium ensemble of systems (characterized by the distribution f_0, at time $t = 0$), to an externally imposed time-dependent shear stress, $-S_{xy}(t)$. For the Nosé–Hoover feedback mechanism, the external field is the mean shear stress, and it appears explicitly in the equations of motion (Hood *et al.*, 1987). This is in contrast to the more difficult Gaussian case (Brown and Clarke, 1986). For the Gaussian feedback mechanism, the numerical value of the constraint variable does not usually appear explicitly in the equations of motion. This is a natural consequence of the differential nature of the Gaussian feedback scheme.

The linear response of an arbitrary phase variable $B(\Gamma)$ to an applied time-dependent external field is given by:

$$\langle B(t)\rangle = \langle B(0)\rangle - \int_0^t \mathrm{d}s \int \mathrm{d}\Gamma B(\Gamma)\exp[-i\mathscr{L}_0(t - s)]i\Delta L(s)f_0(\Gamma), \qquad (6.110)$$

where $i\mathscr{L}_0$ is the equilibrium (Nosé–Hoover thermostatted) f-Liouvillean and $i\Delta\mathscr{L}(s) = i\mathscr{L}(s) - i\mathscr{L}_0$, where $i\mathscr{L}(s)$ is the full field-dependent thermostatted f-Liouvillean. It only remains to calculate $i\Delta\mathscr{L}(s)f_0$. Using the equations of motion and the equilibrium distribution function obtained previously, we see that:

$$\begin{aligned}
i\Delta\mathscr{L}(s)f_0 &= \left(\dot{\Gamma}\cdot\frac{\partial}{\partial\Gamma} + \dot{\xi}\cdot\frac{\partial}{\partial\xi} + \dot{\gamma}\cdot\frac{\partial}{\partial\gamma}\right)f_0 + f_0\left(\frac{\partial}{\partial\Gamma}\cdot\dot{\Gamma} + \frac{\partial}{\partial\xi}\cdot\dot{\xi} + \frac{\partial}{\partial\gamma}\cdot\dot{\gamma}\right) \\
&= -\beta\big(\dot{H}_0 + \dot{\gamma}\gamma Q_\gamma + \dot{\xi}\xi Q_\xi\big)f_0 - 3N\xi f_0 \\
&= \beta V S_{xy}(t)\gamma(\Gamma)f_0.
\end{aligned}$$

$$\qquad (6.111)$$

Here we make explicit reference to the phase dependence of γ, and the explicit time dependence of the external field $S_{xy}(t)$. The quantity $-S_{xy}(t)V\gamma(\Gamma)$ is the adiabatic derivative of the extended internal energy, $E = H_0 + \frac{1}{2}Q_\gamma\gamma^2$.

Combining these results the linear response of the phase variable B is:

$$\langle B(t)\rangle = \langle B(0)\rangle - \beta V\int_0^t \mathrm{d}s\langle B(t - s)\gamma\rangle_0 S_{xy}(s). \qquad (6.112)$$

In order to compute the shear viscosity of the system, we need to calculate the time dependence of the thermodynamic force and flux which appear in the defining constitutive relation for shear viscosity. Because of the presence of the Nosé–Hoover relaxation time, controlled by the parameter Q_γ, the actual shear stress in the system $-P_{xy}(\Gamma)$, does not match the externally imposed shear stress $S_{xy}(t)$, instantaneously. To compute the shear viscosity we need to know the precise relation between P_{xy} and γ, not that between S_{xy} and the shear rate. The two quantities of interest are easily computed from Equation (6.112):

$$\langle \gamma(t) \rangle = -\beta V \int_0^t ds \langle \gamma(t-s)\gamma \rangle_0 S_{xy}(s), \qquad (6.113)$$

$$\langle P_{xy}(t) \rangle = -\beta V \int_0^t ds \langle P_{xy}(t-s)\gamma \rangle_0 S_{xy}(s). \qquad (6.114)$$

Fourier–Laplace transforming, we obtain the frequency-dependent linear response relations:

$$\langle \tilde{\gamma}(\omega) \rangle = -\tilde{\chi}_{\gamma\gamma}(\omega)\tilde{S}_{xy}(\omega), \qquad (6.115)$$

$$\langle \tilde{P}_{xy}(\omega) \rangle = -\tilde{\chi}_{P_{xy}\gamma}(\omega)\tilde{S}_{xy}(\omega), \qquad (6.116)$$

where the Fourier–Laplace transform of $\chi(t)$ is defined to be:

$$\tilde{\chi}_{AB}(\omega) = \int_0^\infty dt \exp[-i\omega t]\chi_{AB}(t) = -\beta V \int_0^\infty dt \exp[-i\omega t]\langle A(t)B \rangle_0. \qquad (6.117)$$

The linear constitutive relation for the frequency-dependent shear viscosity (Section 2.4) is:

$$\tilde{P}_{xy}(\omega) \equiv -\tilde{\eta}(\omega)\gamma(\omega), \qquad (6.118)$$

so that the frequency-dependent viscosity is:

$$\tilde{\eta}(\omega) \equiv -\frac{\tilde{\chi}_{P_{xy}\gamma}(\omega)}{\tilde{\chi}_{\gamma\gamma}(\omega)}. \qquad (6.119)$$

This expression shows that the complex frequency-dependent shear viscosity is given by ratio of two susceptibilities. However, these two different time-correlation functions can be related by using the Nosé–Hoover equation of motion (6.101):

$$\dot{\chi}_{\gamma\gamma}(t) = \beta V \langle \dot{\gamma}(t)\gamma \rangle_0 = \frac{\beta V^2}{Q_\gamma} \langle P_{xy}(t)\gamma \rangle_0 = \frac{V}{Q_\gamma}\chi_{P_{xy}\gamma}(t). \qquad (6.120)$$

In the frequency domain, this relation becomes:

$$\frac{V}{Q_\gamma}\tilde{\chi}_{P_{xy}\gamma}(\omega) = -\chi_{\gamma\gamma}(t=0) + i\omega\tilde{\chi}_{\gamma\gamma}(\omega) = \frac{V}{Q_\gamma} + i\omega\tilde{\chi}_{\gamma\gamma}(\omega). \qquad (6.121)$$

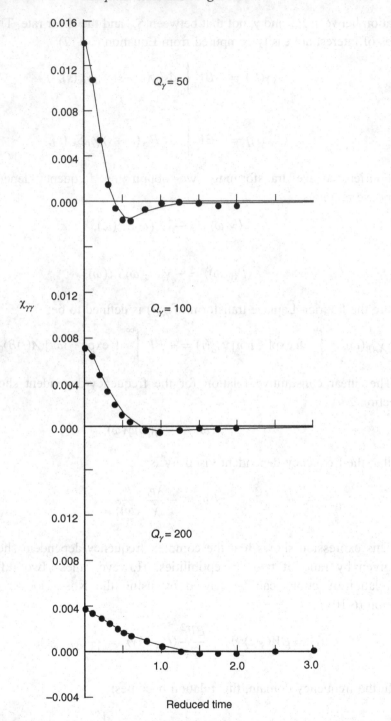

Figure 6.15 A test of Equation (6.121), for the Lennard–Jones triple-point fluid

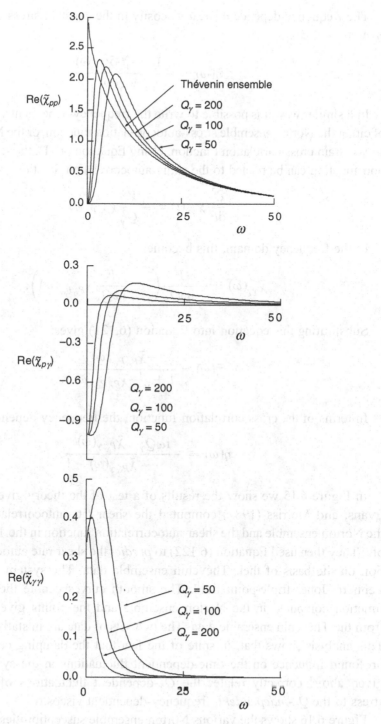

Figure 6.16 The various Norton ensemble susceptibilities as a function of frequency. The system is the Lennard–Jones triple point

The frequency-dependent shear viscosity in the constant-stress ensemble can be written as:

$$\tilde{\eta}(\omega) = -\frac{1 + \frac{i\omega Q_\gamma}{V}\tilde{\chi}_{\gamma\gamma}(\omega)}{\tilde{\chi}_{\gamma\gamma}(\omega)}. \qquad (6.122)$$

In a similar way, it is possible to write the frequency-dependent viscosity in terms of either the Norton ensemble stress-autocorrelation function, or the Norton ensemble stress-strain cross-correlation function. Using Equation (4.11), the stress-autocorrelation function can be related to the strain-autocorrelation function using the relation:

$$\frac{\mathrm{d}^2}{\mathrm{d}t^2}\chi_{\gamma\gamma}(t) = -\frac{V^2}{Q_\gamma^2}\chi_{P_{xy}P_{xy}}(t). \qquad (6.123)$$

In the frequency domain, this becomes:

$$\tilde{\chi}_{\gamma\gamma}(\omega) = -\frac{V}{i\omega Q_\gamma}\left(1 + \frac{V}{i\omega Q_\gamma}\tilde{\chi}_{P_{xy}P_{xy}}(\omega)\right). \qquad (6.124)$$

Substituting this equation into Equation (6.121) gives:

$$\tilde{\eta}(\omega) \equiv \frac{-\tilde{\chi}_{P_{xy}P_{xy}}(\omega)}{1 + \frac{V}{i\omega Q_\gamma}\tilde{\chi}_{P_{xy}P_{xy}}(\omega)}. \qquad (6.125)$$

In terms of the cross-correlation function, the frequency-dependent viscosity is

$$\tilde{\eta}(\omega) \equiv -\frac{i\omega Q_\gamma}{V}\frac{\tilde{\chi}_{P_{xy}\gamma}(\omega)}{\tilde{\chi}_{P_{xy}\gamma}(\omega) - 1}. \qquad (6.126)$$

In Figure 6.15 we show the results of a test of the theory given above. Hood, Evans, and Morriss (1987) computed the shear-rate autocorrelation function in the Norton ensemble and the shear-autocorrelation function in the Thévenin ensemble. They then used Equation (6.122) to *predict* the shear rate autocorrelation function on the basis of their Thévenin ensemble data. The system studied was the Lennard–Jones triple-point fluid. The smooth curves denote the autocorrelation function computed in the Norton ensemble and the points give the predictions from the Thévenin ensemble data. The two sets of data are in statistical agreement. This analysis shows that, in spite of the fact that the damping constant Q_γ has a profound influence on the time-dependent fluctuations in the system, the theory given above correctly relates the Q_γ-dependent fluctuations of shear rate and stress to the Q_γ-*independent*, frequency-dependent viscosity.

Figure 6.16 shows the various Norton ensemble susceptibilities as a function of frequency for the same system.

7

Nonlinear response theory

7.1 Kubo's form for the nonlinear response

In Chapter 6 we saw that nonequilibrium molecular dynamics leads inevitably to questions regarding the nonlinear response of systems. Here we begin a discussion of this subject. It is not widely known that in Kubo's original paper (Kubo, 1957), he not only presented results for adiabatic linear response theory, but that he also included a formal treatment of the adiabatic nonlinear response. The reason why this fact is not widely known is that, like many treatments of nonlinear response theory that followed, his formal results were exceedingly difficult to translate into a useful, experimentally verifiable form. This difficulty can be traced to three sources. Firstly, his results are not easily transformable into explicit representations that involve the evaluation of time-correlation functions of explicit phase variables. Secondly, if one wants to study nonequilibrium steady states, the treatment of thermostats is mandatory. His theory did not include such effects. Thirdly, his treatment gave a power-series representation of the nonlinear response. We now believe that for most transport processes, such expansions do not exist.

We will now give a presentation of Kubo's perturbation expansion for the nonequilibrium distribution function, $f(t)$. Consider an N-particle system evolving under the following dynamics:

$$\dot{\mathbf{q}}_i = \frac{\mathbf{p}_i}{m} + \mathbf{C}_i(\Gamma)F_e,$$

$$\dot{\mathbf{p}}_i = \mathbf{F}_i - \mathbf{D}_i(\Gamma)F_e.$$

(7.1)

The terms $\mathbf{C}_i(\Gamma)$ and $\mathbf{D}_i(\Gamma)$ describe the coupling of the external field F_e to the system. In this discussion, we will limit ourselves to the case where the field is *switched on* at time zero, and thereafter remains at the same steady value. The *f*-Liouvillean is given by:

$$i\mathscr{L} = \dot{\Gamma} \cdot \frac{\partial}{\partial \Gamma} + \frac{\partial}{\partial \Gamma} \cdot \dot{\Gamma} = i\mathscr{L}_0 + i\Delta\mathscr{L}, \tag{7.2}$$

where $i\mathscr{L}_0$ is the equilibrium Liouvillean and $i\Delta\mathscr{L}$ is the field dependent perturbation which is a linear function of F_e. The Liouville equation is:

$$\frac{\partial}{\partial t}f(t) = -i\mathscr{L}f(t). \tag{7.3}$$

To go beyond the linear response, Kubo assumed that $f(t)$ could be expanded as a power series in the external field, F_e:

$$f(t) = f_0 + f_1(t) + f_2(t) + f_3(t) + f_4(t) + \cdots, \tag{7.4}$$

where, $f_i(t)$ is of ith order in the external field F_e. The assumption that $f(t)$ can be expanded in a power series about $F_e = 0$ may seem innocent, but it is not. This assumption rules out any functional form containing noninteger powers of the field, such as F_e^α, where α is noninteger. Substituting Equation (7.4) for $f(t)$, and the expression for $i\mathscr{L}$, into the Liouville Equation (7.3), and equating terms of the same order in F_e, we find an infinite sequence of partial differential equations to solve:

$$\frac{\partial}{\partial t}f_i(t) + i\mathscr{L}_0 f_i(t) = -i\Delta\mathscr{L}f_{i-1}(t), \tag{7.5}$$

where $i \geq 1$. The solution to this series of equations can be written as

$$f_i(t) = -\int_0^t ds \, \exp[-i\mathscr{L}_0(t-s)] \, i\Delta\mathscr{L} \, f_{i-1}(s). \tag{7.6}$$

To prove that this is correct differentiate both sides to obtain Equation (7.5). Recursively substituting Equation (7.6) into Equation (7.4), we obtain a power series representation of the distribution function:

$$f(t) = f(0) + \sum_{i=1}^{\infty}(-1)^i \int_0^t ds_i \int_0^{s_i} ds_{i-1} \ldots$$

$$\int_0^{s_2} ds_1 \exp[-i\mathscr{L}_0(t-s)]\Delta\mathscr{L} \ldots \exp[-i\mathscr{L}_0(s_2-s_1)]\Delta\mathscr{L} f(0). \tag{7.7}$$

Although this result is formally exact, there are a number of difficulties with this approach. The expression for $f(t)$ is a sum of convolutions of operators. In general, the operator $i\Delta\mathscr{L}$ does not commute with the propagator $\exp(-i\mathscr{L}_0 t)$, and no further simplifications of the general result are possible. Further, as we have seen before, there is a strong likelihood that fluxes associated with conserved quantities are nonanalytic functions of the thermodynamic force, F_e. This would mean that the average response of the shear stress, for example, cannot be expanded as a Taylor series about $F_e(=\gamma) = 0$. In Chapter 6 we saw evidence that the shear stress is of the form, $\langle P_{xy} \rangle = -\gamma(\eta_0 + \eta_1 \gamma^{1/2})$ (see Section 6.3). If this is true then $f_2(t) \equiv \frac{1}{2}\gamma^2(\partial^2 f(\gamma)/\partial\gamma^2)_{\gamma=0}$ must be infinite.

7.2 Kawasaki distribution function

An alternative approach to nonlinear response theory was pioneered by Yamada and Kawasaki (1967). Rather than developing power series expansions about $F_e = 0$, they derived a closed expression for the perturbed distribution function. The power of their method was demonstrated in a series of papers in which Kawasaki first predicted the nonanalyticity of the shear viscosity with respect to shear rate (Kawasaki and Gunton, 1973; Yamada and Kawasaki, 1975a). This work predates the first observation of these effects in computer simulations. The simplest application of the Kawasaki method is to consider the adiabatic response of a canonical ensemble of N-particle systems to a steady applied field F_e.

The Liouville equation for this system is:

$$\frac{\partial}{\partial t}f = -i\mathscr{L}f. \qquad (7.8)$$

Here the Liouvillean is field-dependent, defined by the equations of motion (7.1). The formal solution of (7.8) is:

$$f(t) = \exp[-i\mathscr{L}t]f(0). \qquad (7.9)$$

For simplicity we take the initial distribution function $f(0)$, to be canonical, so that $f(t)$ becomes:

$$f(t) = \exp[-i\mathscr{L}t]\frac{\exp[-\beta H_0]}{Z(\beta)}. \qquad (7.10)$$

The *adiabatic* distribution function propagator is the Hermitian conjugate of the phase variable propagator, so in this case $\exp(-i\mathscr{L}t)$ is the negative-time phase variable propagator, $\exp(iL(-t))$. It operates on the phase variable in the numerator, moving time backwards in the presence of the applied field. This implies that:

$$f(t) = \frac{\exp[-\beta H_0(-t)]}{Z(\beta)}. \qquad (7.11)$$

Formally the f-propagator leaves the denominator invariant since it is not a phase variable. The phase dependence of the denominator has been integrated out.

However, since the distribution function must be normalized, we can obviously also write:

$$f(t) = \frac{\exp[-\beta H_0(-t)]}{\int d\Gamma \exp[-\beta H_0(-t)]}. \tag{7.12}$$

This equation is an explicitly normalized version of the so-called *bare* Kawasaki form, Equation (7.11). We will refer to Equation (7.12) as the renormalized Kawasaki distribution function.

Using the equations of motion (7.1) one can write the time derivative of H_0 as the product of a phase variable $J(\Gamma)$ and the magnitude of the perturbing external field, F_e.

$$\dot{H}_0^{ad} = -J(\Gamma)F_e. \tag{7.13}$$

For the specific case of planar Couette flow, we saw in Section 6.2 that $\dot{H}_0^{ad}$ is the product of the shear rate, the shear stress, and the system volume, $-\gamma P_{xy} V$ and thus in the absence of a thermostat we can write:

$$H_0(-t) = H_0(0) - \int_0^t ds \dot{H}_0(s) = H_0(0) + \gamma V \int_0^t ds P_{xy}(-s). \tag{7.14}$$

The *bare* form for the perturbed distribution function at time t is then:

$$f(t) = \exp\left[-\beta \gamma V \int_0^t ds P_{xy}(-s)\right] f(0). \tag{7.15}$$

It is important to remember that the generation of $P_{xy}(-s)$ from $P_{xy}(0)$ is controlled by the field-dependent equations of motion.

A major problem with this approach is that in an adiabatic system the applied field will cause the system to heat up. This process continues indefinitely and a steady state can never be reached. What is surprising is that when the effects of a thermostat are included, the formal expression for the N-particle distribution function remains unaltered, the only difference being that thermostatted, field-dependent dynamics must be used to generate $H_0(-t)$ from $H_0(0)$. This is the next result we shall derive.

Consider an isokinetic ensemble of N-particle systems subject to an applied field. We will assume field-dependent, Gaussian isokinetic equations of motion, Equation (5.66). The f-Liouvillean therefore contains an extra thermostatting term. It is convenient to write the Liouville equation in operator form:

$$\frac{\partial}{\partial t} f(t) = -i \mathscr{L} f(t) = -iL f(t) - f(t)\Lambda = -\frac{\partial}{\partial \Gamma} \cdot (\dot{\Gamma} f) \tag{7.16}$$

The operator $i\mathscr{L}$ is the f-Liouvillean, and iL is the p-Liouvillean. The term Λ, is:

$$\Lambda = \frac{\partial}{\partial \Gamma} \cdot \dot{\Gamma} = -\frac{1}{f}\frac{df}{dt} = -\frac{d}{dt}\ln f, \qquad (7.17)$$

or the phase-space compression factor (Section 3.3). The formal solution of the Liouville equation is given by:

$$f(t) = \exp[-i\mathscr{L}t]f(0) = \exp[-(iL + \Lambda)t]f(0). \qquad (7.18)$$

In the thermostatted case the p-propagator is no longer the Hermitian conjugate of the f-propagator.

We will use the Dyson decomposition derived in Section 3.6, to relate thermostatted p- and f-propagators, assuming that both Liouvilleans have no explicit time dependence. We make a crucial observation, namely that the phase-space compression factor Λ, is a phase variable rather than an operator. Taking the reference Liouvillean to be the adjoint of iL we find:

$$\exp[-iLt - \Lambda t] = \exp[-iLt] - \int_0^t ds\, \exp[-iLt - \Lambda t]\,\Lambda\,\exp[-iL(t - s)]. \quad (7.19)$$

Repeated application of the Dyson decomposition to $\exp[-iLs - \Lambda s]$ on the right-hand side of Equation (7.19) gives:

$$\exp[-iLt - \Lambda t]$$

$$= \sum_{n=0}^{\infty} (-)^n \int_0^t ds_1 \ldots \int_0^{s_{n-1}} ds_n \exp[-iLs_n]\Lambda \exp[-iL(s_{n-1} - s_n)]\Lambda \ldots \exp[-iL(t - s_1)]$$

$$= \sum_{n=0}^{\infty} (-)^n \int_0^t ds_1 \ldots \int_0^{s_{n-1}} ds_n \Lambda(-s_n)\Lambda(-s_{n-1})\ldots\Lambda(-s_1)\exp[-iLt]$$

$$= \exp\left[-\int_0^t ds\,\Lambda(-s)\right]\exp[-iLt]. \qquad (7.20)$$

In deriving the second line of this equation we use the fact that for any phase variable B, $\exp[-iLs]B = B(-s)\exp[-iLs]$. That is, $\exp[-iLs]$ acts on the phase variable B, and then acts on any phase variables on the right-hand side of B. Substituting Equation (7.20) into Equation (7.18) and choosing the iso-kinetic distribution $f(0) = f_T(0) = \delta(K - K_0)\exp[-\beta\Phi]/Z(\beta)$, we obtain:

$$f(t) = \frac{\delta(K - K_0)\exp[-\int_0^t ds\,\Lambda(-s)]\exp[-\beta\Phi(-t)]}{Z(\beta)}. \qquad (7.21)$$

If we change variables in the integral of the phase-space compression factor, and use the identity $\Phi(-t) = \int_0^{-t} ds \dot{\Phi}(s) + \Phi(0)$, we obtain:

$$f(t) = \frac{\delta(K - K_0)\exp[-\beta\Phi(0)]\exp[\int_0^{-t} ds(\Lambda(s) - \beta\dot{\Phi}(s))]}{Z(\beta)}.$$ (7.22)

We know that for the isokinetic distribution, $\beta = 3N/2K$ (see Section 5.2). Since under the isokinetic equations of motion, K is a constant of the motion, we can prove from Equation (5.66), that:

$$\Lambda - \beta\dot{\Phi} = -\beta J F_e.$$ (7.23)

If AIΓ is satisfied the dissipative flux J is defined by Equation (7.13). Substituting Equation (7.23) into Equation (7.22), we find that the thermostatted Kawasaki distribution function can be written as:

$$f_T(t) = f_T(0)\exp\left[\beta\int_0^{-t} ds J(s)F_e\right]$$

$$= f_T(0)\exp\left[-\beta\int_0^{t} ds J(-s)F_e\right].$$ (7.24)

Formally this equation is identical to the adiabatic response (7.15). This is in spite of the fact that the thermostat changes the equations of motion. The adiabatic and thermostatted forms are identical because the changes caused by the thermostat to the dissipation ($\dot{H}_0$), are *exactly* cancelled by the changes caused by the thermostat to the form of the Liouville equation. This observation was first made by Morriss and Evans (1985). Clearly one can renormalize the thermostatted form of the Kawasaki distribution function giving Equation (7.25) as the renormalized form of the isokinetic Kawasaki distribution function:

$$f_{T,n}(t) = \frac{\exp[-\beta\int_0^{t} ds J(-s)F_e]f_T(0)}{\langle\exp[-\beta\int_0^{t} ds J(-s)F_e]\rangle}.$$ (7.25)

As we will see, the renormalized Kawasaki distribution function is very useful for deriving relations between steady-state fluctuations and derivatives of steady-state phase averages. However, it is not efficient for computing nonequilibrium averages themselves. This is because it involves averaging exponentials of extensive integrals. We will now turn to an alternative approach to this problem.

7.3 The transient time-correlation function formalism

The transient time-correlation function formalism (TTCF) provides perhaps the simplest nonlinear generalization of the Green–Kubo relations. A number of authors independently derived the TTCF expression for adiabatic phase averages (Visscher, 1974; Dufty and Lindenfeld, 1979; Cohen, 1983). We will illustrate the derivation for isokinetic planar Couette flow; however, the formalism is quite general and can easily be applied to other systems. The theory gives an exact relation between the nonlinear steady state response and the so-called transient time-correlation functions. We will also describe the links between the TTCF approach and the Kawasaki methods outlined in Section 7.2. Finally, we will present some numerical results which were obtained as tests of the validity of the TTCF formalism.

Following Morriss and Evans (1987), we will give a derivation using the Heisenberg, rather than the customary Schrödinger picture. The average of a phase variable, $B(\Gamma)$, at time t, is:

$$\langle B(t) \rangle = \int d\Gamma B(\Gamma) f(t) = \int d\Gamma f(0) B(\Gamma(t)), \qquad (7.26)$$

where the first equality is the Schrödinger representation and the second equality is the Heisenberg representation. For *time-independent* external fields, differentiating the Heisenberg form with respect to time yields:

$$\frac{d}{dt} \langle B(t) \rangle = \int d\Gamma f(0) \dot{\Gamma} \cdot \frac{\partial}{\partial \Gamma} B(t). \qquad (7.27)$$

In deriving Equation (7.27), we have used the fact that $\dot{B}(t) = iL \exp[iLt]B = \exp[iLt] iLB$. This relies upon the time independence of the Liouvillean, iL. The corresponding equation for the time-dependent case is not true. Integrating Equation (7.27) by parts we see that:

$$\frac{d}{dt} \langle B(t) \rangle = -\int d\Gamma B(t) \frac{\partial}{\partial \Gamma} \cdot (\dot{\Gamma} f(0)). \qquad (7.28)$$

The boundary term vanishes because the distribution function $f(0)$, approaches zero when the magnitude of any component of any particle's momentum becomes infinite, and because the distribution function can be taken to be a periodic function of the particle coordinates. We are explicitly using the periodic boundary conditions used in computer simulations.

Integrating Equation (7.28) with respect to time we see that the nonlinear non-equilibrium response can be written as:

$$\langle B(t) \rangle = \langle B(0) \rangle - \int_0^t ds \int d\Gamma B(t) \frac{\partial}{\partial \Gamma} \cdot (\dot{\Gamma} f(0)). \qquad (7.29)$$

The dynamics implicit in $B(s)$ is, of course, the full field-dependent, thermostatted equations of motion (7.1) and (7.2). For a system subject to a thermostatted shearing deformation, $\dot{\Gamma}$ is given by the thermostatted SLLOD equations of motion (6.35).

If the initial distribution is Gaussian isokinetic it is straightforward to show that, $(\partial/\partial\Gamma) \cdot (f(0)\dot{\Gamma}) = \beta V P_{xy}(\Gamma) f(0)$. If the initial ensemble is canonical then, to first order in the number of particles, $(\partial/\partial\Gamma) \cdot (f(0)\dot{\Gamma}) = \beta V P_{xy}(\Gamma)f(0)$. To show this one writes, (following Section 5.3):

$$\langle B(t)\rangle_c = \left\langle B(0)\rangle_c - \beta\gamma V \int_0^t ds\langle B(s)\left[P_{xy}(0) - P_{xy}^K(0)\frac{\Delta K(0)}{\langle K\rangle_c}\right]\right\rangle_c, \qquad (7.30)$$

where $P_{xy}^K(0)$ is the kinetic part of the pressure tensor evaluated at time zero (compare this with the linear theory given in Section 5.3). Now, we note that $\langle P_{xy}^K(0)\Delta K(0)/\langle K\rangle\rangle_c = 0$. This means that Equation (7.30) can be written as:

$$\langle B(t)\rangle_c = \langle B(0)\rangle_c - \beta\gamma V \int_0^t ds\left\langle\Delta B(s)\left[P_{xy}(0) - P_{xy}^K(0)\frac{\Delta K(0)}{\langle K\rangle_c}\right]\right\rangle_c. \qquad (7.31)$$

As in the linear response case (Section 5.3) we assume, without loss of generality, that $B(\Gamma)$ is extensive. The kinetic fluctuation term involves the average of three zero-mean, extensive quantities and because of the factor $1/\langle K(0)\rangle$, gives only an order one contribution to the average. Thus for both the isokinetic and canonical ensembles, we can write:

$$\langle B(t)\rangle = \langle B(0)\rangle - \beta\gamma V \int_0^t ds\langle\Delta B(s)P_{xy}(0)\rangle. \qquad (7.32)$$

This expression relates the non-equilibrium value of a phase variable B at time t, to the integral of a transient time-correlation function (the correlation between P_{xy} in the equilibrium starting state, $P_{xy}(0)$, and B at time s after the field is turned on). The time-zero value of the transient correlation function is an equilibrium property of the system. For example, if $B = P_{xy}$, then the time-zero value is $\langle P_{xy}^2(0)\rangle$. Under some, but by no means all circumstances, the values of $B(s)$ and $P_{xy}(0)$ will become uncorrelated at long times. If this is the case the system is said to exhibit mixing (see Section 3.4). The transient correlation function will then approach $\langle B(t)\rangle\langle P_{xy}(0)\rangle$, which is zero because $\langle P_{xy}(0)\rangle = 0$.

The adiabatic systems treated by Visscher, 1974, Dufty and Lindenfeld (1979), and Cohen (1983) do not exhibit mixing because in the absence of a thermostat, $\frac{d}{dt}\langle B(t)\rangle$ does not, in general, go to zero at large times. Thus the integral of the associated transient correlation function does not converge. This presumably means that the initial fluctuations in adiabatic systems are remembered forever.

If AIΓ (Section 5.3) is satisfied, the result for the general case is:

$$\langle B(t) \rangle = \langle B(0) \rangle - \beta F_e \int_0^t ds \langle \Delta B(s) J(0) \rangle. \qquad (7.33)$$

We can use recursive substitution to derive the Kawasaki form for the nonlinear response from the transient time-correlation formula, Equation (7.33). The first step in the derivation of the Kawasaki representation is to rewrite the TTCF relation using iL to denote the phase-variable Liouvillean, and $-i\mathcal{L}$ to denote its nonHermitian adjoint, the f-Liouvillean. Thus $\dot{B} \equiv iLB$ and $\partial f / \partial t = -i\mathcal{L}f$. Using this notation, Equation (7.33) can be written as:

$$\langle B(t) \rangle = \int d\Gamma B f(0) - \beta \gamma V \int_0^t ds \int d\Gamma f(0) \exp[iLs](B \exp[-iLs]J)$$

$$= \int d\Gamma B f(0) - \beta \gamma V \int_0^t ds \int d\Gamma \exp(-i\mathcal{L}s) f(0) B \exp[-iLs]J, \qquad (7.34)$$

where we have unrolled the first p-propagator onto the distribution function. Equation (7.34) can be written more simply as:

$$\langle B(t) \rangle = \int d\Gamma B f(0) - \beta \gamma V \int_0^t ds \int d\Gamma f(s) B(0) J(-s). \qquad (7.35)$$

Since this equation is true for all phase variables B, the TTCF representation for the N-particle distribution function can be written formally as:

$$f(t) = f(0) - \beta \gamma V \int_0^t ds J(-s) f(s). \qquad (7.36)$$

Recursively substituting for $f(s)$ using Equation (7.36), and inserting the result into Equation (7.35), the TTCF expression for the nonequilibrium average becomes:

$$\langle B(t) \rangle = \int d\Gamma B f(0) - \beta \gamma V \int_0^t ds_1 \int d\Gamma B(0) J(-s_1) f(0)$$

$$+ (\beta \gamma V)^2 \int_0^t ds_1 \int_0^{s_1} ds_2 \int d\Gamma B(0) J(-s_1) J(-s_2) f(0) + \cdots$$

$$= \int d\Gamma B \exp\left[-\beta \gamma V \int_0^t ds J(-s) \right] f(0). \qquad (7.37)$$

This is precisely the Kawasaki form of the thermostatted nonlinear response. This expression is valid for both the canonical and isokinetic ensembles. It is also valid for the canonical ensemble when the thermostatting is carried out using the Nosé–Hoover thermostat.

One can, of course, also derive the TTCF expression for phase averages from the Kawasaki expression by differentiating Equation (7.37) with respect to time, and then reintegrating. A simple integration of Equation (7.38) with respect to time yields the TTCF relation, Equation (7.33). We have thus proved the formal equivalence of the TTCF and Kawasaki representations for the nonlinear thermostatted response.

Comparing the transient time-correlation expression for the nonlinear response with the Kawasaki representation, we see that the difference simply amounts to a time shift. In the transient time-correlation form, it is the dissipative flux J which is evaluated at time zero, whereas in the Kawasaki form, the response variable B is evaluated at time zero. For equilibrium or steady-state time-correlation functions, the stationarity of averages means that such time shifts are essentially trivial. For transient response-correlation functions, there is, of course, no such invariance principle, consequently the time-translation transformation is accordingly more complex.

The computation of the time-dependent response using the Kawasaki form, Equation (7.32), directly is very difficult. The errors associated with the sampling of the initial ensemble, together with the inaccuracy of the trajectory and the numerical evaluation of the extensive Kawasaki integrand, combine and are magnified by the exponential. This exponential is then multiplied by the phase variable $B(0)$. In contrast, the calculation of the response using the transient correlation expression, Equation (7.33), is, as we shall see, far easier.

It is trivial to see that, in the linear regime, both the TTCF and Kawasaki expressions reduce to the usual Green–Kubo expressions. The equilibrium time-correlation functions that appear in Green–Kubo relations are generated by the field-free thermostatted equations. In the TTCF formulae the field is *turned on* at $t = 0$.

The coincidence at small fields, of the Green–Kubo and transient correlation formulae means that unlike direct NEMD, the TTCF method can be used at small fields. This is impossible for direct NEMD because in the small-field limit, the signal-to-noise ratio goes to zero. The signal to noise ratio for the transient correlation function method becomes equal to that of the equilibrium Green–Kubo method. The transient correlation function method forms a bridge between the Green–Kubo method, which can only be used at equilibrium, and direct NEMD, which is the most efficient strong-field method. Because a field is required to generate TTCF correlation functions, their calculation using a molecular dynamics still requires a *nonequilibrium* computer simulation to be performed.

It is also easy to see that, at short times, there is no difference between the linear and nonlinear stress response. It takes time for the nonlinearities to develop. The way to see this is to expand the transient time-correlation function in a power series in γt. The coefficient of the first term in this series is just $V\langle P_{xy}^2 \rangle / k_B T$, the infinite frequency shear modulus G_∞. Since this is an equilibrium property, its value is unaffected by the shear rate and is thus the same in both the linear and nonlinear cases. If we look at the response of a quantity like the pressure, whose linear response is zero, the leading term in the short time expansion is quadratic in the shear rate and in time. The linear response, of course, is the first to appear.

7.4 Trajectory mappings

To calculate transient time-correlation functions, it is necessary to generate an ensemble of equilibrium starting states. To make this process as efficient as possible, we exploit the symmetries of the initial distribution function. In a simulation of a single field-free trajectory, we take ensemble members every N_e timesteps, but the phase point selected Γ can be used to generate other ensemble members which have the same probability, but which generate different field-dependent trajectories. To do this we use a group of phase space mappings (Morriss and Evans, 1989).

This process is quite general, but in this section we illustrate the idea by considering the SLLOD equations for planar Couette flow. We write the phase point Γ, as $(\mathbf{q}, \mathbf{p}) = (\mathbf{x}, \mathbf{y}, \mathbf{z}, \mathbf{p}_x, \mathbf{p}_y, \mathbf{p}_z)$ where each of the components $\mathbf{x}, \mathbf{y}, \mathbf{z}, \mathbf{p}_x, \mathbf{p}_y, \mathbf{p}_z$ is itself an N-dimensional vector. The four mappings that we introduce are $\mathbf{M}^I$ (the identity), $\mathbf{M}^T$ (time-reversal), $\mathbf{M}^Y$ (y-reflection) and $\mathbf{M}^K$ (Kawasaki) defined by:

$$\Gamma^I = \mathbf{M}^I[\Gamma] = \left(\mathbf{x}, \mathbf{y}, \mathbf{z}, \mathbf{p}_x, \mathbf{p}_y, \mathbf{p}_z\right).$$

$$\Gamma^T = \mathbf{M}^T[\Gamma] = \left(\mathbf{x}, \mathbf{y}, \mathbf{z}, -\mathbf{p}_x, -\mathbf{p}_y, -\mathbf{p}_z\right).$$

$$\Gamma^Y = \mathbf{M}^Y[\Gamma] = \left(\mathbf{x}, -\mathbf{y}, \mathbf{z}, \mathbf{p}_x, -\mathbf{p}_y, \mathbf{p}_z\right). \tag{7.38}$$

$$\Gamma^K = \mathbf{M}^K[\Gamma] = \left(\mathbf{x}, -\mathbf{y}, \mathbf{z}, -\mathbf{p}_x, \mathbf{p}_y, -\mathbf{p}_z\right).$$

The time-reversal map changes the sign of all the momenta. The y-reflection map changes the sign of $\mathbf{y}$ and $\mathbf{p}_y$, and the Kawasaki map is the combination of $\mathbf{M}^T$ and $\mathbf{M}^Y$, so $\mathbf{M}^K = \mathbf{M}^T \mathbf{M}^Y$. The probability of each of these states occurring within the equilibrium canonical or isokinetic distribution is the same, because the Hamiltonian H_0 is invariant or the kinetic and potential energies are each separately invariant. The notation Γ^T means the phase point generated from Γ by the map $\mathbf{M}^T$, so that $\Gamma^T \equiv (\mathbf{x}^T, \mathbf{y}^T, \mathbf{z}^T, \mathbf{p}_x^T, \mathbf{p}_y^T, \mathbf{p}_z^T) = (\mathbf{x}, \mathbf{y}, \mathbf{z}, -\mathbf{p}_x, -\mathbf{p}_y, -\mathbf{p}_z)$ (Table 7.1).

The time-reversal mapping perhaps requires some more explanation. The time evolution of phase variable $B(\Gamma)$ is given by $B(t) = \exp[iLt]B(\Gamma)$. Time reversal can be achieved using the inverse propagator $\exp[iLt]$, so that $\exp[-iLt]B(t) = B(\Gamma)$. This *retraces* the original trajectory. The time-reversal

Table 7.1 *The group multiplication table for mappings*

×	$\mathbf{M}^I$	$\mathbf{M}^T$	$\mathbf{M}^Y$	$\mathbf{M}^K$
$\mathbf{M}^I$	$\mathbf{M}^I$	$\mathbf{M}^T$	$\mathbf{M}^Y$	$\mathbf{M}^K$
$\mathbf{M}^T$	$\mathbf{M}^T$	$\mathbf{M}^I$	$\mathbf{M}^K$	$\mathbf{M}^Y$
$\mathbf{M}^Y$	$\mathbf{M}^Y$	$\mathbf{M}^K$	$\mathbf{M}^I$	$\mathbf{M}^T$
$\mathbf{M}^K$	$\mathbf{M}^K$	$\mathbf{M}^Y$	$\mathbf{M}^T$	$\mathbf{M}^I$

mapping is different. Take $\Gamma(t)$ and apply the time-reversal mapping $\mathbf{M}^T$. Then apply *forward* time evolution $\exp(iLt)$, and then apply $\mathbf{M}^T$ again. Thus:

$$\mathbf{M}^T \exp[iLt]\mathbf{M}^T \Gamma(t) = \mathbf{M}^T \exp[iLt]\mathbf{M}^T \exp[iLt]\Gamma(0)$$

$$= \exp[-iLt]\Gamma(t) = \Gamma(0).$$

It is important to realize that this process does not lead to a *retracing* of the original trajectory, as everywhere along the *return* path the momenta are the opposite sign to those of the *forward* path. These results will be derived in more detail later.

For the SLLOD algorithm, the most important property of the mappings is their effect on the value of the shear stress P_{yx} (Γ). The effect on P_{yx} (Γ) is *at most* a sign change, and the sum $P_{yx}(\Gamma^I) + P_{yx}(\Gamma^T) + P_{yx}(\Gamma^Y) + P_{yx}(\Gamma^K) = 0$. For example, for the time-reversal map:

$$\mathbf{M}^T[P_{yx}(\Gamma)] = \mathbf{M}^T \left[\sum_{i=1}^{N} \left(\frac{p_{yi}p_{xi}}{m} + y_i F_{xi} \right) \right]$$

$$= \sum_{i=1}^{N} \left(\frac{(-p_{yi})(-p_{xi})}{m} + y_i F_{xi} \right) = P_{yx}. \tag{7.39}$$

Similarly:

$$\mathbf{M}^Y[P_{yx}(\Gamma)] = \sum_{i=1}^{N} \left(\frac{p_{yi}(-p_{xi})}{m} + (-y_i)F_{xi} \right) = -P_{yx}(\Gamma). \tag{7.40}$$

The general result is that:

$$\mathbf{M}^X B(\Gamma) = p_B^X B(\Gamma), \tag{7.41}$$

where $p_B{}^X$ is the parity operator (Table 7.2) for phase variable B under the operation of mapping X.

The advantage of using mappings is that the transient response formula, Equation (7.32), requires the time integral of the transient time-correlation function $\langle \Delta B(t) P_{yx}(0) \rangle$. As $t \to \infty$, $\langle \Delta B(t) P_{yx}(0) \rangle \to \langle \Delta B(\infty) \rangle \langle P_{yx}(0) \rangle$, so constructing the time-correlation function from mapped initial states ensures that $\langle P_{yx}(0) \rangle \equiv 0$

Table 7.2 *Mapping parities operators for some phase variables*

Mapping	Shear stress	Pressure	Energy
$\mathbf{M}^I$	1	1	1
$\mathbf{M}^T$	1	1	1
$\mathbf{M}^Y$	-1	1	1
$\mathbf{M}^K$	-1	1	1

precisely! Choosing a Γ, then generating $\Gamma^I, \Gamma^T, \Gamma^Y, \Gamma^K$ ensures that the average of the shear stress is zero. This eliminates the statistical difficulties at long time associated with small nonzero values of the average of $P_{yx}(0)$.

Dynamics

The second aspect of the mappings is to understand their subsequent dynamics. We need each mapped point to lead to different dynamics to achieve the most efficient sampling. The first step is to understand the effect of the mapping on the phase-variable Liouvillean $iL(\Gamma)$. To do this we need to know how the equations of motion transform. We will discuss the transformation of the field-dependent equations of motion under the mapping. This will require an extension of the mappings to include the field itself. We begin with the time-reversal mapping $\mathbf{M}^T$, and simply state the results for other mappings. In what follows, the mapping operator $\mathbf{M}^T$ operates on all functions and operators (which depend upon Γ and γ) to its right. The mapping of the Hamiltonian equations of motion gives:

$$\mathbf{M}^T \dot{\mathbf{q}}_i = \mathbf{M}^T \left[\frac{\mathbf{p}_i}{m} \right] = \frac{\mathbf{p}_i^T}{m} = \frac{-\mathbf{p}_i}{m} = -\dot{\mathbf{q}}_i,$$

$$\mathbf{M}^T \dot{\mathbf{p}}_i = \mathbf{M}^T [\mathbf{F}_i(\mathbf{q})] = \mathbf{F}_i(\mathbf{q}^T) = \mathbf{F}_i(\mathbf{q}) = \dot{\mathbf{p}}_i,$$

(7.42)

so the mapping of the propagator is:

$$\mathbf{M}^T iL(\Gamma) = \mathbf{M}^T \left[\dot{\Gamma} \cdot \frac{\partial}{\partial \Gamma} \right] = \mathbf{M}^T \left[\sum_{i=1}^{N} \left(\dot{\mathbf{q}}_i \cdot \frac{\partial}{\partial \mathbf{q}_i} + \dot{\mathbf{p}}_i \cdot \frac{\partial}{\partial \mathbf{p}_i} \right) \right]$$

$$= \sum_{i=1}^{N} \left(\dot{\mathbf{q}}_i^T \cdot \frac{\partial}{\partial \mathbf{q}_i^T} + \dot{\mathbf{p}}_i^T \cdot \frac{\partial}{\partial \mathbf{p}_i^T} \right)$$

$$= \sum_{i=1}^{N} \left(-\dot{\mathbf{q}}_i \cdot \frac{\partial}{\partial \mathbf{q}_i} + \dot{\mathbf{p}}_i \cdot \frac{\partial}{\partial (-\mathbf{p}_i)} \right) = -iL(\Gamma).$$

(7.43)

With this result it is easy to prove the properties of the time-reversal mapping given previously.

The operation of the mapping $\mathbf{M}^T$ on the SLLOD equations of motion is:

$$\mathbf{M}^T \dot{\mathbf{q}}_i = \mathbf{M}^T \left[\frac{\mathbf{p}_i}{m} + \mathbf{n}_x \gamma y_i \right] = \frac{\mathbf{p}_i^T}{m} + \mathbf{n}_x \gamma^T y_i^T = \frac{-\mathbf{p}_i}{m} + \mathbf{n}_x(-\gamma) y_i = -\dot{\mathbf{q}}_i \quad (7.44)$$

where the choice of $\mathbf{M}^T(\mathbf{q}, \mathbf{p}, \gamma) = (\mathbf{q}, -\mathbf{p}, -\gamma)$ for the field-dependent time-reversal mapping makes the action of the mapping a simple sign change. The equation of motion for the momenta then gives:

$$\mathbf{M}^T \dot{\mathbf{p}}_i = \mathbf{M}^T \left[\mathbf{F}_i(\mathbf{q}) - \mathbf{n}_x \gamma p_{yi} - \alpha \mathbf{p}_i \right] = \mathbf{F}_i(\mathbf{q}^T) - \mathbf{n}_x \gamma^T p_{yi}^T - \alpha^T \mathbf{p}_i^T$$

$$= \mathbf{F}_i(\mathbf{q}) - \mathbf{n}_x \gamma p_{yi} - \alpha \mathbf{p}_i = \dot{\mathbf{p}}_i. \quad (7.45)$$

Notice also that the thermostatting variable α changes sign as:

$$\alpha^T = \mathbf{M}^T \alpha = \mathbf{M}^T \left[\frac{\sum_{i=1}^{N} \left(\mathbf{F}_i \cdot \mathbf{p}_i - \gamma p_{xi} p_{yi} \right)}{\sum_{i=1}^{N} \mathbf{p}_i^2} \right] = -\alpha. \quad (7.46)$$

With this choice for the mapping of the field, the Liouvillean is again a simple sign change:

$$\mathbf{M}^T iL(\Gamma, \gamma) = -iL(\Gamma, -\gamma). \quad (7.47)$$

Following the same steps for the mappings $\mathbf{M}^Y$ and $\mathbf{M}^K$, the results are (Table 7.3):

$$\mathbf{M}^Y iL(\Gamma, \gamma) = iL(\Gamma^Y, \gamma^Y) = iL(\Gamma, -\gamma), \quad (7.48)$$

$$\mathbf{M}^K iL(\Gamma, \gamma) = iL(\Gamma^K, \gamma^K) = -iL(\Gamma, \gamma). \quad (7.49)$$

Using the results obtained in this section, it easy to show that the following four time evolutions of the phase variable B yield identical values. That is:

$$\exp\left[iL(\Gamma, \gamma)t \right] B(\Gamma) = p_B^T \exp[-iL(\Gamma^T, -\gamma^T)t] B(\Gamma^T)$$

$$= p_B^Y \exp[iL(\Gamma^Y, -\gamma^Y)t] B(\Gamma^Y)$$

$$= p_B^K \exp[-iL(\Gamma^K, \gamma^K)t] B(\Gamma^K). \quad (7.50)$$

Notice that these four time evolutions involve changing the sign of the time and/or the sign of the field. If we consider the phase variable $P_{xy}(\Gamma)$, the time evolution leads to a negative average value at long time, and where a single sign change is made in the propagator, the parity operator is -1. As each of the time evolutions in Equation (7.50) represents a different mathematical form for the same trajectory, the stabilities are also the same.

The third equality can be used to interpret the propagation of the dissipative flux in the Kawasaki exponent (7.15). Here a negative time evolution is contained in the integral:

$$-\beta \int_0^t ds\ \gamma P_{yx}(-s, \gamma, \Gamma)V.$$

Using the Kawasaki mapping and Equation (7.50), the negative time evolution can be transformed to an equivalent positive time evolution from a different initial space point. To do this, consider:

$$P_{yx}(-s, \Gamma, \gamma) = \exp[-iL(\Gamma, \gamma)s]P_{yx}(\Gamma) = \exp[iL(\Gamma^K, \gamma^K)s]p_{P_{yx}}^K P_{yx}(\Gamma^K)$$

$$= -\exp[iL(\Gamma^K, \gamma^K)s]P_{yx}(\Gamma^K) = -P_{yx}(s, \Gamma^K, \gamma^K)$$

$$= -P_{yx}(s, \Gamma^K, \gamma). \tag{7.51}$$

The last equality follows from the fact that $\gamma^K = \gamma$. So $P_{yx}(-s, \gamma, \Gamma)$ is equivalent (apart from the sign of the parity operator) to the propagation of P_{yx} forward in time, with the same γ, but starting from a different phase point Γ^K. The probability of this new phase point Γ^K in the canonical (or isothermal) distribution is the same as the original Γ. Therefore the Kawasaki distribution function can be written as:

$$f(t, \Gamma, \gamma) = \exp\left[-\beta\gamma V \int_0^t dsP_{yx}(-s, \Gamma, \gamma)\right]f(0, \Gamma, 0)$$

$$= \exp\left[+\beta\gamma V \int_0^t dsP_{yx}(s, \Gamma^K, \gamma)\right]f(0, \Gamma^K, 0). \tag{7.52}$$

In this form, the sign of the exponent itself changes as well as the sign of the time evolution. At sufficiently large time $P_{yx}(s, \Gamma^K, \gamma)$ approaches the steady-state value $\langle P_{yx}(s, \gamma)\rangle$, regardless of the initial phase point. The negative time evolution with positive external field from Γ is equivalent to positive time evolution with positive field from Γ^K.

Numerical results for the transient time-correlation function

To use the trajectory mappings to improve the efficiency of TTCF calculations, we need the four initial phase points to produce different evolutions of the shear stress. To see that this happens, we only need to consider the initial value of the shear stress and its initial rate of change. If $P_{yx}(\Gamma) = P_{yx}^0$, it is easy to find the values for other initial points using the parities listed in Table 7.2. The rate of change of the shear stress is given by $iL(\Gamma, \gamma)P_{yx}(\Gamma) = \dot{P}^0_{yx}$. For the other mapped points, using the results of Tables 7.2 and 7.3, we find that:

$$iL(\Gamma^T, \gamma)P_{yx}(\Gamma^T) = -iL(\Gamma, -\gamma)P_{yx}(\Gamma) = \dot{P}^1_{yx}, \tag{T}$$

$$iL(\Gamma^Y, \gamma)P_{yx}(\Gamma^Y) = iL(\Gamma, -\gamma)(-)P_{yx}(\Gamma) = -iL(\Gamma, -\gamma)P_{yx}(\Gamma) = \dot{P}^1_{yx}, \tag{Y}$$

$$iL(\Gamma^K, \gamma)P_{yx}(\Gamma^K) = -iL(\Gamma, \gamma)(-)P_{yx}(\Gamma) = iL(\Gamma, \gamma)P_{yx}(\Gamma) = \dot{P}^0_{yx}. \tag{K}$$

From Table 7.4 it is clear that although Γ^I and Γ^T have the same initial value of P_{yx}, the initial rates of change are different, therefore the trajectories are different. Exactly the same result is true for Γ^Y and Γ^K. Therefore each of the mapped initial points leads to a different trajectory.

Computer simulations of 256 WCA particles at the Lennard–Jones triple point (Evans and Morriss, 1988b) used the trajectory mappings to generate four starting states for the nonequilibrium trajectories. The results for the shear stress $\langle P_{yx}(t) \rangle$ in Figure 7.1 show excellent agreement between the TTCF prediction (T), the direct simulation (D) and the long-time steady-state stress computed by conventional NEMD (SS). We show the Green–Kubo prediction for the stress (GK) to compare the linear and nonlinear responses, and the intrinsic nonlinear response is generated at comparatively late times. The response is essentially linear until the stress overshoot time ($t \sim 0.3$). The total nonlinear response converges far more rapidly than does the linear GK response. The linear GK response has

Table 7.3 *Summary of phase-space mappings for SLLOD*

Mapping	$\mathbf{M}^X(\Gamma, \gamma)$	$\mathbf{M}^X iL(\Gamma, \gamma)$
$\mathbf{M}^I$	$(\mathbf{q}, \mathbf{p}, \gamma)$	$iL(\Gamma, \gamma)$
$\mathbf{M}^T$	$(\mathbf{q}, -\mathbf{p}, -\gamma)$	$-iL(\Gamma, -\gamma)$
$\mathbf{M}^Y$	$(\mathbf{x}, -\mathbf{y}, \mathbf{z}, \mathbf{p}_x, -\mathbf{p}_y, \mathbf{p}_z, -\gamma)$	$iL(\Gamma, -\gamma)$
$\mathbf{M}^K$	$(\mathbf{x}, -\mathbf{y}, \mathbf{z}, -\mathbf{p}_x, \mathbf{p}_y, -\mathbf{p}_z, \gamma)$	$-iL(\Gamma, \gamma)$

Table 7.4 *Trajectories obtained from each mapped phase point*

Mapping	$P_{yx}(\Gamma)$	$\dot{P}_{yx}(\Gamma)$
$\mathbf{M}^I$	P^0_{yx}	$\dot{P}^0_{yx}$
$\mathbf{M}^T$	P^0_{yx}	$\dot{P}^1_{yx}$
$\mathbf{M}^Y$	$-P^0_{yx}$	$\dot{P}^1_{yx}$
$\mathbf{M}^K$	$-P^0_{yx}$	$\dot{P}^0_{yx}$

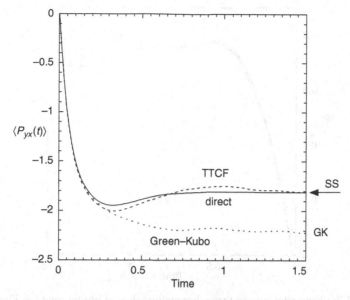

Figure 7.1 The shear stress $\langle P_{yx}(t)\rangle$ in the three-dimensional WCA system. D direct simulation, TTCF prediction; GK Green–Kubo prediction. SS is the long-time steady-state shear stress

obviously not relaxed to its steady-state limiting value at $t \sim 1.5$. This is presumably because of long-time tail effects which predict that the linear response relaxes very slowly as $t^{-1/2}$, at long times.

The shear-induced increase in pressure (shear dilatancy) is intrinsically nonlinear and is zero for Newtonian fluids. The Green–Kubo formulae predict that there is no coupling of the pressure and the shear stress because the equilibrium correlation function, $\langle \Delta p(t)P_{yx}(0)\rangle$, is exactly zero at all times. Figure 7.2 compares the direct and transient correlation function values of the shear dilatancy, $(\Delta p = p - p_0)$ for the WCA fluid. The agreement between the direct and transient correlation function expression at $\gamma = 1$ is impressive. This is a test of the theory for an *entirely* nonlinear effect and is thus a more convincing check on the validity of the TTCF formalism. The TTCF predictions are in statistical agreement with the results from direct simulation. We include a direct calculation of the steady-state pressure shift from conventional NEMD. It is apparent that $t = 1.5$ is sufficient for convergence of the TTCF integral. Note that the initial slope of the pressure response is zero. This contrasts with the initial slope of the shear stress response which is G_∞. This is in agreement with the predictions of the transient time-correlation formalism made in Section 7.3. Figure 7.1 clearly shows that at short time the stress is controled by linear response mechanisms. It takes time for the nonlinearities to develop but paradoxically perhaps, convergence to the steady-state asymptotic values is ultimately much faster in the nonlinear, large-field regime.

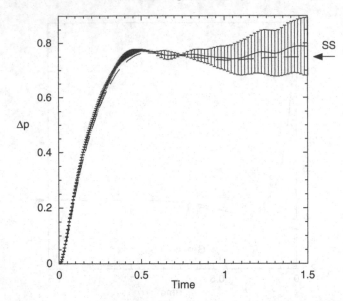

Figure 7.2 Shear dilatancy in three dimensions. For abbreviations see Figure 7.1. Error bars shown on TTCF results

At $\gamma = 1$, conventional NEMD is clearly the most efficient means of establishing the steady-state response. For example, under precisely the same conditions, after 54×10^6 timesteps, the TTCF expression for P_{yx} is accurate to $\pm 0.05\%$, but the directly averaged transient response is accurate to $\pm 0.001\%$. Because time is not wasted in establishing the steady-state from each of 60 000 time origins, conventional steady-state NEMD needs only 120×10^3 timesteps to obtain an uncertainty of ± 0.0017. If we assume that errors are inversely proportional to the square root of the run length, then the relative uncertainties for a 54×10^6 timestep run would be ± 0.05, ± 0.001, and ± 0.00008 for the TTCF, the directly averaged transient response, and for conventional NEMD, respectively. Steady-state NEMD is about 600 times more accurate than TTCF for the same number of timesteps. On the other hand, the transient correlation method has a computational efficiency which is similar to that of the equilibrium Green–Kubo method. An advantage of the TTCF formalism is that it models the rheological problem of stress growth (Bird *et al.*, 1977), not simply steady shear flow, and we can observe the associated effects, such as stress overshoot, and the time development of normal stress differences. Whereas shear stress and shear dilatancy exhibit a simple overshoot before relaxing to their final steady-state values, the normal stress differences $P_{yy} - P_{zz}$ and $P_{xx} - P_{yy}$ (see Figure 7.3), show two maxima before achieving their steady-state values (SS). Again it is apparent that $t = 1.5$ is sufficient time for a complete relaxation to the steady state.

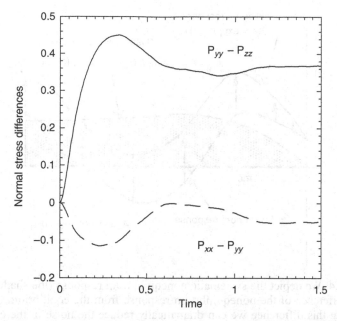

Figure 7.3 Transient responses for the normal stress differences $P_{yy} - P_{zz}$ and $P_{xx} - P_{yy}$ for the three-dimensional WCA system at a reduced shear rate of unity.

Over the years a number of numerical comparisons have been made between the Green–Kubo expressions and the results of NEMD simulations. The work we have just described takes this comparison one step further. It compares NEMD simulation results with the thermostatted, nonlinear generalization of the Green–Kubo formulae. It provides convincing numerical evidence for the usefulness and correctness of the TTCF formalism, which is the natural thermostatted, nonlinear generalization of the Green–Kubo relations.

7.5 Differential response functions

Often we are interested in the intermediate regime where the Green–Kubo method cannot be applied and where, because of noise, direct NEMD is very inefficient. We have just seen how the TTCF method may be applied to strong fields. It is also the most efficient method for treating fields of intermediate strength. Before we demonstrate the application of TTCFs to the small-field response, we will describe an early method that was used to calculate the intermediate field response – the subtraction, or differential, response method. The idea behind this method is simple. By considering a sufficiently small field, the systematic (or field-induced) response will be swamped by the natural (essentially equilibrium) fluctuations in the system. However, it is clear that for short times and small applied fields, there will be a high

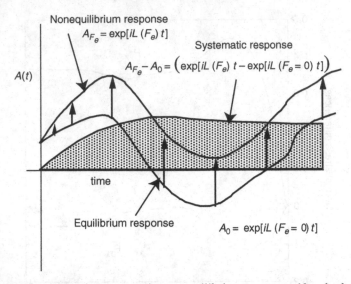

Figure 7.4 We depict the systematic nonequilibrium response (the shaded curve) as the difference of the nonequilibrium response from the equilibrium response. By taking this difference we can dramatically reduce the noise in the computed systematic nonequilibrium response. To complete this calculation one averages this difference over an ensemble of starting states

degree of correlation in the transient response computed with, and without, the external field, (see Figure 7.4).

If we compute $A(t)$ for two trajectories which start at the same phase point Γ, one with the field on and the other with the field off, we might see what is depicted in Figure 7.4. Ciccotti and Jacucci (1975) and Ciccotti *et al.* (1976, 1979) realized that, for short times, the noise in $A(t)$ computed for the two trajectories will be highly correlated. They used this idea to reduce the noise in the response computed at small applied fields.

To use their *subtraction method*, one performs an equilibrium simulation ($F_e = 0$) from which one periodically initiates nonequilibrium calculations ($F_e \neq 0$). The general idea is shown in Figure 7.5. The phases $\{\Gamma_i\}$ are initial points from which one calculates the difference of the response in a phase variable with and without the applied field. The *systematic* or nonequilibrium response is calculated from the equation:

$$\langle A(t; F_e) \rangle = \langle A(t; F_e) \rangle - \langle A(t; 0) \rangle$$

$$= \frac{1}{N} \sum_{i=1}^{N} (\exp[iL(F_e)t] - \exp[iL(0)t]) A(\Gamma_i). \tag{7.53}$$

For many years this was the only method of calculating the small-field nonequilibrium response. It suffers from a major problem, however. For the method to

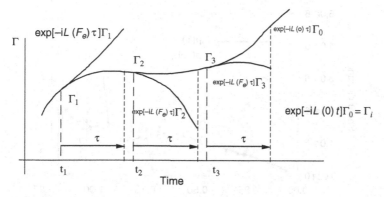

Figure 7.5 Illustration of the subtraction method

work, the noise in the value of $A(t)$ computed with and without the field, must be highly correlated. Otherwise the equilibrium fluctuations will completely swamp the desired response. Now, the noise in the two responses will only be correlated if the two systems remain sufficiently close in phase space. The Lyapunov instability (Section 3.4) will work against this driving the two systems apart exponentially. This can be expected to lead to an exponential growth of noise with time. This is illustrated in Figures 7.6 and 7.7 in which the TTCF (T), and Subtraction techniques (sub), are compared for the 256 particle WCA system considered in Section 7.4.

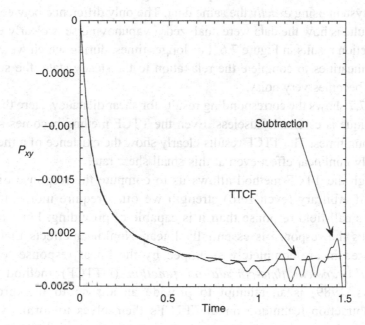

Figure 7.6 Shear stress for the three-dimensional WCA system at a shear rate of $\gamma = 10^{-3}$

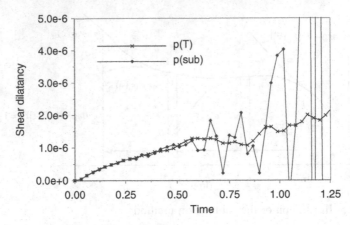

Figure 7.7 Shear dilatancy for the three-dimensional WCA system at a shear rate
of $\gamma = 10^{-3}$. (sub) subtraction; (T) TTCF

Figure 7.6 shows the shear stress for the three-dimensional WCA system at the
comparatively small shear rate of $\gamma = 10^{-3}$. At this field strength, conventional
steady-state NEMD is swamped by noise. However, the subtraction technique
can be used to substantially improve the statistics. It is important to note that
both the subtraction and TTCF techniques are based on an analysis of the transient
response of systems. The results compared in Figure 7.6 were computed for *exactly*
the same system using *exactly* the same data. The only difference between the two
sets of results is how the data were analyzed. Lyapunov noise is clearly evident in
the subtraction results in Figure 7.6. For longer times, during which we expect the
slow nonlinearities to complete the relaxation to the steady state, the subtraction
technique becomes very noisy.

Figure 7.7 shows the corresponding results for shear dilatancy. Here the subtraction technique is essentially useless. Even the TTCF method becomes somewhat
noisy at long times. The TTCF results clearly show the existence of a measurable,
intrinsically nonlinear effect even at this small shear rate.

Although the TTCF method allows us to compute the response of systems
to fields of arbitrary (even zero) strength we often require more information
about the small-field response than it is capable of providing. For example, at
small fields the response is essentially linear. Nonlinear effects that we may
be interested in are completely swamped by the linear response terms. The
differential transient time-correlation function (DTTCF) method (Morriss
and Evans, 1989) is an attempt to provide an answer to this problem. It
uses a subtraction technique on the TTCFs themselves to formally subtract
the linear response.

In the DTTCF method, we consider the difference between $B(s)$ evaluated with and without the external field, starting from the same initial phase point. From the transient correlation function expression this gives:

$$\langle (B(t, \gamma) - B(0, 0)) \rangle$$

$$= -\beta \gamma V \int_0^t ds \langle (B(s, \gamma) - B(s, 0) + B(s, 0)) P_{yx} \rangle$$

$$= -\beta \gamma V \int_0^t ds \langle (B(s, \gamma) - B(s, 0)) P_{yx} \rangle - \beta \gamma V \int_0^t ds \langle B(s, 0) P_{yx} \rangle. \quad (7.54)$$

In this equation $B(s, \gamma)$ is generated from $B(0)$ by the thermostatted field-dependent propagator, $B(s, 0)$, on the other hand, is generated by the zero-field thermostatted propagator. The last term is the integral of an equilibrium time-correlation function. This integral is easily recognized as the *linear*, Green–Kubo response. The first term on the right-hand side is the integral of a differential transient time-correlation function (DTTCF), and is the intrinsically nonlinear response. The left-hand side is termed the direct differential, or subtraction average.

There are two possible cases: the first, in which B has a nonzero linear response term; and the second, where the linear response is identically zero. If B is chosen to be P_{yx}, the third term in Equation (7.54) is the Green–Kubo expression for the response of the shear stress $-\eta(0)\gamma$, where $\eta(0)$ is the zero-shear-rate shear viscosity. The definition of the shear-rate-dependent viscosity, $\eta(\gamma) = -\langle P_{yx} \rangle / \gamma$ gives:

$$\eta(\gamma) - \eta(0) = \beta V \int_0^t ds \langle (P_{yx}(s, \gamma) - P_{yx}(s, 0)) P_{yx} \rangle, \quad (7.55)$$

as the intrinsically nonlinear part of the shear viscosity. As $s \to \infty$ the differential transient time-correlation function (using the mixing assumption) becomes $\langle (P_{yx}(s, \gamma) - P_{yx}(s, 0)) \rangle \langle P_{yx}(0) \rangle = \langle P_{yx}(s, \gamma) \rangle \langle P_{yx}(0) \rangle$. This is zero because $\langle P_{yx}(0) \rangle = 0$. On the other hand $\langle P_{yx}(s, \gamma) \rangle$ is clearly nonzero, which means that the use of our trajectory mappings will improve the statistics as $s \to \infty$.

To obtain four different shearing trajectories, we used the four different initial points $\Gamma^I, \Gamma^T, \Gamma^Y, \Gamma^K$ and we could show that this resulted in different trajectories. Here however, we need the difference between the shear stress evolved from the initial phase point under SLLOD equations of motion, and the shear stress evolved with $\gamma = 0$. The rate of change of the shear stress is given in Table 7.4, and we find that:

$$iL(\Gamma^I, 0) P_{yx}(\Gamma) = \dot{P}_{yx}^0(0), \quad \text{(I)}$$

$$iL(\Gamma^T, 0) P_{yx}(\Gamma^T) = -iL(\Gamma, 0) P_{yx}(\Gamma) = -\dot{P}_{yx}^0(0), \quad \text{(T)}$$

$$iL(\Gamma^Y, 0)P_{yx}(\Gamma^Y) = -iL(\Gamma, 0)P_{yx}(\Gamma) = -\dot{P}^0_{yx}(0), \qquad \text{(Y)}$$

$$iL(\Gamma^K, 0)P_{yx}(\Gamma^K) = iL(\Gamma, 0)P_{yx}(\Gamma) = \dot{P}^0_{yx}(0). \qquad \text{(K)}$$

Clearly there are only two different equilibrium $\gamma = 0$ time evolutions; the remaining two can be obtained from a simple sign change. In practice, a single cycle of the numerical evaluation of a differential transient time-correlation function will involve the calculation of four field-dependent trajectories and two field-free trajectories, yielding four starting states.

The use of the symmetry mappings implies some redundancies in the various methods of calculating the response. In particular the *direct* response of $P_{yx}(t)$ is exactly equal to the *direct differential* response for all time. This means that the contribution from the field-free time evolutions is exactly equal to zero. This is easy to see as there are only two different time evolutions; those corresponding to $\exp[iLt]$ and $\exp[-iLt]$, and for P_{yx} each comes with a positive and negative parity operator. Therefore these two responses exactly cancel for all values of time.

The second redundancy of interest is that the *transient* response of the pressure $p(t)$ is exactly equal to the *differential transient* response for all values of time. This implies that the contribution to the *equilibrium* time correlation function $\langle p(t)P_{yx} \rangle$ from a single sampling of Γ is exactly zero. Clearly this equilibrium time correlation is zero when the full ensemble average is taken, but the result we prove here is that the mappings ensure that $\sum p(t)P_{yx}$ is zero for each starting state Γ for all values of t. The contribution from the field-free trajectories is:

$$\sum_{\alpha \in \{I,T,Y,K\}} p(t, \Gamma^\alpha, \gamma^\alpha = 0)P_{yx}(\Gamma^\alpha)$$

$$= P_{yx}(\Gamma)\Big\{ \exp[iLt]p^I_{P_{yx}}p^I_p + \exp[-iLt]p^T_{P_{yx}}p^T_p$$

$$+ \exp[iLt]p^Y_{P_{yx}}p^Y_p + \exp[-iLt]p^K_{P_{yx}}p^K_p \Big\}p(\Gamma) = 0 \qquad (7.56)$$

Again the product of parities ensures that the two field-free time evolutions $\exp[iLt]$, and $\exp[-iLt]$ occur in cancelling pairs. Therefore the field-free contribution to the differential time-correlation function is exactly zero and the differential transient results are identical to the transient correlation function results.

The DTTCF method suffers from the same Lyapunov noise characteristic of all differential or subtraction methods. In spite of this problem, Morriss and Evans (1989) were able to show, using the DTTCF method, that the intrinsically nonlinear response of three-dimensional fluids undergoing shear flow is given by the classical Burnett form. This is at variance with the nonclassical behavior predicted by mode coupling theory. However, nonclassical behavior can only be expected in the large system limit. The classical behavior observed by Morriss and Evans (1987), is presumably the small shear-rate, asymptotic response for *finite* periodic systems.

Numerical results for the Kawasaki representation

A direct numerical test (Evans and Morriss, 1990a) of the Kawasaki representation of the nonlinear isothermal response showed that phase averages calculated using the explicitly normalized Kawasaki distribution function agree with those calculated directly from computer simulation. The system considered was planar Couette flow using the isothermal SLLOD algorithm. The primary difficulty in using the Kawasaki expression in numerical calculations arises because it involves calculating an extensive exponential. For an N-particle system we would have to average quantities of the order of $\exp[2N]$ to determine the viscosity. Using a small number of particles and a low density, where the viscosity is smaller than at the triple point, can reduce these difficulties. For small systems it is necessary to take into consideration terms of order, $1/N$, in the definition of the temperature, $T = (\sum p_i^2/m)/(dN - d - 1)$, and the shear stress, $P_{yx}V = (dN - d)/(dN - d - 1) \sum (p_{yi}p_{xi}/m + \frac{1}{2}y_{ij}F_{xij})$.

The phase-space integral of the Kawasaki distribution function $f(t)$, Equation (7.24), is:

$$Z(t) = \int d\Gamma f(\Gamma,t) = \int d\Gamma f(\Gamma,0) \exp\left(-\beta F_e \int_0^t ds J(-s)\right). \qquad (7.57)$$

$Z(0)$ is the phase integral of the initial equilibrium distribution function which is equal to unity, since $f(0)$ is the normalized. The numerical results in Figure 7.8 for $Z(t)$, calculate the Kawasaki normalization exactly written in Equation (7.57).

Consider the rate of change of $Z(t)$ after the external field is switched on. Using the Liouville equation we can show that:

$$\frac{dZ(t)}{dt} = \int d\Gamma f(0) \frac{\partial}{\partial t} \exp\left[-\beta F_e \int_0^t ds J(-s)\right]$$
$$= -\beta F_e \int d\Gamma f(t) J(-t) = -\beta F_e \int d\Gamma f(0) J(0) = 0. \qquad (7.58)$$

The last equality is a consequence of the Schrödinger–Heisenberg equivalence (Section 3.3). This implies that the Kawasaki distribution function is normalized at all times. This is not observed in the numerical results in Figure 7.8, but rather $Z(t)$ typically decreases with time. The explanation of this apparent paradox is in the sampling of the initial distribution function $f(0)$. Probable initial conditions in the equilibrium ensemble do not lead to trajectories for which the exponential factor is large, but rather the initial conditions which do

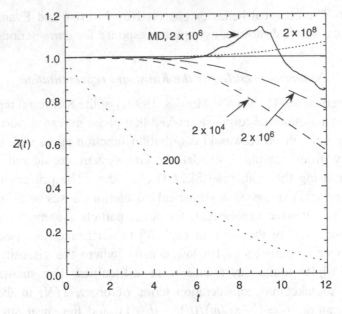

Figure 7.8 The normalization of the Kawasaki distribution for two WCA disks at $T = 1, \rho = 0.39685$ at a color field of one. One result (MD) is molecular dynamics while the others are Monte Carlo calculations

lead to $-F_e \int_0^t \mathrm{d}sJ(-s)$ being large and positive have low probability in the equilibrium ensemble. This means that the initial phase points that lead to probable values of $\Gamma(t)$ in the steady state are highly improbable in the initial distribution. It is the difficulty of sampling the important regions of the initial distribution! Because of this sampling problem it has been suggested that the Kawasaki distribution function should be *explicitly normalized* in numerical calculations (Morriss and Evans, 1987). The renormalized average of the shear stress is then:

$$\langle P_{yx}(t)\rangle = \frac{\int \mathrm{d}\Gamma P_{yx}(\Gamma)f(0)\exp[-\beta F_e \int_0^t \mathrm{d}sJ(-s)]}{\int \mathrm{d}\Gamma f(0)\exp[-\beta F_e \int_0^t \mathrm{d}sJ(-s)]}. \tag{7.59}$$

This appears to work well numerically, but it is clear that the renormalized Kawasaki formalism should not be used as a routine means of computing transport coefficients. It is, however, a very important theoretical tool. It was of crucial importance to the development of nonlinear response theory and it provides an extremely useful basis for subsequent theoretical derivations. As we will see in Chapter 9, the renormalized Kawasaki formalism, in contrast to the TTCF formalism, is very useful in providing an understanding of fluctuations in nonequilibrium steady states.

7.6 The van Kampen objection to linear response theory

Having explored some of the fundamentals of nonlinear response theory, we are now in a better position to comment on one of the early criticisms of linear response theory. In an oft-cited paper (van Kampen, 1971), criticized linear response theory on the basis that microscopic linearity, which is assumed in linear response theory, is quite different from the macroscopic linearity manifest in linear constitutive relations. van Kampen correctly noted that to observe linear *microscopic* response (of individual particle trajectories) over macroscopic time (seconds, minutes, or even hours) requires external fields which are orders of magnitude smaller than those for which linear macroscopic behavior is actually observed. Therefore, so the argument goes, the theoretical justification of, the Green–Kubo relations for linear transport coefficients, is suspect.

In order to explain his assertion that *linearity of microscopic motion is entirely different from macroscopic linearity*, van Kampen considered a system composed of electrons which move freely through a conductor, apart from occasional collisions with impurities. An imposed external electric field, F_e, accelerates the particles between collisions. The distance an electron moves in a time t under the influence of the field is $\frac{1}{2}t^2(eF_e/m)$. In order for the induced current to be linear one requires that $\frac{1}{2}t^2(eF_e/m) << d$, the mean spacing of the impurities. Taking $d \sim 10^{-8}$ m and t to be a macroscopic time, say 1 s, we see that the field must be less than $\sim 10^{-16}$ V m^{-1}! As a criticism of the derivation of linear response theory, this calculation implies that for linear response theory to be valid, trajectories must be subject to a linear perturbation over macroscopic times–the time taken for experimentalists to make sensible measurements of the conductivity. This, however, is incorrect.

The linear response theory expression for the conductivity, $\sigma (\equiv J/F_e)$ is:

$$\sigma = \beta V \int_0^\infty dt \langle J(t)J(0) \rangle_{eq}. \tag{7.60}$$

Now, it happens that in three-dimensional systems the integral of the equilibrium current-autocorrelation function converges rapidly. (In two-dimensional systems this is expected not to be so.) The integral, in fact, converges in microscopic time, a few collision times in the above example. Indeed, if this were not so one could never use equilibrium molecular dynamics to compute transport coefficients from the Green–Kubo formulae. Molecular dynamics is based on the assumption that transport coefficients for simple fluids can be computed from simulations which only follow the evolution systems for $\sim 10^{-10}$ s. These times are sufficient to ensure convergence of the Green–Kubo correlation functions for all the Navier–Stokes transport coefficients. If we require microscopic linearity over 10^{-10} s (rather than van Kampen's 1 s), then we see that the microscopic response will be linear for fields less than about $10^4 \, \text{V m}^{-1}$, not an unreasonable number. It simply does not matter that for times longer than those characterizing the relaxation of the relevant GK correlation function, the motion is perturbed in a nonlinear fashion. In order for linear response theory to yield correct results for linear transport coefficients, linearity is only required for times characteristic of the decay of the relevant correlation functions. These times are microscopic.

We used nonequilibrium molecular dynamics simulation of shear flow in an atomic system to explore the matter in more detail (Morriss *et al.*, 1989). A series of simulations were performed with and without an imposed shear rate γ, to measure the actual separation of phase-space trajectories as a function of the shear rate. The phase-space separation is defined to be

$$d(t, \gamma) = \left[(\Gamma(t, \gamma) - \Gamma(t, 0))^2 \right]^{1/2}, \tag{7.61}$$

where $\Gamma(t, \gamma)$ is the phase-space position at time t under shear rate γ. In measuring the separation of phase-space trajectories, we imposed the initial condition that at time zero the equilibrium and nonequilibrium trajectories start from *exactly* the same point in phase space, $d(0, \gamma) = 0, \quad \forall \gamma$. We used the *infinite checkerboard* convention for defining the Cartesian coordinates of a particle in a periodic system. This eliminates trivial discontinuities in these coordinates. We reported the ensemble average of the phase-space separation, averaged over an equilibrium ensemble of initial phases.

As we have seen, the linear response computed from these equations is given precisely by the Green–Kubo expression for the shear viscosity. Before we begin to analyze the phase-separation data, we need to review some of the relevant features of Lennard–Jones triple-point rheology, for $N = 256$. First, from Section 6.3, this fluid is shear thinning and the shear-rate-dependent shear viscosities are set out in Table 7.5. Note that for reduced shear rates, $\gamma < 10^{-2}$, the fluid is effectively Newtonian with a viscosity which varies at most, by less than $\sim 4 \, \%$ of its zero shear value. We should also remember that the GK equilibrium time-correlation function, whose integral gives the shear viscosity (Table 7.6), has decayed to less than $1 \, \%$ of its zero time value at a time $t = 2.0$.

The viscosity is the time integral of this correlation function and converges relatively slowly due to the presence of the slowly decaying $t^{-3/2}$ long time tail. Here again there is some uncertainty. If one believes that *enhanced long time tail* phenomena (Section 6.3), are truly asymptotic and persist indefinitely then one finds that the viscosity converges to within $\sim 13 \%$ of its asymptotic value at $t = 1$ and to within $\sim 5 \, \%$ of the asymptotic value at $t = 10$. (If we map our simulation onto the standard Lennard–Jones representation of argon, $t = 1$ corresponds

Table 7.5 *Shear-rate-dependent shear viscosities for the Lennard–Jones triple point*

γ	η	Nonlinearity (%)
1.0	2.17 ± 0.03	37
0.1	3.04 ± 0.03	12
0.01	3.31 ± 0.08	~ 4
0.0	3.44 ± 0.2	0

to a time of 21.6 ps.) If *enhanced long time tails* are not asymptotic, then the GK integrand for the shear viscosity converges to within $\sim 5\%$ of its infinite time value by $t = 2$. The only important observation that concerns us here is that the GK estimate for the shear viscosity is determined in *microscopic* time, a few hundreds of picoseconds at the very most, for argon. This observation was omitted from van Kampen's argument. We call the range of times required to ensure, say, 5% convergence of the GK expression for the viscosity, the *GK time window*.

From Figure 7.9, at a shear rate of 10^{-7}, the phase-space separation increases very rapidly initially and then slows to an exponential increase with time. The same pattern is followed at a shear rate of 10^{-5}, except that the initial rise is even more rapid than for a shear rate of 10^{-7}. Remember that at $t = 0$ the phase-space separations start from zero, and therefore the logarithm of the $t = 0$ separations is $-\infty$ for all shear rates. For shear rates greater than 10^{-5}, we notice that at long times the phase separation is a constant independent of time. We see an extremely rapid initial rise, followed by an exponential increase with a slope which is independent of shear rate, followed at later times by a plateau. The plateau is easily understood because at constant kinetic energy the $3N$ components of the momenta lie on the surface of a $3N$-dimensional sphere of radius $r_T = \sqrt{3Nmk_BT}$. The arrow marked on Figure 7.9 shows when the logarithm of the separation is equal to this radius. The maximum separation of phase points within the momentum subspace is $2r_T$, so exponential separation must end at $d(t, \Gamma) = r_T$.

Table 7.6 *Green–Kubo equilibrium stress correlation function for shear viscosity*

t	$\langle P_{yx}(t)P_{yx}(0)\rangle$	% of $\langle P_{yx}(0)^2\rangle$
0.0	24.00	100
0.1	7.17	29
1.0	0.26	1
2.0	0.09	0.3

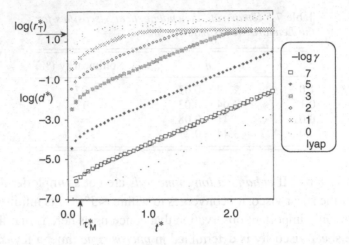

Figure 7.9 Logarithm of the ensemble average of the phase-space separation plotted as a function of reduced time for various values of the imposed shear rate, γ. The shear rates employed were: $\gamma = 1.0$, 10^{-1}, 10^{-2}, 10^{-3}, 10^{-5}, 10^{-7}. Note that for the standard Lennard–Jones argon representation, these shear rates correspond to shear rates of 4.6×10^{11} to 4.6×10^5 Hz. It will be clear from the present results that no new phenomena would be revealed at shear rates less than $\gamma = 10^{-4}$

Between the plateau and the initial (almost vertical) rise is an exponential region. As can be see from the graph, the slope of this rise is virtually independent of shear rate. The slope is related to the largest positive equilibrium Lyapunov exponent. After the initial separation due to the external field, the rate of phase-space separation is governed by Lyapunov instability. The fact that the two trajectories employ slightly different dynamics is a second-order consideration. The Lyapunov exponents are known to be insensitive to the magnitude of the perturbing external field for field strengths less than 10^{-2}.

This conjecture regarding the role played by the Lyapunov exponent in the separation of equilibrium and nonequilibrium trajectories which start from a common phase origin is easily verified numerically. The slope of this Lyapunov curve is essentially identical to the exponential portions of the shear-rate driven curves. The time constants for the exponential portions of the curves are given in Table 7.7. At this stage we see that even at our smallest shear rate, the trajectory separation is exponential in time. It may be thought that this exponential separation in time supports van Kampen's objection to linear response theory. Surely exponentially diverging trajectories imply nonlinearity? The assertion turns out to be false.

In Figure 7.10 we see the field dependence of the phase separations in more detail by plotting the ratio of the separations to the separation observed for a

Table 7.7 *Exponential time constants for phase separation in the triple-point Lennard–Jones fluid under shear*

Time constant	Reduced shear rate	
1.715 ± 0.002	0.0	Lyapunov
1.730 ± 0.002	10^{-7}	Shear induced
1.717 ± 0.002	10^{-5}	Shear induced
1.708 ± 0.012	10^{-3}	Shear induced
1.715 ± 0.002	10^{-2}	Shear induced

field, $\gamma = 10^{-7}$. If the ensemble-averaged trajectory response is linear, then each of the curves in Figure 7.10 will be an equispaced horizontal line. One can see immediately that within the GK time window, $t < 2.0$, all the separations are linear in the field except for the largest two shear rates $\gamma = 1.0,\ 0.1$. We should expect that all shear rates exhibiting linearity within the GK time window should correspond to those systems exhibiting macroscopic linear behavior (i.e. those

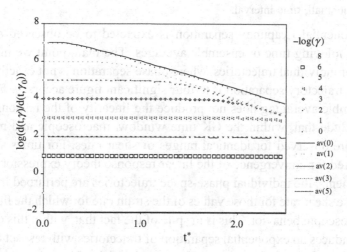

Figure 7.10 We plot the ratio of the phase-space separations as a function of shear rate and time. The ratios are computed relative to the separation at a reduced shear rate of 10^{-7}. Curves denoted by "av" are ensemble averages. Those not so denoted give the results for individual phase trajectories. Since the integrals of the Green–Kubo correlation functions converge to within a few percent by a reduced time of ~ 1.5, we see that the trajectory separation is varying linearly with respect to strain rate for reduced shear rates less than ~ 2. This is precisely the shear rate at which direct nonequilibrium molecular dynamics shows a departure of the computed shear viscosity from linear behavior

which are Newtonian). Those exhibiting microscopic nonlinearity within the GK time window should display nonNewtonian macroscopic behavior. Comparing Table 7.5 with Figure 7.9, this is exactly what is seen.

Although systems at shear rates $\gamma = 10^{-2}$ and 10^{-4}, do exhibit a nonlinear growth in the phase-space separation, it occurs at times which are so late that it cannot possibly effect the numerical values of the shear viscosity. These nonlinearities occur outside the GK time window.

A possible objection to these conclusions might be: since we are computing ensemble averages of the phase-space separations, it might be the averaging process which ensures the observed microscopic linearity. Individual trajectories might still be perturbed nonlinearly with respect to the shear rate. This, however, is not the case. In Figure 7.10 the symbols plotted represent the phase-space separation induced in single trajectories. For all shear rates, a common phase origin is used. We did not average over the time-zero phase origins of the systems. What we see is a slightly noisier version of the ensemble-averaged results. Detailed analysis of the unaveraged results reveals that:

(1) for $\gamma < 10^{-2}$ linearity in shear rate is observed for individual trajectories; and
(2) the exponential behavior in time is only observed when $d(\gamma, t)$ is averaged over some finite, but small, time interval.

The exponential Lyapunov separation is expected to be observed *on average* either by employing time or ensemble averages. The main point we make here is that even for individual trajectories, where phase separation is not exactly exponential in time, trajectory separation is to four significant figure accuracy, *linear* in the field. Ensemble averaging does not produce the linearity of the response.

We conclude that, within the GK time window, macroscopic and microscopic linearities are observed for identical ranges of shear rates. For times shorter than those required for convergence of the linear response theory expressions for transport coefficients, the individual phase-space trajectories are perturbed linearly with respect to the shear rate for those values of the strain rate for which the fluid exhibits linear macroscopic behavior. This is in spite of the fact that, within this domain, the shear rate induces an exponential separation of trajectories with respect to time. We believe that many people have assumed an exponential trajectory separation in time implies an exponential separation with respect to the magnitude of the external field. This work shows that, within the GK time window, the dominant microscopic behavior in fluids which exhibit linear macroscopic behavior is linear in the external field, but exponential in time. For intermediate times the phase separation takes the form:

$$d = A_\gamma \exp[t/\tau_L],$$
(7.62)

where the *Lyapunov time*, τ_L, is the inverse of the largest equilibrium Lyapunov exponent. We can explain why the phase separation exhibits this functional form and, moreover, we can make a rough calculation of the absolute magnitude of the coefficient, A_γ. We know that the exponential separation of trajectories only begins after a time which is roughly the Maxwell relaxation time τ_M, for the fluid. Before the particles sense their mutual interactions, the particles are freely streaming with trajectories determined by the initial values of $(\dot{\mathbf{q}}, \dot{\mathbf{p}})$. After this initial motion, the particles will have coordinates and momenta as follows:

$$\mathbf{q}_i(t) = \mathbf{q}_i(0) + \left(\frac{\mathbf{p}_i(0)}{m} + \hat{\mathbf{x}}\gamma\, y_i(0)\right)t,$$

$$\mathbf{p}_i(t) = \mathbf{p}_i(0) + \left(\mathbf{F}_i(0) - \hat{\mathbf{x}}\gamma\, p_{yi}(0)\right)t. \tag{7.63}$$

When this approximation breaks down, approximately at the Maxwell relaxation time, $\tau_M \equiv \eta/G$, the phase separation $d(\tau_M, \gamma)$ is given by:

$$d(\tau_M, \gamma) = \gamma\tau_M \sqrt{\sum_{i=1}^{N}\left(y_i^2(0) + p_{yi}^2(0)\right)}. \tag{7.64}$$

For our system this distance is:

$$d(\tau_M, \gamma) = \gamma\tau_M\left\{\frac{N^{5/3}}{3n^{2/3}} + NT\right\}^{1/2} \sim 8.7\gamma, \tag{7.65}$$

where n is the number density (number of particles per unit volume) and we have used $\tau_M = 0.137$ as the Maxwell time. After this time the phase separation can be expected to grow as:

$$d(\gamma, t) \sim d(\gamma, \tau_M)\exp\left[\frac{t}{\tau_L + O(\gamma^2)}\right], \tag{7.66}$$

where τ_L is the inverse of the largest zero-shear rate Lyapunov exponent. For fields less than $\gamma = 10^{-2}$, the equilibrium Lyapunov time dominates the denominator of the above expression. This explains why the slopes of the curves in Figure 7.9 are independent of shear rate. Furthermore by combining Equations (7.65) and (7.66), we see that in the regime where the shear rate corrections to the Lyapunov exponents are small, the phase separation takes the form given by Equation (7.62), with the coefficient, $A_\gamma \sim 8.7$. Equation (7.66) is plotted, for a reduced shear rate of 10^{-7}, as a dashed line in Figure 7.10. It is in reasonable agreement with the results. The results for other shear rates are similar. The greatest uncertainty in the prediction is the estimation of the precise time at with Lyapunov behavior begins.

7.7 Time-dependent response theory

In this section, we extend the nonlinear response theory developed in Sections 7.2 and 7.3 to describe the response of classical, many-body systems to time-dependent external fields. The resulting formalism is applicable to both adiabatic and thermostatted systems. The results are then related to a number of known special cases: time-dependent linear response theory, and time-independent nonlinear response theory, as described by the transient time-correlation approach and the Kawasaki response formula. The analysis parallels perturbation treatments of quantum field theory (Raimes, 1972; Parry, 1973). We will give a brief outline of the time-dependent case. The main difference between the two types of nonlinear response is that time-ordered exponentials are required for the time-dependent propagators. The algebraic properties of time-ordered exponentials has no counterpart in the time-independent case, and limit the mathematical forms of the nonlinear time-dependent response. In the time-independent case we have two forms: the Kawasaki and the TTCF forms. Here we will meet yet another, and of these three forms only *one* is applicable in the time-dependent case. This development follows the work of Evans and Morriss (1988a).

When a system is subject to time-dependent external fields, there are two time dependences in the system, one associated with the time of the phase position $\Gamma(t)$ and the other is associated with the explicit time-dependence of the field $F_e(t)$ (Holian and Evans, 1985). We define the *p*-propagator $U_R(0, t)$ to be the operator which advances functions of phase Γ forward in time from 0 to t. That is:

$$\Gamma(t) = U_R(0, t)\Gamma(0). \tag{7.67}$$

The operator $U_R(0, t)$ operates on all functions of phase located to its right. The equations of motion for the system at time t, which are themselves a function of phase Γ, are given by:

$$\dot{\Gamma}(\Gamma(t), t) = U_R(0, t)\dot{\Gamma}(\Gamma(0), t). \tag{7.68}$$

The notation $\dot{\Gamma}(\Gamma(t), t)$ implies that the derivative should be calculated on the current phase $\Gamma(t)$, using the current field $F_e(t)$. On the other hand $\dot{\Gamma}(\Gamma(0), t)$ implies that the derivative should be calculated on the initial phase $\Gamma(0)$ using the current field $F_e(t)$. The p-propagator $U_R(0, t)$ has no effect on the explicit time contained in the field.

The total time derivative of a phase function $B(\Gamma)$ is:

$$\frac{d}{dt}B(\Gamma(t)) = \dot{\Gamma}(\Gamma(t), t) \cdot \frac{\partial B(\Gamma)}{\partial \Gamma}\bigg|_{\Gamma=\Gamma(t)} = U_R(0, t)\dot{\Gamma}(\Gamma, t) \cdot \frac{\partial B(\Gamma)}{\partial \Gamma}\bigg|_{\Gamma} \quad (7.69)$$

$$= U_R(0, t)iL(t)B(\Gamma) = \frac{\partial}{\partial t}U_R(0, t)B(\Gamma),$$

where we have introduced the time-dependent p-Liouvillean, $iL(t) \equiv iL(\Gamma, t)$, which acts on functions of the *initial* phase Γ, but contains the external field at the *current* time. The partial derivative of B with respect to initial phase Γ is simply another phase function, so that the propagator $U_R(0, t)$ advances this phase function to time t. In writing the last line of Equation (7.69), we have used the fact that the p-propagator is an explicit function of time (as well as phase), and that, when written in terms of the p-propagator $\dot{B}(\Gamma(t))$ must *only* involve the partial time derivative of the p-propagator. Equation (7.69) implies that the p-propagator $U_R(0, t)$ satisfies an operator equation of the form:

$$\frac{\partial}{\partial t}U_R(0, t) = U_R(0, t)iL(t), \quad (7.70)$$

where the order of the two operators on the right-hand side is crucial (as they do not commute), since the propagator $U_R(0, t)$ contains sums of products of $iL(s_i)$ at different times s_i and s_j, and $iL(s_i)$ and $iL(s_j)$ do not commute unless $s_i = s_j$. The formal solution of this operator equation is:

$$U_R(0, t) = \sum_{n=0}^{\infty} \int_0^t ds_1 \int_0^{s_1} ds_2 \ldots \int_0^{s_{n-1}} ds_n iL(s_n) \ldots iL(s_2)iL(s_1). \quad (7.71)$$

Notice that the p-Liouvilleans are *right ordered* in time (latest time to the right). As Liouvilleans do not commute, this time ordering is fixed. The integration limits imply that $t > s_1 > s_2 > \ldots > s_n$, so that the time arguments of the p-Liouvilleans in the expression for $U_R(0, t)$ increase as we move from the left to the right.

The inverse theorem

For $t > 0$, it is obvious that the inverse of $U_R(0, t)$, which we write as $U_R(0, t)^{-1}$, should be the propagator that propagates backwards in time from t to 0. From

Equation (7.71) we can write:

$$U_R(0, t)^{-1} = \sum_{n=0}^{\infty} \int_t^0 ds_1 \int_t^{s_1} ds_2 \ldots \int_t^{s_{n-1}} ds_n \, iL(s_n) \ldots iL(s_2) iL(s_1)$$

$$= 1 + \int_t^0 ds_1 iL(s_1) + \int_t^0 ds_1 \int_t^{s_1} ds_2 iL(s_2) iL(s_1) + \ldots \quad (7.72)$$

The first two terms are trivial, so considering the third term it can be shown that:

$$\int_t^0 ds_1 \int_t^{s_1} ds_2 iL(s_2) iL(s_1) = \int_0^t ds_2 \int_0^{s_2} ds_1 iL(s_2) iL(s_1),$$

and therefore Equation (7.72) becomes:

$$U_R(0, t)^{-1} = 1 - \int_0^t ds_1 iL(s_1) + \int_0^t ds_2 \int_0^{s_2} ds_1 iL(s_2) iL(s_1) - \ldots$$

$$= \sum_{n=0}^{\infty} (-)^n \int_0^t ds_1 \int_0^{s_1} ds_2 \ldots \int_0^{s_{n-1}} ds_n \, iL(s_1) iL(s_2) \ldots iL(s_n). \quad (7.73)$$

As s_1 and s_2 are dummy labels, we can interchange them to obtain the final result. The integration limits now imply that the Liouvilleans are left ordered. Comparing this expression with the definition of $U_R(0, t)$ there are two differences, the time ordering *and* the minus signs. We now define the left-ordered operator $U_L(0, t) = U_R(0, t)^{-1}$ to be equal to the right-hand side of Equation (7.73). Using this definition, it can be shown that $U_L(0, t)$ satisfies the operator equation:

$$\frac{\partial}{\partial t} U_L(0, t) = -iL(t) U_L(0, t). \quad (7.74)$$

The associative law and composition theorem

The action of the p-propagator $U_R(0, t)$ is to advance the phase Γ of a phase variable forward in time from 0 to t, so $U_R(0, t)B(\Gamma) = B(t)$. This must be equivalent to advancing time from 0 to s, then advancing time from s to t, or $B(t) = U_R(s, t)[U_R(0, s)B(\Gamma)] \neq U_R(s, t)U_R(0, s)B(\Gamma)$, whenever $0 < s < t$. The problem is that the operator product $U_R(s, t)U_R(0, s)$ is no longer a right-ordered operator. The correct operator equation is:

$$U_R(0, t) = U_R(0, s)U_R(s, t), \quad (7.75)$$

as can be shown directly from Equation (7.71). Apart from the present discussion, we will always write operators in a form which reflects the mathematical rather than the causal ordering. The f-propagators are causally ordered.

The distribution function

The Liouville equation for a system subject to a time-dependent external field is:

$$\frac{\partial}{\partial t} f(t) = -\frac{\partial}{\partial \Gamma} \cdot \left(\dot{\Gamma}(\Gamma,t) f(\Gamma,t) \right) = -i\mathscr{L}(t) f(t), \qquad (7.76)$$

where $i\mathscr{L}(t)$ is the time dependent f-Liouvillean. This equation describes how the density of phase points changes with time at a *fixed* point in phase space. The distribution function propagator $U_R^\dagger(0, t)$ which advances the time dependence of $f(\Gamma,0)$ from 0 to t, is:

$$f(\Gamma,t) = U_R^\dagger(0, t) f(\Gamma,0). \qquad (7.77)$$

In this equation $U_R^\dagger(0, t)$ is the adjoint of $U_R(0, t)$. It is therefore closely related to $U_L(0, t)$ except that the Liouvilleans appearing in Equation (7.73) are replaced by their adjoints $i\mathscr{L}(s_i)$. Combining Equation (7.77) with the Liouville Equation (7.76) we find that $U_R^\dagger(0, t)$ satisfies the following equation of motion:

$$\frac{\partial}{\partial t} U_R^\dagger(0, t) = -i\mathscr{L}(\Gamma,t) U_R^\dagger(0, t). \qquad (7.78)$$

The formal solution to this operator equation is:

$$U_R^\dagger(0, t) = \sum_{n=0}^{\infty} (-)^n \int_0^t ds_1 \int_0^{s_1} ds_2 \ldots \int_0^{s_{n-1}} ds_n i\mathscr{L}(s_1) i\mathscr{L}(s_2) \ldots i\mathscr{L}(s_n). \qquad (7.79)$$

Here the f-Liouvilleans are *left time ordered*. This is opposite to the time ordering in the p-propagator $U_R(0, t)$, but the definition of $U_R^\dagger(0, t)$ is consistent with the definition of $U_L(0, t)$.

A common notation in quantum mechanics is to refer to the phase and distribution-function propagators as *right-* and *left-ordered exponentials* ($\exp_R$ and $\exp_L$) respectively. To exploit this notational simplification, we introduce the time-ordering operators T_R and T_L. The operator T_R simply reorders a product of operators so that the time arguments increase from left to right. In this notation we write the p-propagator $U_R(0, t)$ as:

$$U_R(0, t) = \exp_R\left[\int_0^t ds\ iL(s)\right] = T_R \exp\left[\int_0^t ds\ iL(s)\right]$$

$$= T_R \sum_{n=0}^{\infty} \frac{1}{n!} \int_0^t ds_1 \int_0^t ds_2 \dots \int_0^t ds_n iL(s_n) \dots iL(s_2)iL(s_1). \quad (7.80)$$

Considering the term with two integrals in more detail we see that:

$$T_R\frac{1}{2!}\int_0^t ds_1 \int_0^t ds_2 iL(s_2)iL(s_1)$$

$$= \frac{1}{2!}T_R\left\{\int_0^t ds_1 \int_0^{s_1} ds_2 iL(s_2)iL(s_1) + \int_0^t ds_1 \int_{s_1}^t ds_2 iL(s_2)iL(s_1)\right\}$$

$$= \frac{1}{2!}T_R\left\{2\int_0^t ds_1 \int_0^{s_1} ds_2 iL(s_2)iL(s_1)\right\} = \int_0^t ds_1 \int_0^{s_1} ds_2 iL(s_2)iL(s_1).$$

Repeating this exercise for all the higher-order terms in the sum in Equation (7.80), we see that the time-ordering operator effectively removes all the factorials in the normal exponential Taylor series. Using the same arguments for the f-propagator gives:

$$U_R^\dagger(0, t) = \exp_L\left[-\int_0^t ds\ i\mathscr{L}(s)\right] = T_L \exp\left[-\int_0^t ds\ i\mathscr{L}(s)\right]. \quad (7.81)$$

The use of time-ordered exponentials leads to many simplifications.

We can compute the average of a phase variable B at time t using either the Schrödinger or Heisenberg representations and these must give the same result. The Heisenberg representation:

$$\langle B(t)\rangle = \int d\Gamma B(\Gamma(t))f(\Gamma,0) = \int d\Gamma f(\Gamma,0) U_R(0, t)B(\Gamma), \quad (7.82)$$

while the Schrödinger representation is:

$$\langle B(t)\rangle = \int d\Gamma B(\Gamma)f(\Gamma,t) = \int d\Gamma B(\Gamma)U_R^\dagger(0, t)f(\Gamma,0). \quad (7.83)$$

The Dyson equation is useful for deriving relationships between propagators, as we saw for time-independent systems in Section 3.6. For two arbitrary

p-Liouvilleans, the most general form of the Dyson equation is:

$$U_R(0, t) = U_{R0}(0, t) + \int_0^t ds U_R(0, s)(iL(s) - iL_0(s))U_{R0}(s, t)$$

$$= U_{R0}(0, t) + \int_0^t ds U_{R0}(0, s)(iL(s) - iL_0(s))U_R(s, t). \quad (7.84)$$

Both Liouvilleans $iL(s)$ and $iL_0(s)$ may be time dependent. The corresponding equations for left-ordered propagators are:

$$U_R^\dagger(0, t) = U_{R0}^\dagger(0, t) - \int_0^t ds U_R^\dagger(0, s)(i\mathscr{L}(s) - i\mathscr{L}_0(s))U_{R0}^\dagger(s, t)$$

$$= U_{R0}^\dagger(0, t) - \int_0^t ds U_{R0}^\dagger(0, s)(i\mathscr{L}(s) - i\mathscr{L}_0(s))U_R^\dagger(s, t). \quad (7.85)$$

An important application of this result is to the case where the difference between two Liouvilleans is a phase variable rather than an operator. We consider:

$$i\mathscr{L}(s) - iL(s) = \Lambda(\Gamma, s) = \frac{\partial}{\partial \Gamma} \cdot \dot{\Gamma}(s). \quad (7.86)$$

If the operator $iL(s)$ generates the propagator $U_R(0, t)$, and the operator $iL(s)$ generates the propagator $U_{R0}(0, t)$, then, using Equation (7.84), it can be shown that:

$$U_R(0, t) = \exp\left[\int_0^t ds \Lambda(\Gamma(s), s)\right] U_{R0}(0, t). \quad (7.87)$$

Response theory

Consider an equilibrium ensemble of systems with initial distribution function f_0 at $t < 0$, to which an external time-dependent field $F_e(t)$ is applied at $t = 0$. We assume that the equilibrium system has evolved under the influence of the Gaussian isokinetic Liouvillean iL_0, which has no explicit time dependence. The equilibrium distribution could be the canonical or the isokinetic distribution. The equations of motion for the system can be written as:

$$\mathbf{q}_i = \frac{1}{m}\mathbf{p}_i + \mathbf{C}_i(\Gamma)F_e(t),$$

$$\dot{\mathbf{p}}_i = \mathbf{F}_i + \mathbf{D}_i(\Gamma)F_e(t) - \alpha(\Gamma, t)\mathbf{p}_i. \quad (7.88)$$

The terms $\mathbf{C}_i(\Gamma)$ and $\mathbf{D}_i(\Gamma)$ couple the external field $F_e(t)$ to the system. We assume that the AIΓ holds. The dissipative flux is defined in the usual way:

$$iL^{ad}(s)H_0 \equiv -J(\Gamma)F_e(s). \tag{7.89}$$

The response of an arbitrary phase variable $B(\Gamma)$ can be written as:

$$\langle B(t)\rangle = \int d\Gamma f(\Gamma) \exp_R\left[\int_0^t ds\, iL(s)\right]B(\Gamma) = \int d\Gamma f(\Gamma)U_R(0,t)B(\Gamma). \tag{7.90}$$

In this equation $iL(s)$ is the p-Liouvillean for the field-dependent Gaussian thermo-statted dynamics at $t > 0$. If we use the Dyson decomposition of the field-dependent p-propagator (Equation 7.84) in terms of the equilibrium thermostatted propagator we find that:

$$\langle B(t)\rangle = \langle B(0)\rangle + \int_0^t ds \int d\Gamma f(\Gamma)\, U_{R0}(0,s)(iL(s) - iL_0(s))U_R(s,t)B(\Gamma). \tag{7.91}$$

By successive integrations, we unroll U_{R0} propagator onto the distribution function:

$$\langle B(t)\rangle = \langle B(0)\rangle + \int_0^t ds \int d\Gamma (U_{R0}^\dagger(0,s)f(\Gamma))\,(iL(s) - iL_0(s))U_R(s,t)B(\Gamma). \tag{7.92}$$

However, $U_{R0}^\dagger$ is the equilibrium f-propagator and it has no effect on the equilibrium distribution function f_0, so:

$$\langle B(t)\rangle = \langle B(0)\rangle + \int_0^t ds \int d\Gamma f(\Gamma)\,(iL(s) - iL_0(s))U_R(s,t)B(\Gamma). \tag{7.93}$$

We can now unroll the Liouvilleans to attack the distribution function rather than the phase variables. The result is:

$$\langle B(t)\rangle = \langle B(0)\rangle - \int_0^t ds \int d\Gamma[(i\mathscr{L}(s) - i\mathscr{L}_0(s))f(\Gamma)]\,U_R(s,t)B(\Gamma). \tag{7.94}$$

Obviously only the operation of the field-dependent Liouvillean needs to be con-sidered. Provided AIΓ is satisfied, we know that $i\mathscr{L}_0 f_0 = -\partial f/\partial t = 0$, so:

$$(i\mathscr{L}(s) - i\mathscr{L}_0)f_0 = i\mathscr{L}(s)f_0 = \beta f_0 J(\Gamma)F_e(s). \tag{7.95}$$

For either the canonical or Gaussian isokinetic ensembles therefore:

$$\langle B(t) \rangle = \langle B(0) \rangle - \beta \int_0^t ds \int d\Gamma f_0(\Gamma) JF_e(s) \, U_R(s, t) B(\Gamma). \qquad (7.96)$$

Thus far the derivation has followed the same procedures used for the time-dependent linear response and time-independent nonlinear response. The operation of $U_R(s, t)$ on $B(\Gamma)$, however, presents certain difficulties. No simple meaning can be attached to $U_R(s, t)B(\Gamma)$. We can now use the composition and the inverse theorems to break up the incremental p-propagator $U_R(s, t)$. Using Equations (7.75):

$$U_R(s, t) = U_R^{-1}(0, s)U_R(0, t). \qquad (7.97)$$

Substituting this result into Equation (7.96) we find:

$$\langle B(t) \rangle = \langle B(0) \rangle - \beta \int_0^t ds \int d\Gamma f_0(\Gamma) JF_e(s) \, U_R^{-1}(0, s)U_R(0, t)B(\Gamma). \qquad (7.98)$$

Using the inverse theorem (7.74), and integrating by parts we find:

$$\langle B(t) \rangle = \langle B(0) \rangle - \beta \int_0^t ds \int d\Gamma F_e(s)B(t) \exp_R\left[\int_0^s ds_1 i\mathscr{L}(s_1)\right] Jf_0(\Gamma), \qquad (7.99)$$

where, after unrolling $U_R^{-1}(0, s)$, we attack B with $U_R(0, t)$, giving $B(t)$. As it stands, the exponential in this equation has the right time ordering of a p-propagator, but the argument of the exponential contains an f-Liouvillean. We obviously have some choices here. We choose to use Equation (7.87) to rewrite the exponential in terms of a p-propagator. This equation gives:

$$\exp_R\left[\int_0^s ds_1 i\mathscr{L}(s_1)\right] = \exp_R\left[\int_0^s ds_1 \Lambda(s_1)\right] U_R(0, s), \qquad (7.100)$$

where $\Lambda(s_1) = -3N\alpha(\Gamma(s_1), s_1) + O(1)$, and $\alpha(\Gamma(s_1), s_1)$ is the Gaussian isokinetic multiplier required to maintain a fixed kinetic energy. Substituting these results into Equation (7.99), using the fact that, $iL(s)H_0(\Gamma) = -J(\Gamma)F_e(s) - 3Nk_BT\alpha(\Gamma, s)$ gives:

$$\langle B(t) \rangle = \langle B(0) \rangle - \beta \int_0^t ds_1 \left\langle B(t)J(s_1) \exp\left[\beta \int_0^{s_1} ds_2 J(s_2)F_e(s_2)\right]\right\rangle F_e(s_1). \qquad (7.101)$$

This equation is the fundamental result for time-dependent nonlinear response theory. It must be remembered that all time evolution is governed by the field-dependent thermostatted equations of motion implicit in the Liouvillean, $iL(t)$.

We have described a consistent formalism for the nonlinear response of many-body systems to time-dependent external perturbations. This theory reduces to the standard results of linear response theory in the linear regime and can be used to derive the Kawasaki form of the time-independent nonlinear response. It also is easy to show that our results lead to the transient time-correlation function expressions for the time-independent nonlinear case. It may be thought that we can move freely between the various forms of nonlinear response theory, which each have different time arguments for B and J. However for the time-dependent non-linear case only our new form, Equation (7.101), seems to be valid. One can develop a Kawasaki version of the nonlinear response to time-dependent fields, but it is found that the resulting expression is not very useful. It, like the corresponding transient correlation form, involves convolutions of incremental propagators, Liouvilleans, and phase variables which have no directly interpretable meaning. None of the operators in the convolution chains commute with one another and the resulting expressions are intractable and formal.

8

Dynamical stability

8.1 Introduction

In the previous chapter we have developed a theory which can be applied to calculate the nonlinear response of an arbitrary phase variable to an applied external field. We have described several different representations for the N-particle, non-equilibrium distribution function, $f(\Gamma, t)$: the Kubo representation (Section 7.1), which is only useful from a formal point of view; and two related representations, the transient time-correlation function formalism (Section 7.3) and the Kawasaki representation (Section 7.2), both of which can be applied to obtain useful results. We now turn our interest towards thermodynamic properties, which are not simple phase averages, but rather are functionals of the distribution function itself. We will consider the entropy and free energy of nonequilibrium steady states. At this point, it is useful to recall the connections between equilibrium statistical mechanics, the thermodynamic entropy (Gibbs, 1902) and Boltzmann's famous H-theorem (Boltzmann, 1964). Gibbs pointed out that, *at equilibrium*, the entropy of a classical N-particle system can be calculated from the relation:

$$S(t) = -k_B \int d\Gamma f(\Gamma) \ln f(\Gamma),$$ (8.1)

where $f(\Gamma)$ is a time-independent equilibrium distribution function. Boltzmann calculated the *nonequilibrium* entropy of gases in the low density limit, and showed that for the single-particle distribution of velocities obtained from the irreversible Boltzmann equation, the entropy of a gas at equilibrium is greater than that of any nonequilibrium gas with the same number of particles, volume, and energy. Furthermore he showed that the Boltzmann equation predicts a *monotonic* increase in the entropy of an isolated gas as it relaxes towards equilibrium. These results are the content of his H-theorem (Huang, 1963). They are in accord with our intuition that the increase in entropy is the driving force behind the relaxation to equilibrium.

One can use the reversible Liouville equation to calculate the change in the entropy of a dense many-body system. Suppose we consider a Gaussian isokinetic system subject to a time-independent external field F_e. We expect that the entropy of a nonequilibrium steady state will be finite and less than that of the corresponding equilibrium system with the same energy. From Equation (8.1) we see that:

$$\dot{S}(t) = -k_B \int d\Gamma [1 + \ln f(\Gamma)] \frac{\partial f}{\partial t}. \tag{8.2}$$

Using successive integrations by parts, one finds, for an N-particle system in three dimensions:

$$\dot{S}(t) = -k_B \int d\Gamma f \, \dot{\Gamma} \cdot \frac{\partial}{\partial \Gamma} [1 + \ln f(\Gamma)] = -k_B \int d\Gamma \, \dot{\Gamma} \cdot \frac{\partial}{\partial \Gamma} f(\Gamma)$$

$$= k_B \int d\Gamma f(\Gamma) \frac{\partial}{\partial \Gamma} \cdot \dot{\Gamma} = -3Nk_B \langle \alpha(t) \rangle. \tag{8.3}$$

Now, for any nonequilibrium steady state, the average of the Gaussian multiplier α is positive. The external field does work on the system which must be removed by the thermostat. This means that the Liouville equation predicts that the Gibbs entropy (Equation 8.1) diverges to negative infinity! After the decay of initial transients (Equation 8.3) shows the rate of decrease of the entropy is constant. This paradoxical result was derived by Evans (1985). If there is no thermostat, the Liouville equation predicts that the Gibbs entropy of an arbitrary system, satisfying AIΓ and subject to an external dissipative field, is constant! This result was known to Gibbs.

Gibbs went on to show that if one computes a coarse-grained entropy, by limiting the resolution with which we compute the distribution function, then the coarse-grained entropy based on Equation (8.1) obeys a generalized H-theorem. Gibbs showed that the coarse-grained entropy cannot decrease. We shall return to the question of coarse graining in Section 10.2.

The reason for the divergence in Equation (8.3) is not difficult to find. Consider a small region of phase space, $d\Gamma$, at $t = 0$, when the field is turned on. If we follow the phase trajectory of a point originally within $d\Gamma$, the local relative density of ensemble points in phase space about $\Gamma(t)$ can be calculated from the Liouville equation:

$$\frac{1}{f(t)} \frac{df(t)}{dt} = 3N\alpha(t). \tag{8.4}$$

If the external field is sufficiently large, we know that there will be some trajectories along which the multiplier $\alpha(t)$ is positive for *all* time. For such trajectories, Equation (8.4) predicts that the local density of the phase-space distribution function must diverge in time, towards positive infinity. The distribution function of a steady state will be singular at long times. One way in which this could happen would be for the distribution function to evolve into a space of lower dimension than the *ostensible* 6-N dimensions of phase space. If the dimension of the phase space which is *accessible* to nonequilibrium steady states is lower than the ostensible dimension, the volume of accessible phase space (as computed from within the ostensible phase space) will be zero. If this were so, the Gibbs entropy of the system (which occupies zero volume in ostensible phase space) would be minus infinity.

At this stage these arguments are not at all rigorous. We have yet to define what we mean by a continuous change in the dimension. In the following sections we will show that a reduction in the dimension of accessible phase space is a universal feature of nonequilibrium steady states. The phase-space trajectories are chaotic and separate exponentially with time, and for nonequilibrium systems, the accessible steady-state phase space is a strange attractor whose dimension is less than that of the initial equilibrium phasespace. Before we start a detailed analysis, it is instructive to consider two generic problems from dynamical systems theory – the quadratic map and the Lorenz model. The first is a discrete mapping and the other is a *flow* in continuous time. This will introduce many of the concepts needed later to quantitatively characterize nonequilibrium steady states.

8.2 Chaotic dynamical systems

It was long thought that the complex behavior of nonlinear systems of many degrees of freedom was inherently different to that of simple mechanical systems. It is now known that simple one-dimensional nonlinear systems can indeed show very complex behavior and that this behavior is, in many senses, typical of other chaotic systems. For example, the family of quadratic maps $f_\mu(x) = \mu x(1 - x)$ demonstrates many of these features (see Devaney, 1986; Schuster, 1988; Ott, 2002; Sprott, 2003). The connection between a discrete mapping and the solution of a system of ordinary differential equations (a flow) is clear when we realize that the numerical solution involves the construction of an iterative mapping that approximates the flow. All numerical schemes, such as Runge–Kutta, transform a set of ordinary differential equations into a discrete map f where $\Gamma(n\Delta) = f^n(\Gamma(0))$ is the result of evolving the initial point $\Gamma(0)$ forward in time an amount $t = n\Delta$, where Δ is the time step. Here f^n means n applications of the map f.

There is an important difference between difference equations and similar differential equations (or flows), for example if we consider the differential equation:

$$\frac{\mathrm{d}x}{\mathrm{d}t} = \mu x(1 - x), \tag{8.5}$$

the solution can easily be obtained as:

$$x(t) = \frac{x_0 \exp[\mu t]}{1 - x_0 + x_0 \exp[\mu t]} \tag{8.6}$$

where $x_0 = x(t = 0)$ is the initial condition. The trajectory for this system is now quite straightforward to understand. The solution of the quadratic-map difference equation is a much more difficult problem, which is still not completely understood.

The quadratic map

The quadratic map is defined by the equation:

$$x_{n+1} = f_\mu(x_n) = \mu x_n(1 - x_n). \tag{8.7}$$

If we iterate this mapping for $\mu = 4$, starting with a random number in the interval between zero and one, then we obtain dramatically different behavior depending upon the initial value of x. Sometimes the values repeat; other times they do not; and usually they wander aimlessly about in the range zero to one. Initial values of x which are quite close together can have dramatically different iterates. This unpredictability or sensitive dependence on initial conditions is a property familiar in statistical-mechanical simulations of higher-dimensional systems. If we change the map to $x_{n+1} = 3.839 x_n(1 - x_n)$, then a random initial value of x leads to a repeating cycle of three numbers (0.149888..., 0.489172..., 0.959299...). This mapping includes a set of initial values which behave just as unpredictably as those in the $\mu = 4$ example, but due to round-off error we don't see this randomness.

Before we look at the more complicated behavior, we consider some simpler properties of the family of quadratic maps (Figure 8.1). We begin with some definitions; x_1 is a fixed point of the map f if $f(x_1) = x_1$. x_1 is a periodic point, of period n, if $f^n(x_1) = x_1$, where f^n represents n applications of the mapping f. Clearly a fixed point is a periodic point of period one. The fixed point at x_1 is stable if $|f'(x_1)| < 1$. We consider the quadratic map $f_\mu(x)$ on the interval $0 < x < 1$, as a function of the parameter μ.

Region 0: $0 < \mu < 1$

Here the map $f_\mu(x)$ has only one fixed point $x = 0$. $f'_\mu(0) = \mu < 1$ so that in this region the fixed point at $x = 0$ is attracting or stable.

Region 1: $1 < \mu < 3$

Here $f_\mu(x)$ has two fixed points $x = 0$ and $x_p = (\mu - 1)/\mu$. The fixed point $x = 0$ is repelling or unstable while $|f'_\mu(x_p)| = |2 - \mu| < 1$, so that x_p is an attracting or stable fixed point.

Region 2: $3 < \mu < 1 + \sqrt{6}$

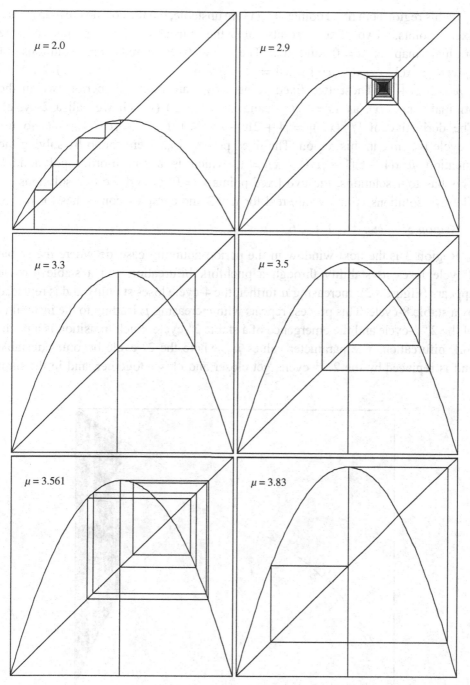

Figure 8.1 The iterates of the quadratic map for particular values of the parameter μ. The horizontal axis is x_n and the vertical axis is x_{n+1}. For $\mu = 2$ and 2.9 there is a single stable fixed point. For $\mu = 3.3$ there is a stable 2-cycle; for $\mu = 3.5$ a stable 4-cycle and for $\mu = 3.561$ a stable 8-cycle. The value $\mu = 3.83$ is in the period 3 window

In this region both fixed points of $f_\mu(x)$ are unstable, but the composite map f_μ^2 has fixed points. Two fixed points are the unstable fixed points of the original map at $x = 0$ and x_p. The new fixed points are solutions of $f_\mu^2(x) = \mu^2 x(1-x)[1-\mu x(1-x)] = x$, given by $x_\pm = (\mu+1)/2\mu\{1 \pm \sqrt{\mu - 3/\mu + 1}\}$. These two fixed points of f_μ^2 are points of period two in the original map $f_\mu(x)$, as $x_+ = f_\mu(x_-)$ and $x_- = f_\mu(x_+)$ (which we call a 2-cycle). The derivative of $|f_\mu^{2'}(x_\pm)| = |4 + 2\mu - \mu^2| < 1$ for $3 < \mu < 1 + \sqrt{6}$ so the 2-cycle is stable in this region. The fixed points of f_μ^2 were found by solving the equation $\mu^2 x(1-x)[1-\mu x(1-x)] = x$, which is a fourth-order polynomial. This has four solutions; the two fixed points $x = 0$, x_p and the two solutions $x_\pm$. The two solutions x_+ and x_- are real for $\mu > 3$ and complex conjugates for $\mu < 3$.

Region 3, 4, etc: $1 + \sqrt{6} < \mu < \mu_\infty$

Region 3 is the next window in the period-doubling cascade where the stable 2-cycle loses its stability through a pitchfork bifurcation and a stable 4-cycle appears (Figure 8.2). Increasing μ further, the 4-cycle loses stability and is replaced by a stable 8-cycle. This process repeats with increasing μ leading to the instability of the 2^{n-1}-cycle and the emergence of a stable 2^n-cycle. Each transition is a pitchfork bifurcation. The parameter values μ_n, where the 2^n-cycle becomes unstable and is replaced by the 2^{n+1}-cycle, get closer and closer together, and in the limit

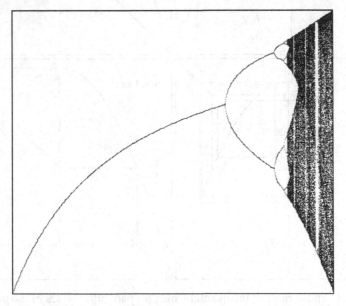

Figure 8.2 The iterates of the quadratic map as a function of the parameter μ. The horizontal axis is the parameter $1 \le \mu \le 4$, and the vertical axis is the iterate $0 \le x_n \le 1$

as $n \to \infty$, μ_n approaches $\mu_\infty = 3.5699456\ldots$ for the quadratic map. Notice that in region m, the order of the polynomial we need to solve to find the various fixed points is 2^m, and this leads to at least one m-cycle. The critical values μ_n, where the 2^n-cycle loses its stability and a 2^{n+1}-cycle is created, obey a very famous scaling relation discovered by Feigenbaum (1978):

$$\lim_{n \to \infty} \frac{\mu_n - \mu_{n-1}}{\mu_{n+1} - \mu_n} = \delta, \tag{8.8}$$

where for unimodal maps with a quadratic maximum $\delta = 4.6692016091\ldots$ Interestingly, for a unimodal map with a quartic maximum the value changes to $\delta = 7.28\ldots$

The Chaotic Region: $\qquad \mu_\infty < \mu < 4$

Here stable periodic and chaotic regions are densely interwoven. Chaos here is characterized by sensitive dependence on the initial value x_0. Close to every value of μ where there is chaos, there is a different value of μ where there is a stable periodic orbit, that is, the mapping displays sensitive dependence on the parameter μ (Figure 8.3). The windows of period three, five, and six are examples. From the

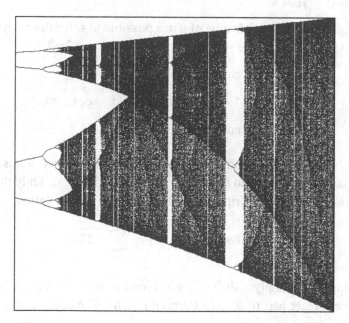

Figure 8.3 The iterates of the quadratic map as a function of the parameter μ. This is an expanded version of Figure 8.2 to include more detail in the chaotic region. The horizontal axis is the parameter $3.5 < \mu < 4$, and the vertical axis is the iterate $0 \le x_n \le 1$. The windows of period three (at about $\mu = 3.83$), period 5 (at about $\mu = 3.74$), and period 6 (at about 3.63) are clearly visible

mathematical perspective, the sequence of cycles in a unimodal map is completely determined by the Sarkovskii theorem (Sarkovskii, 1964). If $f(x)$ has a point x which leads to a cycle of period p, then it must have a point x' which leads to a q-cycle for every $q \leftarrow p$ where p and q are elements of the following sequence (here we read $\leftarrow$ as *precedes*):

$$1 \leftarrow 2 \leftarrow 4 \leftarrow 8 \leftarrow 16 \leftarrow 32 \leftarrow \ldots \leftarrow 2^m \leftarrow \ldots$$

$$\ldots 2^m.9 \leftarrow 2^m.7 \leftarrow 2^m.5 \leftarrow 2^m.3 \leftarrow \ldots$$

$$\vdots$$

$$\ldots 2^2.9 \leftarrow 2^2.7 \leftarrow 2^2.5 \leftarrow 2^2.3 \leftarrow \ldots$$

$$\ldots 2^1.9 \leftarrow 2^1.7 \leftarrow 2^1.5 \leftarrow 2^1.3 \leftarrow \ldots$$

$$\ldots 9 \leftarrow 7 \leftarrow 5 \leftarrow 3 \leftarrow \ldots$$

(8.9)

This theorem applies to values of x at a fixed parameter μ, but says nothing about the stability of the cycle or the range of parameter values for which it is observed.

Region ∞^-: $\mu = 4$

Surprisingly for this special value of μ it is possible to solve the mapping exactly (Kadanoff, 1983). Making the substitution $x_n = \frac{1}{2}(1 - \cos 2\pi\theta_n)$:

$$\begin{aligned}
x_{n+1} &= \tfrac{1}{2}(1 - \cos 2\pi\theta_{n+1}) \\
&= 4 \times \tfrac{1}{2}(1 - \cos 2\pi\theta_n)\big(1 - \tfrac{1}{2}(1 - \cos 2\pi\theta_n)\big) \\
&= \tfrac{1}{2}(1 - \cos 4\pi\theta_n).
\end{aligned}$$

(8.10)

A solution is $\theta_{n+1} = 2\theta_n \bmod 1$, or $\theta_n = 2^n \theta_0 \bmod 1$. Since x_n is related to $\cos(2\pi\theta_n)$, adding an integer to θ_n leads to the same value of x_n. Only the fractional part of θ_n has significance. Writing θ_n in binary (base two) notation:

$$\theta_n = 0.a_1a_2a_3a_4a_5\cdots = \sum_{i=1}^{\infty} a_i 2^{-i},$$

(8.11)

thus the mapping is simply a shift of the decimal point one place to the right and removing the integer part of θ_{n+1} (a Bernoulli shift). The equivalent mapping is:

$$f(0.a_1a_2a_3a_4a_5\cdots) = 0.a_2a_3a_4a_5\cdots$$

(8.12)

It is easy to see that any finite precision approximation to the initial starting value θ_0 consisting of N digits will lose all of its significant digits in N iterations.

We can also discuss the evolution of a distribution of initial values under the operation of a map $x' = f(x)$. If the distribution consists of a delta function at x_0, then $\rho_0(x) = \delta(x - x_0)$. After one iteration of the map, the distribution $\delta(x - x_0)$ evolves to $\rho_1(x) = \delta(x - f(x_0))$. This evolution can be written as:

$$\delta(x - f(x_0)) = \int_0^1 dy\,\delta(x - f(y))\delta(y - x_0). \tag{8.13}$$

The evolution of an arbitrary density $\rho_n(x)$ (perhaps constructed from the normalized sum of infinitely many delta functions) in one step of the map, satisfies an equation of the form:

$$\rho_{n+1}(x) = \int_0^1 dy\,\delta(x - f(y))\rho_n(y). \tag{8.14}$$

To construct an invariant or steady-state distribution $\rho(x)$ that is independent of the number of iterations n, we can imagine iterating Equation (8.14) until $\rho(x)$ satisfies:

$$\rho(x) = \int_0^1 dy\,\delta(x - f(y))\rho(y) \tag{8.15}$$

The solution to this equation is not unique as $\rho(x) = \delta(x - x^*)$, where x^* is any fixed point of the map, is always a solution. However, in general there is a physically relevant solution which typically corresponds to iterating Equation (8.14) with an initially uniform distribution. This is called the *natural measure*. Note that the set of fixed points is measure zero in the interval [0, 1] so the probability of choosing to start from a fixed point x^* is zero.

To define the natural measure we consider covering the space with a grid of cubes and then looking at the fraction of time a typical orbit spends in that cube. If these frequencies are the same for all initial conditions (except for a set of Lebesgue measure zero, such as the periodic points) then the frequencies give the natural measure for the grid of cubes. For a typical point x_0, the natural measure of the ith cube is:

$$\mu_i = \lim_{T \to \infty} \frac{\eta(C_i, x_0, T)}{T}, \tag{8.16}$$

where $\eta(C_i, x_0, T)$ is the amount of time the orbit beginning at x_0 spends in cube C_i in the time interval $0 \le t \le T$. In Figure 8.4 we show the natural measure for the quadratic map at $\mu = 3.65$. The natural measure contains a number of peaks which we will see (in Section 8.3) are, in fact, fractional power law singularities of the underlying probability distribution.

For the parameter value $\mu = 4$, the quadratic map becomes $\theta_{n+1} = 2\theta_n \bmod 1$, and it is easy to see that the continuous loss of information about any finite

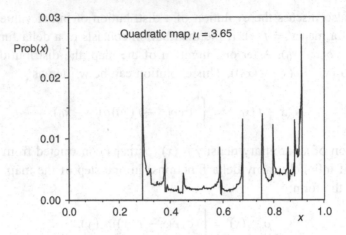

Figure 8.4 The natural measure for the quadratic map at $\mu = 3.65$

precision initial point means that the invariant measure is uniform on $[0, 1]$ (that is $g(\theta) = 1$). From the change of variable $x = \frac{1}{2}(1 - \cos 2\pi\theta)$ it is easy to see that x is a function of θ, $x = q(\theta)$ (but θ is not a function of x). If $x_1 = q(\theta_1)$, then the number of counts in the distribution function histogram bin centered at x_1 with width dx_1, is equal to the number of counts in the bins centered at θ_1 and $1 - \theta_1$ with widths $d\theta_1$. That is:

$$f(x_1) = \frac{g(\theta_1) + g(1 - \theta_1)}{|dx/d\theta|}. \tag{8.17}$$

It is then straightforward to show that the natural invariant measure as a function of x is given by:

$$f(x)dx = \frac{1}{\pi} \frac{dx}{\sqrt{x(1 - x)}}. \tag{8.18}$$

This has inverse square-root singularities at $x = 0$ and $x = 1$. In Figure 8.5 we present the invariant measure for the quadratic map at $\mu = 4$. Note the two square root singularities at $x_0 = 0$ and $x_0 = 1$.

Region ∞: $\mu > 4$

Here the maximum of $f_\mu(x)$ is greater than one so some of the interval $[0, 1]$ is mapped outside $[0, 1]$. Once the iterate leaves the interval $[0, 1]$ it does not return. Similarly, the map $f_\mu^2(x)$ has two maxima, both of which are greater than one. If I is the interval $[0, 1]$, and A_1 is the region of I mapped out of I by $f_\mu(x)$, A_2 the region of

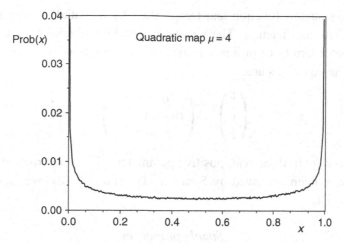

Figure 8.5 The distribution of iterates for the quadratic map at $\mu = 4$. When correctly normalized, this natural measure agrees well with Equation (8.15)

I mapped out of I by $f_\mu^2(x)$, etc., then the trajectory wanders the interval defined by $I - (A_1 \cup A_2 \cup \ldots)$. It can be shown that this set is a Cantor set.

In this example, a seemingly very simple iterative equation has very complex behavior as a function of initial value x_0 and parameter μ. As μ is increased, the stable fixed point becomes unstable and is replaced by stable 2^n-cycles (for $n = 1, 2, 3, \ldots$), until chaotic behavior develops at $\mu_\infty > 3.5699456\ldots$ (about 3.5699456). For $\mu_\infty > 3.5699456\ldots$, the behavior of the quadratic map shows sensitive dependence upon the parameter μ, with an infinite number of islands of periodic behavior immersed in a sea of chaos. This system is not atypical, and a wide variety of nonlinear problems show similar behavior. We will now consider a simple model for atmospheric dynamics which has had a dramatic impact in the practical limitations of weather forecasting.

The Lorenz model

Consider two flat plates, separated by a liquid layer. The lower plate is heated and the fluid is assumed to be two-dimensional and incompressible. In three-dimensions this is termed a Rayleigh–Benard cell. A coupled set of nonlinear field equations must be solved in order to determine the motion of the fluid between the plates (the continuity equation, the Navier–Stokes equation, and the energy equation). These equations are simplified by introducing the stream function in place of the two velocity components. Saltzmann (1961) and Lorenz (1963) proceed by making the field equations dimensionless and then representing the dimensionless stream function and temperature by a spatial Fourier series (with time-dependent coefficients). The resulting equations obtained by Lorenz are a three-parameter family of ordinary differential equations which can have chaotic solutions. The equations are:

$$
\begin{pmatrix} \dot{x} \\ \dot{y} \\ \dot{z} \end{pmatrix} = \begin{pmatrix} -\sigma\,(x-y) \\ (r-z)x - y \\ xy - bz \end{pmatrix},
\tag{8.19}
$$

where σ, r and b are three real, positive parameters. The properties of the Lorenz equations have been reviewed by Sparrow (1982) and below we summarize the principle results.

Simple properties

(1) Symmetry: the Lorenz equations are symmetric with respect to the mapping $(x, y, z) \rightarrow (-x, -y, z)$.

(2) The z-axis is invariant. All trajectories which start on the z-axis remain there and move toward the origin. All trajectories which rotate around the z-axis do so in a clockwise direction (when viewed from above the $z = 0$ plane). This can be seen from the fact that if $x = 0$, then $\dot{x} > 0$ when $y > 0$, and $\dot{x} < 0$ when $y < 0$.

(3) Existence of a bounded attracting set of zero volume, that is the existence of an attractor. The divergence of the flow, is given by:

$$
\frac{\partial \dot{x}}{\partial x} + \frac{\partial \dot{y}}{\partial y} + \frac{\partial \dot{z}}{\partial z} = -(1 + b + \sigma).
\tag{8.20}
$$

The volume element V is contracted by the flow into a volume element $V\exp[-(1 + b + \sigma)t]$ in time t. We can show that there is a bounded region E, such that every trajectory eventually enters E and remains there forever. There are many possible choices of Lyapunov function which defines the surface of the bounded region E. One simple choice is $V = rx^2 + \sigma y^2 + \sigma(z - 2r)^2$. Differentiating with respect to time and substituting the equations of motion gives:

$$
\frac{dV}{dt} = -2\sigma(rx^2 + y^2 + bz^2 - 2brz).
\tag{8.21}
$$

Another choice of Lyapunov function is $E = r^2x^2 + \sigma y^2 + \sigma(z - r(r-1))^2$ for $b \le r + 1$. This shows that there exists a bounded ellipsoid, and together with the negative divergence shows that there is a bounded set of zero volume within E, towards which all trajectories tend.

(4) Fixed points: the Lorenz equations have three fixed points; one at the origin, and the other two at $C_1 = \left(-\sqrt{b(r-1)}, -\sqrt{b(r-1)}, r-1\right)$ and $C_2 = \left(\sqrt{b(r-1)}, \sqrt{b(r-1)}, r-1\right)$.

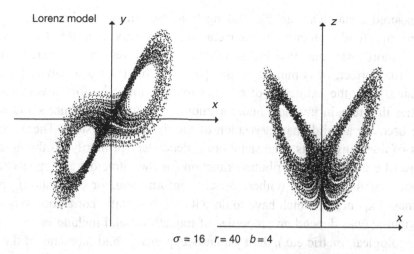

Figure 8.6 The iterates of the Lorenz model for a typical set of parameters which leads to chaotic behavior. The iterates are the values obtained at the end of each fourth order Runge-Kutta step

(5) Eigenvalues for linearized flow about the origin are:

$$\lambda_1 = -b$$

$$\lambda_2 = \frac{-(\sigma + 1) - \sqrt{(\sigma + 1)^2 - 4\sigma(1 - r)}}{2}$$

$$\lambda_3 = \frac{-(\sigma + 1) + \sqrt{(\sigma + 1)^2 - 4\sigma(1 - r)}}{2}$$

(8.22)

(6) Stability

$0 < r < 1$, the origin is stable.

$r > 1$, the origin is non-stable. Linearized flow about the origin has two negative and one positive, real eigenvalues.

$1 < r < \frac{470}{19}$, C_1 and C_2 are stable. All three eigenvalues of the linearized flow about C_1 and C_2, have negative real part. For $r > 1.346(\sigma = 10, b = \frac{8}{3})$ there is a complex conjugate pair of eigenvalues.

$r > \frac{470}{19}$ C_1 and C_2 are non-stable. Linearized flow about C_1 and C_2 has one negative real eigenvalue and a complex conjugate pair of eigenvalues with positive real part.

Again we have a nonlinear system which is well behaved for small values of the parameter r, but for $r > \frac{470}{19}$ chaotic behavior begins. Typical iterates of the Lorenz model are shown in Figure 8.6.

8.3 The characterization of chaos

The experimental measurement of the onset and development of chaos in dissipative physical systems is often accompanied by some arbitrariness in the choice of the measured dynamical variable. Taking fluid systems as an example, one can measure the fluid velocity, its temperature, heat flux etc. Rarely does one measure more than one variable simultaneously. Moreover, one rarely knows what is the correct, or complete, phase space in which the dissipative dynamics take place. Thus the extraction of relevant information calls for measurement of quantities that remain invariant under a smooth change of coordinates and which can be used for a valid characterization of the dynamical system. There are two classes of these invariants. The static ones, dependent primarily on the invariant measure (the underlying distribution function for the attractor) and appear as the dimension of the attractor (either capacity, information, or correlation) and as other mass exponents, which have to do with various static correlation functions. The dynamic ones depend on properties of trajectories and include various entropies (topological, metric etc.), the Lyapunov exponents, and moments of the fluctuations in the Lyapunov exponents. Here we present a short review of the theory of these invariants and the interrelations between them.

Studies of simple dissipative systems have shown that if we begin with a Euclidian space of initial phase positions, then as time passes, transients relax, some modes may damp out, and the point in phase space that describes the state of the system approaches an *attractor*. In this process, it is common for the number of degrees of freedom to be reduced, and hence the dimension of the system is lowered. This change in dimension is a continuous process and to describe such systems we have to generalize the concept of dimension, (Farmer, 1982; Farmer *et al.*, 1983). We distinguish three intuitive notions of dimension: direction, capacity, and measurement. These lead to the definitions of: topological dimension (Hurewicz and Wallman, 1948); capacity dimension (Mandelbrot, 1983); and information dimension (Balatoni and Renyi, 1956). As we will see the capacity and information dimensions allow the dimension to be a continuous, positive variable.

The capacity and information dimensions

The capacity dimension of an attractor can be defined by the following construction. Let $b(\varepsilon)$ be the minimum number of balls of diameter ε needed to cover the attractor. The capacity dimension is defined by the limit:

$$D_F = \lim_{\varepsilon \to 0} \frac{\ln b(\varepsilon)}{|\ln \varepsilon|}. \tag{8.23}$$

As the length scale ε is reduced, the number of balls required to cover the attractor increases. As $b(\varepsilon)$ is a positive integer, its logarithm is positive. The term $\ln \varepsilon$ is negative as soon as the length scale ε is less than one (in the appropriate units); the dimension is a positive real quantity.

To obtain the information dimension we suppose an observer makes an isolated measurement of the coarse-grained probability distribution function p_i. Coarse graining implies a length scale ε for the observation, and an associated number of cells $N(\varepsilon)$. The discrete entropy $S(\varepsilon)$ as a function of the length scale is given by:

$$S(\varepsilon) = -\sum_{i=1}^{N(\varepsilon)} p_i \ln p_i. \tag{8.24}$$

Notice that $S(\varepsilon)$ is positive as for each i, $-p_i \ln p_i$ is positive. The information dimension D_I is then defined by:

$$D_I = \lim_{\varepsilon \to 0} \frac{S(\varepsilon)}{|\ln \varepsilon|}. \tag{8.25}$$

This dimension is a property of any distribution function as nothing in the definition is specific to attractors, or to some underlying dynamics.

If all the $N(\varepsilon)$ elements have the same probability, then $S(\varepsilon) = \ln N(\varepsilon)$. Further if $b(\varepsilon)$ is a minimal covering, then a smaller covering can be formed by removing the overlapping parts of circles so that $\ln b(\varepsilon) \geq \ln N(\varepsilon) = S(\varepsilon)$. It is then straightforward to see that the fractal dimension is an upper bound on the information dimension. (We will generalize this result later.) From a computational point of view, it is easier to tabulate the steady-state distribution function and calculate D_I, rather than to attempt to identify the attractor and construct a covering to calculate D_F.

Correlation dimension

The correlation dimension D_C introduced by Grassberger and Procaccia (1983a; 1983b) is a scaling relation on the correlation function $C(\varepsilon)$:

$$C(\varepsilon) = \frac{1}{N^2} \sum_{i \neq j} \theta(\varepsilon - |\Gamma_i - \Gamma_j|) \tag{8.26}$$

where $\theta(x)$ is the Heaviside step function. $C(\varepsilon)$ is the correlation integral which counts the number of pairs of points whose distance of separation $|\Gamma_i - \Gamma_j|$ is less than ε. The correlation dimension is:

$$D_C = \lim_{\varepsilon \to 0} \lim_{N \to \infty} \frac{\ln C(\varepsilon)}{|\ln \varepsilon|}. \tag{8.27}$$

It has been argued that the correlation dimension can be calculated numerically, more easily and more reliably than either the capacity or information dimension.

Generalized dimensions

In a series of papers (Grassberger, 1983; Hentschel, and Procaccia, 1983; Halsey *et al.*, 1986) have shown that the concept of dimension can be generalized further. They introduce a generating function D_q which provides an infinite spectrum of dimensions depending upon the value of the parameter q. We will show that all previous dimensions correspond to special values of q. We begin with a discrete probability distribution $p_i(\varepsilon)$ on cells of length scale ε. Summing powers of the p_is over all cells, the generalized dimension D_q is obtained as:

$$D_q = - \lim_{\varepsilon \to 0} \frac{1}{q-1} \frac{1}{|\ln \varepsilon|} \ln\left(\sum_i p_i^q \right). \tag{8.28}$$

There are formal similarities between the D_q and the free energy per particle F_β in the thermodynamic limit:

$$F_\beta = - \lim_{N \to \infty} \frac{1}{\beta N} \ln\left(\sum_i (\exp[-E_i])^\beta \right), \tag{8.29}$$

where the E_i are the discrete energy levels, N is the number of particles, and $\beta = (k_B T)^{-1}$ is the inverse temperature. The analogy is not a strict one, as the probability of state i is $\exp[-\beta E_i]$ rather than simply $\exp[-E_i]$ as implied above. Also the probabilities $p_i \Rightarrow \exp[-E_i]$ are not normalized. This is crucial as normalized probabilities in Equation (8.29) give a zero free energy F_β.

It straightforward to see that D_q gives each of the previously defined dimensions. For $q = 0$, $p_i^q = 1$ for all values of i, so that:

$$D_0 = \lim_{\varepsilon \to 0} \frac{\ln\left(\sum_{i=1}^{N(\varepsilon)} 1 \right)}{|\ln \varepsilon|} = \lim_{\varepsilon \to 0} \frac{\ln N(\varepsilon)}{|\ln \varepsilon|}. \tag{8.30}$$

This is the capacity or Hausdorff dimension defined in Equation (8.23).

For $q = 1$ we use d'Hopitals rule to consider the limit:

$$\lim_{q \to 1} \frac{\ln\left(\sum_i p_i^q \right)}{q-1} = \lim_{q \to 1} \frac{\frac{d}{dq} \ln\left(\sum_i e^{q \ln p_i} \right)}{\frac{d}{dq}(q-1)} = \sum_i p_i \ln p_i = -S(\varepsilon). \tag{8.31}$$

Substituting this limit into the expression for D_q gives:

$$\lim_{q \to 1} D_q = \lim_{\varepsilon \to 0} \frac{S(\varepsilon)}{|\ln \varepsilon|} = D_1. \tag{8.32}$$

This is simply the information dimension. For $q = 2$ it is easy to show that the generalized dimension is the correlation dimension.

The generalized dimension D_q is a non-increasing function of q. To show this we consider the generalized mean $M(t)$ of the set of positive quantities $\{a_1, \ldots, a_n\}$, where p_i is the probability of observing a_i. The generalized mean is defined to be:

$$M(t) = \left(\sum_{i=1}^{n} p_i a_i^t \right)^{1/t}. \qquad (8.33)$$

This reduces to the familiar special cases; $M(1)$ is the arithmetic mean and the limit as $t \to 0$ is the geometric mean. It is not difficult to show that if $a_i = p_i(\varepsilon)$, where the $p_i(\varepsilon)$ are a set of discrete probabilities calculated using a length scale of ε, then the generalized dimension in Equation (8.28) is related to the generalized mean by:

$$D_q = - \lim_{\varepsilon \to 0} \frac{\ln M(q-1)}{|\ln \varepsilon|}. \qquad (8.34)$$

Using a theorem concerning generalized means, namely if $t < s$, then $M(t) \leq M(s)$ (Hardy et al., 1934, p. 26), it follows that if $s > t$ then $D_s \leq D_t$, thus D_q is a non-increasing function of q.

The probability distribution on the attractor

If we consider the quadratic map for $\mu = 4$, the distribution of the iterates shown in Figure 8.5 is characterized by the two singularities at $x = 0$ and $x = 1$. For $\mu = 3.65$, the distribution of iterates shown in Figure 8.4 has a small number of large peaks which also appear to be singularities. It is common to find a probability distribution on an attractor which consists of sets of singularities with differing fractional power-law strengths. The size of these sets of singularities can be calculated from the generalized dimension D_q. To illustrate the connection between the generalized dimensions D_q and the size of the singularity sets, we consider a one-dimensional system whose underlying distribution function is:

$$\rho(x) = \tfrac{1}{2} x^{-1/2} \qquad \text{for} \qquad 0 \leq x \leq 1. \qquad (8.35)$$

First note that, despite the fact that $\rho(x)$ is singular, $\rho(x)$ is integrable on the interval $0 \leq x \leq 1$ and it is correctly normalized. The generalized dimension D_q requires discrete probabilities, so we divide the interval into bins of length $\varepsilon - [0, \varepsilon)$ is bin 0, $[\varepsilon, 2\varepsilon)$ is bin 1, etc. The probability of bin 0 is given by:

$$p_0 = \int_0^\varepsilon dx \tfrac{1}{2} x^{-1/2} = \varepsilon^{1/2}, \qquad (8.36)$$

and, in general, the probability of bin i is given by:

$$p_i = \int_{x_i}^{x_i + \varepsilon} dx \tfrac{1}{2} x^{-1/2} = (x_i + \varepsilon)^{1/2} - x_i^{1/2}, \qquad (8.37)$$

where $x_i = i\varepsilon$. As $(x_i + \varepsilon)^{1/2}$ is analytic for $i \neq 0$, we can expand this term to obtain:

$$p_i = \frac{1}{2}x_i^{-1/2}\varepsilon + O(\varepsilon^2) = \rho(x_i)\varepsilon + O(\varepsilon^2). \tag{8.38}$$

So for $i = 0$ the probability scales with the bin size as $p_i \sim \varepsilon^{1/2}$, but for nonzero values of i, $p_i \sim \varepsilon$. To construct D_q we need to calculate:

$$\sum_{i=0} p_i^q = p_0^q + \sum_{i\neq0} p_i^q = \varepsilon^{q/2} + \varepsilon^{q-1}\sum_{i\neq0}\rho(x_i)^q\varepsilon, \tag{8.39}$$

For small ε we can replace the last sum in this equation by an integral:

$$\sum_{i\neq0}\varepsilon\rho(x_i)^q \cong \int_\varepsilon^1 dx\rho(x)^q = \int_\varepsilon^1 dx\left(\frac{1}{2}x^{-1/2}\right)^q = \frac{\left(1 - \varepsilon^{1-q/2}\right)}{2^q\left(1 - \frac{q}{2}\right)} \tag{8.40}$$

$$= a\left(1 - \varepsilon^{1-q/2}\right)$$

where $a = (\frac{1}{2})^q/(1 - \frac{q}{2})$. Combining this result with that for $i = 0$ we obtain:

$$\sum_{i=0} p_i^q = \varepsilon^{q/2} + \varepsilon^{q-1}a\left(1 - \varepsilon^{1-q/2}\right) = (1 - a)\varepsilon^{q/2} + a\varepsilon^{q-1}. \tag{8.41}$$

The distribution function $\rho(x)$ in Equation (8.35) gives rise to singularities in the discrete probabilities p_i. If the discrete probabilities scale with exponent α_i, so that $p_i \sim \varepsilon^{\alpha_i q}$ and:

$$p_i^q \sim \varepsilon^{\alpha_i q}, \tag{8.42}$$

then α_i can take on a range of values corresponding to different regions of the underlying probability distribution. In particular, if the system is divided into pieces of size ε, then the number of times α_i takes on a value between α' and $\alpha' + d\alpha'$ will be of the form:

$$d\alpha'\rho(\alpha')\varepsilon^{-f(\alpha')}, \tag{8.43}$$

where $f(\alpha')$ is a continuous function. The exponent $f(\alpha')$ reflects the differing dimensions of the sets whose singularity strength is α'. Thus fractal probability distributions can be modeled by interwoven sets of singularities of strength, α each characterized by its own dimension $f(\alpha)$.

In order to determine the function $f(\alpha)$ for a given distribution function, we must relate it to observable properties, in particular we relate $f(\alpha)$ to the generalized dimension D_q. As q is varied, subsets associated with different scaling indices become dominant. Using Equation (8.43), we obtain:

$$\sum_i p_i^q = \sum_i \varepsilon^{\alpha_i q} = \int d\alpha'\rho(\alpha')\varepsilon^{-f(\alpha')}\varepsilon^{\alpha'q}. \tag{8.44}$$

Since ε is very small, the integral will be dominated by the value of α' which makes the exponent $q\alpha' - f(\alpha')$ smallest, provided that $\rho(\alpha')$ is nonzero. The condition for an extremum is:

$$\frac{d}{d\alpha'}(q\alpha' - f(\alpha')) = 0 \quad \text{and} \quad \frac{d^2}{d\alpha'^2}(q\alpha' - f(\alpha')) > 0. \tag{8.45}$$

If $\alpha(q)$ is the value of α' which minimizes $q\alpha' - f(\alpha')$ then $f'(\alpha(q)) = q$ and $f''(\alpha(q)) < 0$. If we approximate the integral in Equation (8.44) by its maximum value, and substitute this into Equation (8.28) then:

$$D_q = \frac{1}{q-1}(q\alpha(q) - f(\alpha(q))), \tag{8.46}$$

so that:

$$f(\alpha) = q\alpha(q) - (q-1)D_q. \tag{8.47}$$

Thus if we know $f(\alpha)$ and the spectrum of α' values we can find D_q. Alternatively, given D_q we can find $\alpha(q)$, since $f'(\alpha) = q$ implies that:

$$\alpha(q) = \frac{d}{dq}((q-1)D_q). \tag{8.48}$$

and knowing $\alpha(q), f(\alpha(q))$ can be obtained.

To calculate the generalized dimension D_q, we need to determine which of the two terms in Equation (8.41) dominate in the limit as $\varepsilon \to 0$. First notice that if $q - 1 > q/2$ then $a\varepsilon^{q-1}$ dominates, whereas if $q - 1 < q/2$ then $(1 - a)\varepsilon^{q/2}$ dominates. In the first case, $q - 1 > q/2 \Rightarrow q > 2$ so:

$$D_q = \lim_{\varepsilon \to 0} \frac{1}{\ln \varepsilon} \frac{1}{q-1} \ln((1-a)\varepsilon^{q/2}) = \frac{q}{2(q-1)}. \tag{8.49}$$

The other case is when $q - 1 < q/2 \Rightarrow q < 2$, and:

$$D_q = \lim_{\varepsilon \to 0} \frac{1}{\ln \varepsilon} \frac{1}{q-1} \ln(b\varepsilon^{q-1}) = 1. \tag{8.50}$$

This is the analytic form for the generalized dimension D_q. From this the singularity spectrum $f(\alpha)$ can also be obtained analytically as:

$$\alpha(q) = \frac{d}{dq}[(q-1)D_q] = \begin{cases} \frac{d}{dq}\left[\frac{q}{2}\right] = \frac{1}{2} & q > 2 \\ \frac{d}{dq}[q-1] = 1 & q < 2 \end{cases} \tag{8.51}$$

and:

$$f(\alpha) = q\alpha(q) - (q-1)D_q = \begin{cases} 0 & q > 2 \\ 1 & q < 2 \end{cases}. \tag{8.52}$$

The graph of $f(\alpha)$ versus α contains just two points: $(\frac{1}{2}, 0)$ and $(1, 1)$, a set of measure zero that scales as $\varepsilon^{1/2}$ and a set of measure one that scales as ε^1. It is also possible to invert this procedure, so that if $f(\alpha)$ is known we can obtain the generalized dimension D_q. For example, if $f(\alpha)$ consists of a single point: $(3, 3)$, then using Equation (8.47) implies $D_q = 3$ for all values of q.

Further generalizations of this approach are possible and Grassberger and Procaccia (Grassberger, 1983; Grassberger and Procaccia, 1983a, 1983b; Procaccia, 1985) and Eckmann and Procaccia (1986) have shown that how to define a range of scaling indices for the dynamical properties of chaotic systems and develop Legendre transforms (Jensen *et al.*, 1987).

Lyapunov exponents

In Section 3.4 we introduced the concept of the Lyapunov exponent as a quantitative measure of the dynamical instability of a system. Here we develop these ideas further, but first we briefly review the methods which can be used to calculate the Lyapunov exponents. The standard method of calculating Lyapunov exponents for dynamical systems is due to Benettin *et al.* (1976; 1978; 1980a; 1980b) and Shimada and Nagashima (1979). They linearize the equations of motion and study the time evolution of a set of orthogonal vectors. To avoid problems with rapidly growing vector lengths they periodically renormalize the vectors using a Gram–Schmidt procedure. This allows one vector to follow the fastest growing direction in phase space, and the second to follow the next fastest direction, while remaining orthogonal to the first vector, etc. The Lyapunov exponents are given by the average rates of growth of each of the vectors.

A variant of this method of calculating Lyapunov exponents developed by Hoover, Posch, and Morriss (Hoover and Posch, 1985, 1987; Morriss, 1988) uses Gauss' principle of least constraint to fix the length of each tangent vector, and to maintain the orthogonality of the set of tangent vectors. The two methods differ in the detail of the constraint forces. In Chapter 6 we used Gauss' principle to change from one ensemble to another (from Thévenin to Norton ensembles) and this application of Gauss' principle exactly parallels that application. In the Benettin method, one monitors the divergence of a pair of trajectories, with periodic rescaling. In the Gaussian scheme we monitor the *force* required to keep two trajectories a fixed distance apart.

Lyapunov dimension

The rate of exponential growth of a vector $\delta x(t)$ is given by the largest Lyapunov exponent. The rate of growth of a surface element $\delta \sigma(t) = \delta x_1(t) \wedge \delta x_2(t)$ is given by the sum of the two largest Lyapunov exponents. (Here $\wedge$ signifies the wedge product.) In general, the exponential rate of growth of a k-volume element is determined by the sum of the largest k Lyapunov exponents $\lambda_1 + \cdots + \lambda_k$. This sum may be positive, implying growth of the k-volume element, or negative implying shrinkage of the k-volume element. The number of Lyapunov exponents is equal the number of independent phase space directions. All of the previous characterizations of chaos that we have considered have led to a single scalar measure of the dimension of the attractor. From a knowledge of the complete spectrum of Lyapunov exponents, Kaplan and Yorke (1979) have conjectured that the effective dimension of an attractor is given by that value of k for which the k-volume element neither grows nor decays. This requires some generalization of the idea of a k-dimensional volume element, as the result is almost always noninteger. The Kaplan and Yorke conjecture is that the Lyapunov dimension can be calculated from:

$$D_L^{KY} = n + \frac{\sum_{i=1}^{n} \lambda_i}{|\lambda_{n+1}|} \tag{8.53}$$

where n is the largest integer for which $\sum_{i=1}^{n} \lambda_i > 0$.

Essentially, the Kaplan–Yorke conjecture corresponds to plotting the sum of Lyapunov exponents $\sum_{i=1}^{n} \lambda_i$ versus n, and the dimension is estimated by finding where the curve intercepts the n-axis by linear interpolation (Figure 8.7).

When the phase-space dimensional contraction is less than one, there is an *exact* limiting relationship between the Kaplan–York dimension and the linear transport coefficient L defined by Equation (6.20) (Evans *et al.*, 2000). This can be solved for L giving an exact expression for the linear transport coefficient and D_{KY}. This is like a Green–Kubo relation for linear transport coefficients:

$$\frac{D_{KY}(F_e)}{2dN - f} = 1 - \frac{LF_e^2}{\lambda_{\max} 2 d \rho k T} + O(F_e^4) \tag{8.54}$$

where d is the spatial dimension, f is number of constants of the motion, and the approximation $|\lambda_{\min}| \sim \lambda_{\max}$ has been used.

There is a second postulated relation between Lyapunov exponents and dimension due to Mori (1980):

$$D_L^M = m_0 + m^+ \left(1 + \frac{|\lambda^+|}{|\lambda^-|} \right), \tag{8.55}$$

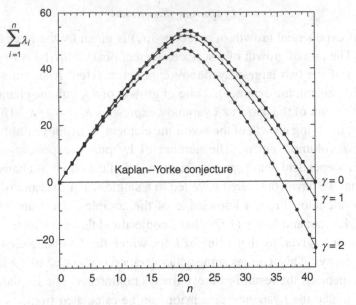

Figure 8.7 The sum of the largest n exponents, plotted as a function of n, for three-dimensional eight-particle Couette flow at three different shear rates $\gamma = 0$, 1, and 2. The Kaplan-Yorke dimension is the n-axis intercept

where m_0 and m^+ are the number of zero and positive exponents, respectively, and $\lambda^{\pm}$ is the mean value of the positive or negative exponents (depending upon the superscript). Farmer (1982) gives a modified form of the Mori dimension which is found to give integer dimensions for systems of an infinite number of degrees of freedom.

8.4 Chaos in planar Couette flow

We have seen in Section 8.2 that in apparently simple dynamical systems, such as the quadratic map and the Lorenz model, a single trajectory or sequence of iterates can have quite unusual behavior. In Section 8.3 we introduced a number of techniques to characterize the dynamical behavior of a system with a strange attractor. Here we will apply those techniques to the SLLOD algorithm for planar Couette flow introduced in Section 6.3. The first difficulty is that to determine the dimension of the attractor, the dimension of the initial phase space must be small enough to make the numerical calculations feasible. To calculate the static dimensions D_q we need to calculate the discrete probability distribution function in phase space. We do this by dividing phase space into boxes of size ε. The number of boxes needed varies as ε^{-6N}, for a $6N$ dimensional phase space. Such calculations quickly become impractical as the phase-space dimension increases. A typical statistical mechanical system has a phase space of $2dN$ dimensions (where d is the spatial dimension, typically 2 or 3) so clearly N must be small, but also N must be large enough to give nontrivial behavior. Surprisingly enough, both of these considerations can be satisfied with $d = 2$ and $N \geq 2$ (Ladd and Hoover, 1985; Morriss, 1985; 1987).

The SLLOD equations of motion for Gaussian isokinetic planar Couette flow are (Equation 6.39):

$$\dot{\mathbf{q}}_i = \frac{\mathbf{p}_i}{m} + \hat{\mathbf{x}}\gamma y_i,$$

$$\dot{\mathbf{p}}_i = \mathbf{F}_i - \hat{\mathbf{x}}\gamma p_{yi} - \alpha \mathbf{p}_i,$$

(8.56)

and the thermostatting multiplier is given by Equation (6.40):

$$\alpha = \frac{\sum(\mathbf{F}_i \cdot \mathbf{p}_i - \gamma p_{xi}p_{yi})}{\sum p_i^2}$$

(8.57)

where γ is the shear rate. The dissipative flux is the shear stress $-P_{xy}(\Gamma)V$, given by:

$$P_{xy}(\Gamma)V = \sum_{i=1}^{N}\left(\frac{p_{xi}p_{yi}}{m} + y_i F_{xi}\right).$$

(8.58)

The shear-rate-dependent viscosity $\eta(\gamma)$ is related to the shear stress in the usual way $\eta(\gamma)\gamma = -\langle P_{xy}\rangle$.

For two-dimensional, two-body, planar Couette flow, the phase space has eight degrees of freedom – $(x_1, y_1, x_2, y_2, p_{x1}, p_{y1}, p_{x2}, p_{y2})$. Using *sliding-brick* periodic boundary conditions with an origin where $\sum_i \mathbf{p}_i = 0$ and $\sum_i y_i = 0$, the centre of mass and total momentum are conserved. As the total kinetic energy is fixed, the accessible phase space is three-dimensional, and we choose the separation of the two particles (x_{12}, y_{12}) and the direction of the momentum of particle one (p_{x1}, p_{y1}) with respect to the x-axis, θ. For $N > 2$ the phase space reduces from $4N$ degrees of freedom to $4N - 5$. The sliding-brick periodic boundary conditions induce an explicit time dependence in the equations of motion (they become *non-autonomous*). As the position of image of particle j' is an explicit function of time, then the force $F_{ij'}$ on particle i is also an explicit function of time. The non-autonomous equations of motion do not have a zero Lyapunov exponent. However, the $4N - 5$ equations can be transformed into $4N - 1$ autonomous equations by the introduction of an extra. In this form there is a zero Lyapunov exponent (see Haken, 1983). The results reported below are for the $4N - 5$ nonautonomous equations of motion so there is no zero Lyapunov exponent.

Information dimension

The first evidence for the existence of a strange attractor in the phase space of the two-dimensional, two-body planar Couette flow system was obtained by Morriss (1987). He calculated the information dimension numerically for two-body planar Couette flow as a function of the shear rate, and found that it dropped steadily, from three, towards a value near two, before dropping dramatically at some critical value of the shear rate to approach one. These results are for a WCA potential at a temperature $T = 1$ and a density $\rho = 0.4$. The sudden change in dimension, from a little greater than two to near one, is associated with the onset of the *string-phase* for this system (see Section 6.4). As we have seen, the string phase is an artifact of the definition of the temperature with respect to an *assumed* streaming velocity profile, so it is likely that this decrease in dimensionality is a pathology of the model system, and not associated with the attractor which is found at intermediate shear rates.

Generalized dimensions

Morriss (1989b) has calculated the generalized dimension D_q and the spectrum of singularities $f(\alpha)$ for the steady-state phase-space distribution function of two-dimensional two-body planar Couette flow (the same system used for the information-dimension calculations). The maximum resolution of the distribution function was 3×2^6 bins in each of the three degrees of freedom. The results showed that, at equilibrium, the discrete probabilities $p_i(\varepsilon)$ scale with the dimension of the phase space, but away from equilibrium the $p_i(\varepsilon)$ scale with a range of indices, extending from the full phase-space dimension to a lower limit which is controlled by the value of the shear rate γ.

The results (Morriss and Kruss, unpublished) near $\gamma = 0$ depend significantly on the grid size. At higher values of γ (say $\gamma = 1$) the values of $f(\alpha)$ above the shoulder in Figure 8.8, are insensitive to grid size. However, the position of the shoulder does change. In the limit $q \to \infty$, the value of D_q and hence $\alpha_{\min}$ for which $f(\alpha) \to 0$, is controlled by the scaling of the most probable p_i in the histogram $p_{\max}$. It is easy to identify $p_{\max}$ and determine its scaling as an independent check on the value of $\alpha_{\min}$. Just as large positive values of q weight the most probable p_i most strongly, large negative values of q weight the least probable p_i most strongly. The least probable p_i corresponds to a bin with only one count (so $p_{\min} = 1/N$), so its accuracy is poor and the calculation of D_q for $q \leq 0.5$ is typically unreliable, and the estimate of the capacity dimension D_0 is also poor.

From Figure 8.8, the value of $f(\alpha)$ is the measure of the set of points on the attractor which scale as ε^α in the discrete distribution function $\{p_i\}$. This implies singularities of the form $|\Gamma - \Gamma_0|^{\alpha-3}$ in the underlying (continuous) phase-space distribution function $f(\Gamma, \gamma)$. At equilibrium, most p_is scale as ε^3, with a very

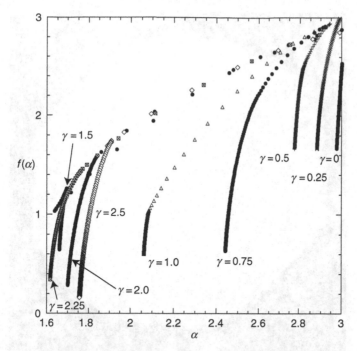

Figure 8.8 The spectrum of phase-space singularities for two-dimensional two-particle planar Couette flow at $T = 1$ and $\rho = 0.4$ as a function of shear rate γ. The function $f(\alpha)$ is the dimension of the set of points on the attractor that scale with exponent α

narrow spread of lower α values. Away from equilibrium the dimension of the set of p_is which scale as ε^3 drops with increasing γ, and the distribution of values of α increases downwards with the lower limit α_{min} depending upon γ. This distribution is monotonic with the appearance of a shoulder at some intermediate value of α.

From the full phase-space distribution function we can investigate the behavior of the various reduced distributions, for example we may consider the coordinate space distribution function $f_2(r, \varphi)$, or the distribution of the momentum angle θ. Each of these reduced distributions is obtained by integrating over the redundant coordinates or momenta. Perhaps the most interesting of these reduced distribution functions is the coordinate space distribution $f_2(x_{12}, y_{12})$, shown in Figure 8.9.

Lyapunov exponents

The complete set of Lyapunov exponents for two-dimensional isokinetic planar Couette flow have been calculated for two, four- and eight-particle systems by Morriss (1988; 1989a). For the two-particle system, values of the dimension have been calculated using both the Mori and Kaplan–Yorke conjectures (Equations (8.53) and (8.54)) along with the generalized dimension. Table 8.1

Figure 8.9 The coordinate space distribution function for the relative position coordinate (x_{12}, y_{12}) for $\gamma = 1.25$, $\rho = 0.4$ and $e = 0.25$. The centre of the plot is the excluded region. The grey scale is black for few counts and increasingly white for the more probable regions. Notice that there is a preference for collisions to occur in the top right-hand side and lower left-hand side, and fewer collisions near $x_{12} = 0$

Table 8.1 *Lyapunov exponents for two-body, two-dimensional isokinetic Couette flow and various estimates of the dimension calculated from them*

γ	λ_1	λ_2	λ_3	D_L^{KY}	D_L^M	D_0	D_1	D_2
0	2.047(2)	0.002(2)	−2.043(2)	3.003	3.00	2.90	2.98	2.98
0.25	2.063(3)	−0.046(2)	−2.1192(3)	2.952	2.90	2.91	2.98	2.98
0.5	1.995(3)	−0.187(4)	−2.242(3)	2.81	2.64	2.91	2.97	2.95
0.75	1.922(4)	−0.388(3)	−2.442(3)	2.62	2.36			
1.0	1.849(5)	−0.63(1)	−2.74(1)	2.445	2.10	2.89	2.90	2.67
1.25	1.807(4)	−0.873(5)	−3.17(1)	2.295	1.89			
1.5	1.800(5)	−1.121(2)	−4.12(5)	2.14	1.68	2.87	2.75	2.29
1.75	1.733(4)	−1.424(3)	−5.63(6)	2.058	1.49			
2.0	1.649(9)	−1.54(1)	−7.36(8)	2.015	1.37	2.80	2.65	2.20
2.25	1.575(3)	−1.60(1)	−9.25(9)	1.981	1.29			

contains a comparison of these the results. For both the Kaplan–Yorke and Mori dimensions, the Lyapunov dimension is found to be a decreasing function of the shear rate. This is consistent with the contraction of phase-space dimension that we have already seen from the numerical evaluated static dimensions D_q. It confirms that the nonequilibrium distribution function is a fractal attractor whose dimension is less than that of the equilibrium phase-space. At $\gamma = 0$, both methods of calculating the Lyapunov dimension agree. However, as soon as the shear rate changes from zero, differences appear. In the Kaplan–Yorke formula (Equation 8.53), the value of n is 2 from $\gamma = 0$, until the magnitude of λ_2 exceeds that of λ_1 (somewhere between $\gamma = 2$ and 2.25). This means that $2 < D_L^{KY} < 3$ in this range. For $\gamma > 2$, $1 < D_L^{KY} < 2$ as long as λ_1 remains positive. The value of λ_3 is irrelevant as soon as $|\lambda_2| > \lambda_1$. Then, as λ_1 becomes negative, the dimension is equal to zero. The Kaplan-Yorke formula can never give fractional values between zero and one. In the Mori formula the value of λ_3 always contributes to the dimension, and its large negative value tends to dominate the denominator, reducing D_L^M. The transition from $D_L^M > 2$ to $D_L^M < 2$ is somewhere between $\gamma = 1$ and 1.25. Indeed, the Mori dimension is systematically less than the Kaplan–Yorke dimension.

Of the two routes to the Lyapunov dimension, the Kaplan–Yorke method agrees best with the information-dimension results of Table 8.1, whereas the Mori method does not. In particular, the Kaplan–Yorke method and the information dimension both give a change from values greater than two, to values less than two at about $\gamma = 2.25$. There are a number of points to note about these numerical calculations. It can be shown that D_1 is a lower bound for D_0, however, the numerical results for D_0 and D_1 are inconsistent with this as $D_0 < D_1$. We have already remarked that the results for D_q are poor when $q < 0.5$. It has been argued that the capacity (Hausdorff) dimension and Kaplan–Yorke Lyapunov dimension should yield the same result, at least for homogeneous attractors. In this work, we find that D_L^{KY} is significantly lower than D_1 for all values of the shear rate. From a practical point of view, it is much easier to calculate the Lyapunov exponents of systems with many particles, than it is to extend the box-counting algorithms required for the information or generalized dimensions. There is now a large collection of numerical calculations of the full Lyapunov spectrum for many-particle planar Couette flow systems in both two and three dimensions.

Numerical conjugate pairing

In Figure 8.10 we show the Lyapunov spectra for an eight-particle system at $\rho = 0.4$ for a range of values of the shear rate (Morriss, 1989a; 2002). At equilibrium

Dynamical stability

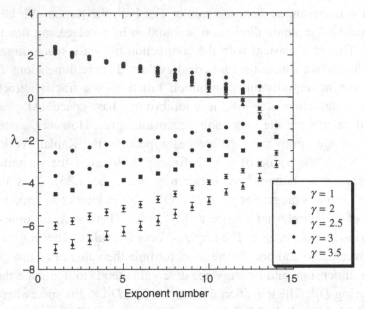

Figure 8.10 The Lyapunov spectrum for an eight-particle two-dimensional Couette flow system using the SLLOD algorithm at $T = 1$ and $\rho = 0.4$. The Lyapunov spectra shifts downwards with increasing shear rate with the largest exponent shifting least. The sum of the exponents is zero at equilibrium and becomes more negative with increasing strain

($\gamma = 0$) one exponent is zero, while the others occur in conjugate pairs $\{\lambda_{-i}, \lambda_i\}$, where $\lambda_{-i} = -\lambda_i$. Physically this symmetry is a consequence of the time reversibility of the equations of motion and the conservation of phase-space volume from the Liouville theorem. As the external field is increased systematic changes occur in the Lyapunov spectrum. The positive branch decreases in magnitude, with the smallest positive exponent decreasing most. The largest positive exponent seems almost independent of the external field. We expect that the strongest trajectory separation depends on the curvature of the particles (the higher the curvature the more defocusing is each collision), and upon the collision frequency (and hence the density). For this reason the insensitivity of the largest exponent is expected. The zero exponent (at $\gamma = 0$) becomes more negative with increasing field, as does the whole negative branch of the Lyapunov spectrum. The change in the negative branch is larger than the change in the positive branch. However, the change in the sum of each pair of conjugate exponents is the same, that is $\lambda_i + \lambda_{-i} = c$, where c is a constant independent of i and related directly to the dissipation. The change in the unpaired exponent is shear-rate dependent. This result has been called the *conjugate pairing* rule for Lyapunov exponents (Evans and Morriss, 1990b) and in the next section we will prove it for a particular class of systems.

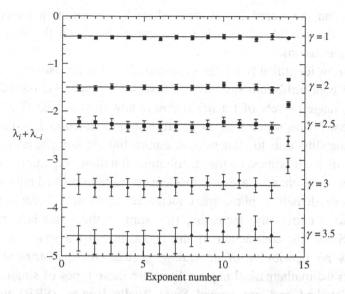

Figure 8.11 The sum of conjugate pairs of Lyapunov exponents for two-dimensional eight-particle planar Couette flow at $T = 1$ and $\rho = 0.4$. The sum of each pair is the same within error bars supporting the conjugate-pairing rule. The shift in the unpaired exponent #13 is different to the shift in a conjugate pair

In Figure 8.11 we show the sum of each conjugate pair of exponents for this system. In summary, the results confirm the dimensional contraction observed previously in two-body, two-dimensional planar Couette flow simulations. The initial phase-space dimension of $D = 2dN - 2d - 1$, contracts with increasing shear rate, and the distribution function is only nonzero on a fractal attractor of dimension less than $2dN - 2d - 1$. Although the results for these systems differ in detail from the generalized-dimension results, the observation of significant dimensional contraction is universal.

If we consider the volume element V_{2dN} where $2dN$ is the phase space dimension of the initial system (d is the spatial dimension and N is the number of particles), then the phase-space compression factor gives the rate of change of phase-space volume (see Equation 3.89), so that the average of the divergence is equal to the sum of the Lyapunov exponents. A careful calculation of the divergence for the SLLOD algorithm, taking into account, the precise number of degrees of freedom gives:

$$\sum_{i-1}^{2dN-2d-1} \lambda_i = -(dN-d-1)\langle\alpha\rangle + \frac{\gamma\langle P_{xy}^K\rangle V}{(dN-d-1)kT}, \qquad (8.59)$$

where $\langle P_{xy}^K\rangle$ is the kinetic contribution to the shear stress. The term involving $\langle P_{xy}^K\rangle$ is order one, whereas the first term is order N, so for many particle systems the second term can be ignored. For systems where $N = 8$ both terms need to be considered and this provides a valuable consistency check on the accuracy of the numerical calculations.

We have now identified two effects associated with the phase-space distribution functions of nonequilibrium systems: the first is dimensional contraction; and the second is a range of sets of fractional power law singularities. The two results are consistent in the sense that, as the distribution function is normalized, the loss of probability due to dimensional contraction is compensated for by the appearance of singularities in the distribution function. Equation (8.4) implies that as a comoving phase-space volume element contains a fixed number of trajectories, the local density in phase space increases indefinitely, because the Lagrangian volume is constantly decreasing (the sum of the Lyapunov exponents is negative). Since the contraction of the accessible phase space is *continuous*, there seems no possibility of generating a steady-state distribution function. However, in the mathematical physics literature these types of singular measures have been studied and are termed Sinai–Ruelle–Bowen (SRB) measures. In Section 8.6 we will examine a series of approximate methods for calculating averages using SRB measures.

8.5 Conjugate pairing of Lyapunov exponents

It is possible to prove the conjugate pairing rule exactly (Dettmann and Morriss, 1996b) for a system with an isokinetic thermostat and forces derivable from a potential Φ for all values of N. There is an important difference between the present result and past statements of the conjugate-pairing rule (Evans *et al.*, 1990). Here we explicitly single out two trivial exponents (equal to zero), which do not pair. These are associated with conservation of kinetic energy and time translation symmetry. These exponents sum to zero so should not be included with the other pairs, which sum to a different constant. Excluding these directions, we define the time evolution in a reduced $(6N - 2)$-dimensional space, in which the pairing is exact.

The isokinetic equations of motion in an external field $\mathbf{F}_{ext}$, where the force is the sum of internal and external parts $\mathbf{F} = \mathbf{F}_{int} + \mathbf{F}_{ext} = -\boldsymbol{\nabla}\Phi$ ($\mathbf{F}_{int}$ is the internal interparticle forces) takes the form:

$$\dot{\mathbf{q}} = \mathbf{p},$$

$$\dot{\mathbf{p}} = -\boldsymbol{\nabla}\Phi - \alpha\mathbf{p} = \mathbf{F} - \alpha\mathbf{p},$$

$$\alpha = -\frac{\mathbf{p} \cdot \boldsymbol{\nabla}\Phi}{\mathbf{p} \cdot \mathbf{p}},$$

$$(8.60)$$

where we have transformed the variables to remove the mass. Here $\mathbf{q} \equiv (\mathbf{q}_1, \ldots, \mathbf{q}_N)$ and $\mathbf{p} \equiv (\mathbf{p}_1, \ldots, \mathbf{p}_N)$ are $3N$-dimensional vectors and Φ is a scalar potential that generates the total force $\mathbf{F}$ (also a $3N$-dimensional vector).

There are two time-dependent matrices which can be used to describe the evolution of a linear perturbation $\delta\Gamma$ in tangent space, and they both depend on the initial phase-space point Γ. These are the infinitesimal and finite evolution matrices, T and L, respectively introduced in Section 7.7, defined by:

$$\dot{\delta\Gamma}(t) = T(t)\delta\Gamma(t) \quad \text{and} \quad \delta\Gamma(t) = L(t)\delta\Gamma(0). \tag{8.61}$$

The matrix T is usually obtained by differentiating the equations of motion, however, we will evaluate it in a restricted subspace of the tangent space, which is slightly more complicated. L can be obtained from T as the solution of $\dot{L}(t) = T(t)L(t)$ with the initial condition $L(0) = I$. The Lyapunov exponents are defined as the logarithms of the eigenvalues of the matrix Λ (Eckmann and Ruelle, 1985), where:

$$\Lambda = \lim_{t \to \infty} \left(L^T(t)L(t) \right)^{1/2t}. \tag{8.62}$$

We introduce a set of comoving basis vectors, which span the tangent space containing $\delta\Gamma$, and rotate with the motion of the trajectory so as to remain perpendicular to the direction of increasing kinetic energy (Fermi–Walker transport [Misner *et al.*, 1970]). This means that the finite time eigenvalues may be different to those obtained with fixed basis vectors, but in the long time limit the results are the same. We introduce $6N - 2$ orthonormal basis vectors, none of which are exactly along the flow, and demand that a perturbation $\delta\Gamma$ be in the space spanned by these vectors. This effectively means that we are taking a Poincaré section, and considering the perturbed point to be the one at which the perturbed trajectory intersects with the $(6N - 2)$-dimensional space spanned by the vectors. In general, the time elapsed along the perturbed trajectory t', runs at an infinitesimally different rate to t. We scale the time so that $\mathbf{p} \cdot \mathbf{p} = 1$, and choose $\mathbf{e}_0 = \mathbf{p}$ as one of the unit vectors. At some initial time, arbitrarily choose $3N - 1$ unit vectors $\mathbf{e}_i$, which, together with $\mathbf{e}_0$, form an orthonormal set in $3N$-space. This set of vectors is used in *both* position and momentum space to form the required basis. The separation of phase space into position and momentum space, while retaining the symplectic structure is what makes this proof possible in the isokinetic case. The perturbations are taken from the $(6N - 2)$-dimensional subspace defined by the two sets of $\mathbf{e}_i$. This ensures that there are no perturbations in the direction of increasing kinetic energy and none directly along the flow (which contains a component of $\mathbf{e}_0$ in position space). Now, the equations of motion in

$6N$-dimensional space may be written as:

$$\frac{d\mathbf{q}}{dt} = \mathbf{p} \qquad \frac{d\mathbf{p}}{dt} = \dot{\mathbf{e}}_0 = \sum_{i=1}^{3N-1} \mathbf{f} \cdot \mathbf{e}_i \mathbf{e}_i, \tag{8.63}$$

where $\mathbf{f} = -\nabla\Phi$. If we choose the unit vectors to have equations of motion $\dot{\mathbf{e}}_i = -\mathbf{f} \cdot \mathbf{e}_i \mathbf{e}_0$ (see Figure 8.12), then the basis remains orthonormal as the time derivatives of $\mathbf{e}_\mu \cdot \mathbf{e}_\nu$ are zero for all $\mu, \nu \in \{0, 3N-1\}$. The perturbed trajectory $(\mathbf{q}', \mathbf{p}')$ is given by:

$$\mathbf{q}' = \mathbf{q} + \sum_{i=1}^{3N-1} \delta q_i \mathbf{e}_i,$$
$$\mathbf{p}' = \mathbf{p} + \sum_{i=1}^{3N-1} \delta p_i \mathbf{e}_i, \tag{8.64}$$

with equations of motion:

$$\frac{d\mathbf{q}'}{dt'} = \mathbf{p}',$$
$$\frac{d\mathbf{p}'}{dt'} = \sum_{i=1}^{3N-1} \mathbf{f}' \cdot \mathbf{e}'_i \mathbf{e}'_i. \tag{8.65}$$

Here, $\mathbf{f}'$ is the value of $\mathbf{f}$ at the perturbed coordinates:

$$\mathbf{f}' = \mathbf{f} + \sum_{i=1}^{3N-1} \delta q_i \nabla_i \mathbf{f}, \tag{8.66}$$

and the $\mathbf{e}'_i$ are new (arbitrary) unit vectors perpendicular to $\mathbf{p}'$. We choose the transported orthonormal set $\mathbf{e}'_i$ as in Figure 8.12. Then:

$$\mathbf{e}'_0 = \mathbf{p}', \qquad \mathbf{e}'_i = \mathbf{e}_i - \delta p_i \mathbf{e}_0. \tag{8.67}$$

Taking the first equation in (8.65) and substituting for $\mathbf{q}'$ and $\mathbf{p}'$ using (8.64) gives:

$$\frac{d\mathbf{q}'}{dt'} = \frac{dt}{dt'}\left(\frac{d\mathbf{q}}{dt} + \sum (\delta\dot{q}_i \mathbf{e}_i + \delta q_i \dot{\mathbf{e}}_i)\right) = \mathbf{p}' = \mathbf{p} + \sum_{i=1}^{3N-1} \delta p_i \mathbf{e}_i. \tag{8.68}$$

Here we have used the chain rule to change from derivatives with respect to t', to derivatives with respect to t. From Equation (8.64) and Figure 8.11, $\mathbf{e}_0 = \mathbf{p}$ and $\dot{\mathbf{e}}_i = -\mathbf{f} \cdot \mathbf{e}_i \mathbf{e}_0$, so this equation can be written as:

$$\frac{dt}{dt'}\left(\mathbf{e}_0 + \sum (\delta\dot{q}_i \mathbf{e}_i - \delta q_i \mathbf{f} \cdot \mathbf{e}_i \mathbf{e}_0)\right) = \mathbf{e}_0 + \sum_{i=1}^{3N-1} \delta p_i \mathbf{e}_i, \tag{8.69}$$

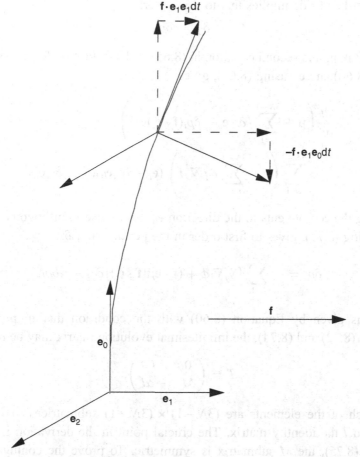

Figure 8.12 The basis vectors for one particle in three dimensions are Fermi–Walker transported along the trajectory

or equivalently:

$$\sum_{i=1}^{3N-1} \left(\frac{dt}{dt'} \delta \dot{q}_i - \delta p_i \right) \mathbf{e}_i = \mathbf{e}_0 \left\{ 1 - \left(1 - \sum_i \delta q_i \mathbf{f} \cdot \mathbf{e}_i \right) \frac{dt}{dt'} \right\}. \tag{8.70}$$

As the basis vector $\mathbf{e}_0$ is orthogonal to all the other basis vectors $\mathbf{e}_i$, both sides of this equation must be zero. This is satisfied only if the coefficients of $\mathbf{e}_i$ and $\mathbf{e}_0$ are both equal to zero. Thus the right-hand side implies that to first order in the perturbation δ:

$$\frac{dt}{dt'} = 1 + \sum_{i=1}^{3N-1} \mathbf{f} \cdot \mathbf{e}_i \delta q_i, \tag{8.71}$$

and the left hand side implies that to first order:

$$\delta \dot{q}_i = \delta p_i. \tag{8.72}$$

Similarly, taking the second equation in (8.65) and substituting for $\mathbf{p}'$ using (8.64), $\mathbf{f}'$ using (8.66) and $\mathbf{e}'_i$ using (8.67), gives:

$$\frac{dt}{dt'}\left(\dot{\mathbf{p}} + \sum_{i=1}^{3N-1} \left(\delta \dot{p}_i \mathbf{e}_i - \delta p_i (\mathbf{f} \cdot \mathbf{e}_i) \mathbf{e}_0 \right) \right)$$

$$= \sum_{i=1}^{3N-1} \left(\mathbf{f} + \sum_{i=1}^{3N-1} \delta q_j \boldsymbol{\nabla}_J \mathbf{f} \right) \cdot (\mathbf{e}_i - \delta p_i \mathbf{e}_0)(\mathbf{e}_i - \delta p_i \mathbf{e}_0). \tag{8.73}$$

Separating the components in the direction $\mathbf{e}_0$ from those in orthogonal directions $\mathbf{e}_i$, and using (8.71), gives to first order in the perturbation δ:

$$\delta \dot{p}_i = -\sum_{j=1}^{3N-1} \left(\nabla_j \nabla_i \phi + (\mathbf{f} \cdot \mathbf{e}_j)(\mathbf{f} \cdot \mathbf{e}_i) \right) \delta q_j - \alpha \delta p_i, \tag{8.74}$$

where α is given by Equation (8.60) with the condition that $\mathbf{p} \cdot \mathbf{p} = 1$. From Equations (8.72) and (8.74), the infinitesimal evolution matrix may be read off as:

$$T = \begin{pmatrix} 0 & I \\ M & -\alpha I \end{pmatrix}, \tag{8.75}$$

where each of the elements are $(3N-1) \times (3N-1)$ submatrices. 0 is the zero matrix and I the identity matrix. The crucial point in the derivation is that from Equation (8.75), the M submatrix is symmetric. To prove the conjugate-pairing rule, we need a generalization of the symplectic eigenvalue theorem (Abraham and Marsden, 1978) which we outline below.

Definition 1

The T matrix is infinitesimally α-symplectic if:

$$T^{\mathrm{T}}J + JT = -\alpha J, \tag{8.76}$$

where:

$$J = \begin{pmatrix} 0 & I \\ -I & 0 \end{pmatrix}. \tag{8.77}$$

Definition 2

The L matrix is globally μ-symplectic if:

$$\mu L^{\mathrm{T}} J L = J. \tag{8.78}$$

Let $K(t) = L^T(t)JL(t)$ then using $\dot{L}(t) = T(t)L(t)$ and Equation (8.75), it follows that:

$$\dot{K}(t) = \dot{L}^T JL + L^T J\dot{L} = L^T T^T JL + L^T JTL = L^T(T^T J + JT)L$$
$$= -\alpha K(t). \tag{8.79}$$

Clearly, an infinitesimally symplectic T matrix generates an L matrix that is globally symplectic. Solving for K, given the initial condition $K(0) = J$, we find that $K(t) = \exp\left[-\int_0^t \alpha(s)\mathrm{d}s\right]J$. This is essentially an application of the techniques of Section 7.7. So μ is given by:

$$\mu = \exp\left[\int_0^t \alpha(s)\mathrm{d}s\right]. \tag{8.80}$$

Theorem: a-symplectic eigenvalue theorem

If matrix M satisfies $aM^T JM = J$ for some finite a, and if χ is an eigenvalue of M, then $(\chi a)^{-1}$ is also an eigenvalue of M.

Proof

We note that, if χ is an eigenvalue of M, then $\det(M - \chi I) = 0$, and M^T has the same eigenvalues as M. The determinant of a product is equal to the product of the determinants and $\det(J) = 1$. Therefore:

$$\det(M - \chi I) = 0 \Rightarrow \det(M^T JM - \chi M^T J) = 0$$
$$\Rightarrow \det(a^{-1}J - \chi M^T J) = 0$$
$$\Rightarrow \det(((a\chi)^{-1}I - M^T)\chi J) = 0 \tag{8.81}$$
$$\Rightarrow \det((a\chi)^{-1}I - M^T) = 0$$
$$\Rightarrow \det(M - (a\chi)^{-1}I) = 0. \quad \text{QED.}$$

Applying this result to $M = L^T L$, where $a = \mu^2$, we find that, for each eigenvalue $\chi > 1$, there is another $(\mu^2\chi)^{-1}$. The sum of the logarithms of this pair of eigenvalues of $L^T L$ is:

$$\ln\chi + \ln(\mu^2\chi)^{-1} = -2\int_0^t \alpha(s)\mathrm{d}s, \tag{8.82}$$

which is clearly independent of which pair of eigenvalues we chose. The eigenvalues of $L^T L$ are the eigenvalues of Λ^{2t}, so that two of the logarithms of the eigenvalues of Λ are λ_+ and λ_-, where:

$$\lambda_+ + \lambda_- = -\lim_{t\to\infty}\frac{1}{t}\int_0^t \alpha(s)\mathrm{d}s = -\langle\alpha\rangle_t. \tag{8.83}$$

The finite t result applies to any segment of trajectory, no matter how small, if the comoving basis set is used. The $t \to \infty$ result is that any pair of Lyapunov exponents (except the trivial zeros) sum to $-\langle\alpha\rangle_t$. An important corollary is that, if there is an invariance in the equations of motion leading to a zero exponent, the conjugate exponent is not zero as in the Hamiltonian case, but $-\langle\alpha\rangle_t$. This proof is valid for any number of particles moving in a potential which may contain both external fields and interactions between the particles. There is still no proof of conjugate pairing for SLLOD dynamics, where conjugate pairing was first observed numerically, and has recently been confirmed within error bars (Morriss, 2002). Indeed, now there is a suggestion that conjugate pairing does not hold exactly for SLLOD (Panja and van Zon, 2002; Frascoli *et al.*, 2006). This point is not yet resolved, although it is likely that the SLLOD algorithm does not satisfy the strong version of conjugate pairing proved above, however, it may be satisfied in the thermodynamic limit (Taniguchi and Morriss, 2002).

8.6 Periodic orbit measures

The distribution function or *density* $\rho(x)$ is the fundamental statistical quantity that gives the probability of observing the system in some particular state x. Here we think of the possible states of the system as being specified by a continuous variable in phase space x (in general, x may be a vector). In the mathematical literature it is more common to find the *measure* $d\mu(x) = \rho(x)dx$ as more fundamental than the density (essentially the measure always exists, but the density can be singular). For the measure we think of the phase space M as being partitioned into regions M_i (this can be considered a coarse graining, in the Gibbsian sense). The *characteristic* function χ_i defined on the coarse graining is very useful for manipulating measures. If x is in region M_i, then $\chi_i = 1$, and otherwise χ_i is zero, so:

$$\chi_i(x) = \begin{cases} 1 & \text{if } x \in M_i \\ 0 & \text{otherwise} \end{cases}. \tag{8.84}$$

Thus the measure of region M_i is $\Delta\mu_i$, and it is obtained from:

$$\Delta\mu_i = \int_M d\mu(x)\chi_i(x) = \int_{M_i} d\mu(x). \tag{8.85}$$

We require the measure be normalized, so adding up the measure of all partitions must give one, and:

$$\sum_i^n \Delta\mu_i = 1 \quad \text{or} \quad \text{equivalently} \quad \int_M d\mu(x) = 1. \tag{8.86}$$

A stationary or *invariant* measure is unchanged by the time evolution of the system, so:

$$\Delta\mu_i(t) = \Delta\mu_i(0), \tag{8.87}$$

for all times and all partitions of the phase space.

We can imagine sprinkling the phase space M with some initial (smooth) distribution of phase points according to some initial density and following the time evolution f^τ of this ensemble of initial conditions. This construction defines the *natural* measure of the system:

$$\bar{\rho}_{x_0}(y) = \lim_{t \to \infty} \frac{1}{t} \int_0^t d\tau \delta(y - f^\tau(x_0)). \tag{8.88}$$

From the knowledge of the Lyapunov spectrum, or more specifically the knowledge of the local stretching rates or Lyapunov numbers for general points, we can construct the natural measure of the system. Here we follow the argument given in Grebogi *et al.* (1988) and Ott (2002). Once the measure is known, it is straightforward to calculate the averages of observables. This is an extension of the idea of classical ensembles to nonequilibrium states, but now the basis for the measure is modern dynamical-systems theory so it is no longer restricted to equilibrium. To illustrate the ideas we consider a two-dimensional hyperbolic invertible map $F(x) = x'$ (here x and x' are vectors).

Hyperbolicity implies that for an arbitrary point x in phase space there exist stable and unstable manifolds (Figure 8.13) denoted by $W^s(x)$ and $W^u(x)$. The *stable manifold* of x is the set of points y such that $\|F^n(x) - F^n(y)\| \to 0$ as $n \to \infty$. That is, the set of points that map together under forward iteration of the map. The *unstable manifold* is the set of points z such that $\|F^n(x) - F^n(z)\| \to 0$ as $n \to -\infty$. That is, the set of points that map together under backward iteration of the map. The stable manifold is generated by forward iteration, the unstable by backward iteration. Under time reversal, the unstable manifold becomes a stable manifold and vice versa.

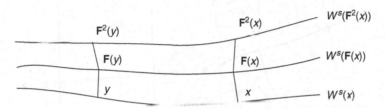

Figure 8.13 An illustration of the evolution of the stable manifold of a generic point x under successive applications of the mapping. Notice that the stable manifold of a generic point moves in space

For a hyperbolic attractor:

(1) Stable and unstable manifolds exist at each point on the attractor. The stable and unstable manifolds are in different directions. The dimensions of the stable and unstable manifolds, d_s and d_u, are the same for all points x on the attractor, and $d_s + d_u = d$ where d is the dimension of the space. This ensures that the space consists of either expanding or contracting directions. There are no neutral directions (or centre manifolds) and hence no zero Lyapunov exponents.

(2) If $DF(x)$ is the Jacobian matrix consisting of the partial derivatives of $F(x)$ evaluated at x, then there exists a constant $K > 1$ such that for all points on the attractor, if a vector v is chosen tangent to the unstable manifold, then $\|DF(x)v\| \geq K\|v\|$ and if v is chosen tangent to the stable manifold, then $\|DF(x)v\| \leq \|v\|/K$. This ensures that nearby points on the same stable manifold approach each other at least as fast as $\exp[-Kn]$.

We imagine that a two-dimensional space will locally consist of approximately parallel unstable manifolds and approximately parallel stable manifolds. Thus we construct a partition of the space into cells C_i which have unstable manifolds as two of the edges and stable manifolds as the other two edges (see Figure 8.14). If the cells are very small, the curvature of the boundaries will be slight, and we can regard them as parallelograms. We now construct a measure on these cells. Consider a cell C_k and a large number of initial conditions sprinkled within the cell according to the natural probability measure on the attractor. In general, the measure on the attractor will be fractal, but the variation of the measure in the direction of the unstable manifold will be stretched and smoothed by the action of the map. Similarly the variation of the measure in the direction of the stable manifold will be reinforced, as the compression forces the same variations on a smaller length scale. Imagine that we iterate each of the initial conditions n times. After n iterates,

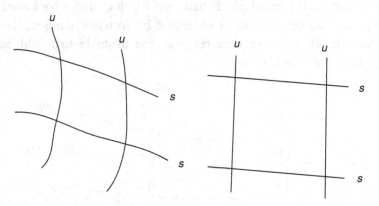

Figure 8.14 The cells C_i constructed from the intersections of stable and unstable manifolds of generic points. On the right is a very small cell which can be considered a parallelogram

a small fraction of the initial conditions may return to the cell C_k. Since we assume the attractor to be ergodic and mixing, this fraction is asymptotically equal to the natural measure of the attractor in the cell, $\mu(C_k)$.

There are periodic points of the n times iterated map $F^n(x_{jn}) = x_{jn}$ and these have special properties (here x_{jn} is the *j*th periodic point of the n-times iterated map). The stable and unstable manifolds of a periodic point are invariant sets and do not move in the space, as do the manifolds of generic points.

In Figure 8.15 the cell C_k is mapped forward in time to the narrow rectangle. The unstable manifold is the long side of the rectangle and the measure is smooth in this direction (Bowen, 1978). Therefore the fraction of the measure that returns to C_k is the ratio of the length of C_k to the length of the rectangle (or the ratio of the shaded length to the total length of the rectangle). This ratio is Λ_1^{-1} where $\Lambda_1 = \exp[\lambda_1 n]$ is the stretching factor and λ_1 is the Lyapunov exponent for the periodic point. For the two-dimensional example we have considered, there is only one stretching factor, but for higher-dimensional hyperbolic systems the same construction is needed in each expanding direction and then the ratio involves the product of all stretching factors.

Figure 8.15 also shows both the forward and backward images of C_k. If the forward and backward images intersect then the intersection must contain a periodic point. Since, for $n \to \infty$, the fraction of initial conditions starting in C_k which return to it is $\mu(C_k)$, we have that the measure of an area S is:

$$\mu(S) = \lim_{n \to \infty} \left(\sum_{\text{fixed points} \in S} \Lambda_1^{-1}(x_{jn}, n) \right). \tag{8.89}$$

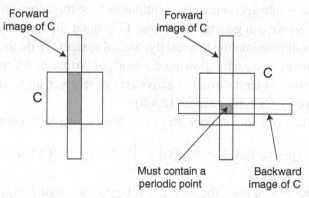

Forward image of C   Forward image of C   C

C

Must contain a periodic point   Backward image of C

Figure 8.15 If C is the initial cell, then the forward image of C is the thin rectangle. Only the shaded area of the forward image remains in the cell C. The intersection of the forward and backward images of C must contain a periodic point of the map

Also note that, as n gets larger, Λ_{1j}^{-1} and Λ_{2j} get exponentially smaller and the number of fixed points in a cell of fixed size C_k grows exponentially. Since we imagine that we can make the partition into cells as small as we wish, (with reasonably smooth boundaries), by a covering of cells, the result above follows. For hyperbolic systems, a partition into cells can always be made to ensure these properties and these are called Markov partitions (Bowen, 1970).

The theory of unstable periodic orbits as a means of deriving stationary measures for axiom-A systems is well developed (Ruelle, 1978; Grebogi *et al.*, 1988; Artuso *et al.*, 1990a). Orbital probability measures, weighted by the expanding eigenvalues of the associated stability matrices converge weakly to the SRB measure (Parry, 1986). The measure constructed in this way can be used to calculate the average properties of the system (Vance, 1992). If $B(x)$ is an arbitrary function of the phase point x then the average of B is:

$$\langle B \rangle = \lim_{n \to \infty} \frac{\displaystyle\sum_{i \in \{\text{fixed points of } F^n\}} \Lambda_{1i}^{-1} \sum_{j=1}^{n} B(x_{ji})}{\displaystyle\sum_{i \in \{\text{fixed points of } F^n\}} n \Lambda_{1i}^{-1}}, \tag{8.90}$$

where for each value of n, $\{x_{ji}\}$ are the points of the n-cycle, and Λ_{1i} is the product of the expanding eigenvalues of the stability matrix.

Time evolution of densities

The result we have just obtained is for maps, and our interest is in dynamical systems that evolve in time. We may approximate a dynamical system as a map by choosing a Poincaré surface of a section and considering the iterates of the map as successive intersections of the continuous time trajectory with the Poincaré surface. Likewise we can generalize the result obtained for maps by assuming that the map is some approximation to a real dynamical system. To do this we will take a more mathematical approach following the work of Artuso and Cristadoro (2004). There is an extensive literature on this active area of research and a good entry point is the web-book by Cvitanovic *et al.* (2005).

In the language of Liouville operators $i\mathscr{L}$, the change in the density is:

$$\rho(x, t) = \exp[-i\mathscr{L}t]\rho(x) = \mathscr{L}[\rho(x)] = \int_M dx_0 \delta(x - f^t(x_0))\rho(x_0, 0). \tag{8.91}$$

In dynamical systems theory, the Perron–Frobenius operator L plays a central role in the time evolution of densities. The invariant density $\rho_I(x)$ is a solution of this equation. If we write Equation (8.91) as an eigenvalue equation $\lambda\rho(x) = L[\rho(x)]$, then the invariant density $\rho_I(x)$ is the solution with eigenvalue equal to one, that is $\lambda = 1$. The Perron–Frobenius operator is a linear operator on a general class of functions – the distributions. The invariant density is a fixed-point function in that function space. The kernel of the Perron–Frobenius is singular, so sophisticated techniques are required to determine the spectral properties of the operator (Baladi, 2000).

If x is a scalar variable, then the integral of the delta function for an arbitrary function of x, $G(x)$ is given by:

$$\int dx \delta(G(x)) = \int dx \frac{\delta(s - s_0)}{G'(s)} = \sum_{\{x_i : G(x_i) = 0\}} \frac{1}{|G'(x_i)|}, \qquad (8.92)$$

where the sum is over all x_is that are zeros of $G(x)$. As $G(x) = x - f^t(x)$, the x_is are fixed points of $f^t(x)$, or solutions of $x_i = f^t(x_i)$. If x is a vector variable then the derivative of $G(x)$ is $G' \Rightarrow \frac{\partial}{\partial x}(x - f^t(x)) = 1 - \partial f^t / \partial x = 1 - \mathbf{J}^t(x)$ so that the time evolution of the density is:

$$\rho(x, t) = \exp[-i\mathscr{L}t]\rho(x) = \sum_{x_0 = f^{-t}(x)} \frac{\rho(x_0)}{|1 - df^t(x_0)/dx_0|}$$

$$= \sum_{x_0 = f^{-t}(x)} \frac{\rho(x_0)}{|1 - \mathbf{J}^t(x_0)|}. \qquad (8.93)$$

We can define a generalized transfer operator where the singular kernel is weighted by a factor $\exp[\beta(x - f^t(x_0)]$ as:

$$[L_\beta \rho](x) = \int_M dx_0 \exp[\beta(x - f^t(x_0)]\delta(x - f^t(x_0))\rho(x_0). \qquad (8.94)$$

An arbitrary function of the phase alone is a *phase variable* and we choose $b(x)$ as a generic function. This function or system *observable* connects each phase point x with a number (or set of numbers). In physical applications the function must be integrable and usually smooth, so that a small variation in x leads to a small variation in b. The phase average of the observable b with respect to a normalized measure $d\mu$ is given by:

$$\langle b \rangle_\mu = \int_M d\mu(x)b(x). \qquad (8.95)$$

Inserting the natural measure we obtain the time average of an observable beginning at the point x_0.

$$\overline{b(x_0)} = \lim_{t \to \infty} \frac{1}{t} \int_0^t d\tau b(f^\tau(x_0)). \qquad (8.96)$$

Moreover, if the dynamical system is ergodic, the time average beginning from almost any initial point x_0 tends to the phase space average:

$$\lim_{t \to \infty} \frac{1}{t} \int_0^t d\tau \, b(f^\tau(x_0)) = \langle b \rangle_\rho, \qquad (8.97)$$

so the left-hand side of Equation (8.97) is independent of x_0. By *almost any trajectory*, we mean all apart from a set of measure zero (for example, the average calculated beginning on a periodic point can be different). For a periodic orbit the time average of b is quite simple:

$$B_p = b_p T_p = \int_0^{T_p} b(f^\tau(x_p)) d\tau, \qquad (8.98)$$

where p is a periodic orbit beginning at x_p, T_p is the period, and B_p is the integral of the observable along the orbit. The contribution from a single cycle of the orbit is then $b_p = B_p / T_p$.

The Liouville operator (or Perron–Frobenius operator) is bounded in the sense that no value grows faster than exponentially, so for $M > 0$ and $\beta \geq 0$, $\| \exp[-i\mathcal{L}t] \| \leq M \exp[\beta t]$ for all $t \geq 0$. It then follows that we can define the resolvent of the Liouville operator:

$$\int_0^\infty dt \, \exp[-st] \exp[-i\mathcal{L}t] = \frac{1}{s + i\mathcal{L}}. \qquad (8.99)$$

For $\mathrm{Re}\, s > \beta$, the resolvent is bounded by $M / (s + \beta)$.

To construct the dynamical average, we consider the evolution of the map f of the phase variable x beginning at some initial value. The flow moves the phase point x along some path in phase space so we can consider the integral $B(t)$ of the observable $b(x)$ along this path (for a mapping the integral is replaced by a sum):

$$B(t) = \int_0^t b(f^\tau(x)) d\tau. \qquad (8.100)$$

Then the average is given by:

$$\langle b \rangle = \lim_{t \to \infty} \int dx \, \frac{1}{t} \int_0^t d\tau \, b(f^\tau(x)) = \lim_{t \to \infty} \int dx \, \frac{B(t)}{t}. \qquad (8.101)$$

It is more convenient to consider a slightly different object, a generating function whose moments provide the averages that we need:

$$\langle e^{\beta B(t)} \rangle = \int_M dx \, \exp[\beta B(x(t))]. \qquad (8.102)$$

As $t \to \infty$, then $B(t) \to \bar{B}t$ and $\langle \exp[\beta B(t)] \rangle \sim \langle \exp[\beta \bar{B}t] \rangle \propto \exp[ts(\beta)]$, so that:

$$s(\beta) = \lim_{t \to \infty} \frac{1}{t} \ln \langle \exp[\beta B(t)] \rangle. \tag{8.103}$$

Now the reason for considering $s(\beta)$ becomes much clearer, as:

$$\left. \frac{\partial s}{\partial \beta} \right|_{\beta=0} = \lim_{\beta \to 0} \lim_{t \to \infty} \frac{1}{t} \frac{\langle B(t) \exp[\beta B(t)] \rangle}{\langle \exp[\beta B(t)] \rangle} = \lim_{t \to \infty} \frac{\langle B(t) \rangle}{t} = \langle b \rangle. \tag{8.104}$$

The derivative of $s(\beta)$ with respect to β is the average $\langle b \rangle$ that we need to calculate. We can look at the fluctuations by going to the second derivative of $s(\beta)$, but then we need to know higher-order contributions to $B(t) \sim \langle b \rangle t + \cdots$

Evolution operators

The shift from studying $\langle b \rangle$ to studying $\langle \exp[\beta B(t)] \rangle$ is the key to what follows. Let:

$$\langle \exp[\beta B(t)] \rangle = \frac{1}{|M|} \int_M dx \int_M dy \delta(y - f^t(x)) \exp[\beta B(t, x)] \tag{8.105}$$

where we have inserted the identity into Equation (8.102). By the identity we mean the following:

$$1 = \int_M dy \delta(y - f^t(x)). \tag{8.106}$$

The treatment of mappings is a little more simple than systems with continuous time evolution (flows) (Artuso and Cristadoro, 2004), so we consider a map $f(x)$. We also assume that the map satisfies two conditions $f(-x) = -f(x)$ and $f(x + n) = n + f(x)$, so that the problem reduces to a map on the unit interval $\hat{f}(\theta) = f(\theta)|_{\mathrm{mod}1}$. The evolution can then be separated into an integer part and a fractional part where:

$$N_{n+1} = N_n + \sigma(\theta_n),$$
$$\theta_{n+1} = \hat{f}(\theta_n), \tag{8.107}$$

where θ_n is the fractional part and N_n is the integer part. The transfer operator on the unit interval can be written as:

$$[L_\beta \rho](\phi) = \int d\theta \exp\left[\beta\left(\hat{f}(\theta) - \theta + \sigma(\theta)\right)\right] \delta\left(\hat{f}(\theta) - \phi\right) \rho(\theta). \tag{8.108}$$

The leading eigenvalue is the inverse of the smallest z satisfying the secular equation:

$$F_\beta(z) = \det(1 - zL_\beta) = 0. \tag{8.109}$$

F_β is called the spectral determinant. Using the formal relation:

$$\det(1 - zL_\beta) = \exp[\operatorname{Tr}\ln(1 - zL_\beta)] = \exp\left[-\sum_{n=1}^{\infty}\frac{z^n}{n}\operatorname{Tr}L_\beta^n\right]. \tag{8.110}$$

Spectral determinants

Define the stretching factor for fixed point n and $\sigma_n(x)$:

$$\operatorname{Tr}L_\beta^n = \int dx\,\delta(f^n(x) - x)\exp[f^n(x) - x] = \sum_{x=f^n(x)}\frac{\exp[\beta\sigma_n(x)]}{|1 - \Lambda_n(x)|}. \tag{8.111}$$

Expand the denominator of Equation (8.111) in a geometric series for $\|\Lambda_n(x)\| < 1$

$$\operatorname{Tr}L_\beta^n = \sum_{k=0}^{\infty}\sum_{x=f^n(x)}\frac{\exp[\beta\sigma_n(x)]}{|\Lambda_n(x)|(\Lambda_n(x))^k}. \tag{8.112}$$

This expansion can only be justified if $|\Lambda_n(x)| > 1$ for every point x and any order n. The next step is to see that this sum is over *all* fixed points, so that a periodic orbit of length p appears every time n is an integer multiple of p. Also all p points along the periodic orbit contribute the same amount.[1]

$$\sum_{n=1}^{\infty}\frac{z^n}{n}\sum_{x=f^{(n)}(x)}\Lambda_n^{-1}(x) = \sum_{\{p\}}\sum_{r=1}^{\infty}\frac{1}{r}z^{n_p r}\Lambda_p^{-r}. \tag{8.113}$$

We convert the sum over periodic points to a sum over *orbits* p, computed with respect to its prime period n_p. Therefore:

$$\det(1 - zL_\beta) = \exp\left[-\sum_{\{p\}}\sum_{r=1}^{\infty}\sum_{k=0}^{\infty}\frac{z^{rn_p}}{r}\frac{\exp[\beta\sigma_{pr}]}{|\Lambda_p|^r\Lambda_p^{kr}}\right]$$

$$= \exp\left[-\sum_{\{p\}}\sum_{r=1}^{\infty}\sum_{k=0}^{\infty}\frac{t_{p,k}^r}{r}\right], \tag{8.114}$$

---

1. As an example, consider the fixed point of $f^{(2)}$ consisting of the two points $\{x_1, x_2\}$. On the left-hand side, at $n = 2$, we have $\frac{z^2}{2}(\Lambda_{12}^{-1} + \Lambda_{21}^{-1}) = z^2\Lambda_{12}^{-1}$. At $n = 4$ both of these points are again periodic points so we have $\frac{z^4}{4}(\Lambda_{1212}^{-1} + \Lambda_{2121}^{-1}) = \frac{z^4}{2}\Lambda_{1212}^{-1}$. The right-hand side at $r = 1$, gives $\Lambda_{12}^{-1}z^{2.1}$ and at $r = 2$ gives $\frac{1}{2}(\Lambda_{12}^{-1})^2z^{2.2}$.

where we have collected together a group of terms as:

$$t_{p,k} = \frac{z^{n_p} \exp[\beta\sigma_p]}{|\Lambda_p|\Lambda_p^k}.$$

(8.115)

The sum over r gives all the repeats of single periodic orbits and this can be resummed exactly[2] to give:

$$\det(1 - zL_\beta) = \exp\left[\sum_{\{p\}}\sum_{k=1}^{\infty} \ln\left(1 - \frac{z^{n_p}\exp[\beta\sigma_p]}{|\Lambda_p|\Lambda_p^k}\right)\right]$$

$$= \prod_{k=0}^{\infty}\prod_{\{p\}}\left(1 - \frac{z^{n_p}\exp[\beta\sigma_p]}{|\Lambda_p|\Lambda_p^k}\right).$$

(8.116)

Thus the spectral determinant is the infinite product of the generalized zeta functions:

$$\zeta_{(k)\beta}^{-1} = \prod_{\{p\}}\left(1 - \frac{z^{n_p}\exp[\beta\sigma_p]}{|\Lambda_p|\Lambda_p^k}\right),$$

(8.117)

and

$$F_\beta(z) = \prod_{k=0}^{\infty}\zeta_{(k)\beta}^{-1}.$$

(8.118)

The leading zero is obtained from the zero-order zeta function $\zeta_{(0)\beta}^{-1}$, called the *dynamical zeta function*:

$$\zeta_{(0)\beta}^{-1} = \prod_{\{p\}}\left(1 - \frac{z^{n_p}\exp[\beta\sigma_p]}{|\Lambda_p|\Lambda_p^0}\right) = \prod_{\{p\}}\left(1 - \frac{z^{n_p}\exp[\beta\sigma_p]}{|\Lambda_p|}\right)$$

$$= \prod_{\{p\}}(1 - t_{p,0}).$$

(8.119)

The zeta function calculations can be rewritten as a perturbative expansion that provides exponentially good estimates (Artuso *et al.*, 1990a; 1990b; Cvitanovic *et al.*, 2005). As a simple illustration, for a complete grammar in the binary alphabet

---

2. The sum of repeated orbits is just the Taylor series for the natural logarithm $\sum_{r=1}^{\infty}\frac{1}{r}t_{p,k}^r = -\ln(1 - t_{p,k})$.

$\sum_{r=1}^{\infty} \frac{1}{r} t_{p,k}^r = -\ln(1 - t_{p,k})$., and using $t_p = t_{p,0} = z^{n_p} \exp[\beta \sigma_p]/\Lambda_p$ (from Equation 8.114), we can expand the infinite product as a series:

$$\zeta_{(0)\beta}^{-1}(z) = \prod_{\{p\}} (1 - t_p) = (1 - t_0)(1 - t_1)(1 - t_{01})(1 - t_{001})(1 - t_{011}).$$

$$= 1 + \sum_{p_1,\ldots,p_k} t_{\{p_1,\ldots,p_k\}} \tag{8.120}$$

$$= 1 - t_0 - t_1 - (t_{01} - t_0 t_1) - (t_{001} - t_0 t_{01})$$

$$- (t_{011} - t_{01} t_1) - \cdots,$$

where $t_{\{p_1,\ldots,p_k\}} = (-)^k t_{p_1} t_{p_2} \ldots t_{p_k}$. The terms t_0 and t_1 are called fundamental terms whereas the other bracketed terms are called curvature corrections. For completely hyperbolic systems with a finite grammar, the curvature corrections are exponentially small so that Equation (8.120) (a power-law expansion of the zeta function) is indeed a perturbative series.

To illustrate the techniques we consider an *oversimplified* example, the mapping on the real line given by:

$$f(x) = \begin{cases} ax & x \in \left[0, \frac{1}{2}\right] \\ a(x-1)+1 & x \in \left[\frac{1}{2}, 1\right] \end{cases}. \tag{8.121}$$

For an arbitrary value of the parameter a, the dynamics are quite complicated and can lead to fractal dependence of the diffusion constant on the slope (Klages and Dorfman, 1995). However if $a = 2m$ where m is an integer, then the map consists of $2m$ complete branches, so the symbolic dynamics are an unrestricted alphabet in $2m$ symbols. Further, the linearity of the map leads to a complete cancellation of the curvature terms. There are $2m$ fixed points and the expanding rate is $\Lambda_i = a$ for every fixed point, and the jumping numbers associated with each branch are $0, 1, \ldots, (a-1), -(a-1), \ldots, -1, 0$. The zeta function for this map is then given by

$$\zeta_{0\{\beta\}}^{-1} = 1 - \frac{2z}{a} \left(1 + \sum_{k=1}^{a/2-1} \cos h(\beta k)\right). \tag{8.122}$$

In general, probability conservation gives $\bar{z}_0(0) = 1$ and the transport properties are obtained by various derivatives with respect to β:

$$\sigma_k(n) = \langle (x_n - x_0)^k \rangle_0 = \frac{\partial^k}{\partial \beta^k} G_n(\beta)\Big|_{\beta=0}$$

$$\sim \frac{\partial^k}{\partial \beta^k} \frac{1}{2\pi i} \int_{a-i\infty}^{a+i\infty} ds \, \exp[sn] \frac{d}{ds} \ln\left[\zeta_{(0)\beta}^{-1} \exp[-s]\right]\Big|_{\beta=0}. \tag{8.123}$$

Determinants tend to have larger analyticity domains because if $\mathrm{Tr}L/(1 - zL) = -\frac{d}{dz}\ln\det(1-zL)$ diverges at a particular value of z, then $\det(1-zL)$ might have an isolated zero there, and a zero of a function is easier to find than a pole. If we are only interested in the leading eigenvalue of L^t, the size of the p cycle neighbourhood can be approximated by $1/|\Lambda_p|^r$, the dominant term in the $rT_p = t \to \infty$ limit, where $\Lambda_p = \prod_e \Lambda_{p,e}$ is the product of the expanding eigenvalues of the Jacobian.

8.7 Positivity of transport coefficients

As an example of the application of the ζ function formalism to calculate the diffusion coefficient, we consider the Lorentz gas (Figure 8.16). The cycle expansion for the average Lyapunov exponent and diffusion coefficient are obtained by considering $\ln\Lambda = \lambda\tau$. The cycle expansion for the average Lyapunov exponent can be written as:

$$\langle\lambda\rangle = \frac{\sum\limits_{\{p_1\dots p_k\}} (-1)^k(\ln\Lambda_{p_1} + \cdots + \ln\Lambda_{p_k})t_{\{p_1\cdots p_k\}}}{\sum\limits_{\{p_1\dots p_k\}} (-1)^k(\tau_{p_1} + \cdots + \tau_{p_k})t_{\{p_1\cdots p_k\}}}, \qquad (8.124)$$

where $\{p_1 \dots p_k\}$ is all possible partitions of the prime symbol string $p_1 \dots p_k$. An arbitrary symbol string may be decomposable into smaller substrings, where all the substrings are themselves prime orbits. In that case k is the number of substrings. The cycle expansion formula for the diffusion coefficient is:

$$D = \frac{1}{2d}\frac{\sum\limits_{p_1\dots p_k} (-1)^k(n_{p_1}^2 + \dots + n_{p_k}^2)t_{\{p_1\dots p_k\}}}{\sum\limits_{p_1\dots p_k} (-1)^k(\tau_{p_1} + \dots + \tau_{p_k})t_{\{p_1\dots p_k\}}}, \qquad (8.125)$$

where for the two-dimensional Lorentz gas $\hat{n}_{p_i}$ is the displacement for the symbol string p_i. There are some results that follow directly from the fact that a periodic orbit representation exists and the fact that the field-dependent equations of motion are time reversible (Dettmann *et al.*, 1997). The simplest illustration of these ideas is obtained using the periodic Lorentz gas in an external field. The Lorentz gas consists of an infinite array of circular scatterers on a triangular lattice, through which a single point particle moves (Cvitanovic *et al.*, 1992; Vance 1992; Lloyd *et al.*, 1995).

When we apply an external field to the Lorentz gas, the response function is the current **J** generated by the field $\mathbf{F}_e$. The current is the systematic drift induced on the wandering particle by the field. If the field is directed along a symmetry axis of the triangular lattice, then the current will be in the same direction as the field. Using the formula (8.90) to calculate the average current, we take $\mathbf{J} \equiv B$.

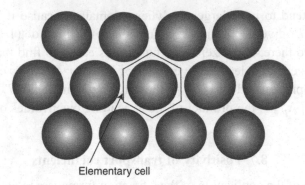

Elementary cell

Figure 8.16 The arrangement of the circular scatterers on a triangular lattice for the Lorentz gas. The central hexagon is the elementary cell that generates the lattice by periodic replication. A single-point particle moves through the lattice colliding with the fixed scatterers

Then in Equation (8.90), we need the sum $\sum_j J(x_{ji})$ along the orbit beginning at x_{0i}, or more correctly for a flow rather than a mapping we need $\Delta x_i = \int_0^{\tau_i} J(x_i(s)) \mathrm{d}s$. Equation (8.90) becomes:

$$\langle J \rangle = \lim_{n \to \infty} \frac{\displaystyle\sum_{i \in \{\text{fixed points of } F^n\}} \Lambda_{1i}^{-1} \Delta x_i}{\displaystyle\sum_{i \in \{\text{fixed points of } F^n\}} \tau_i \Lambda_{1i}^{-1}}. \tag{8.126}$$

For this periodic Lorentz gas, periodic-orbit theory shows that the transport coefficient, in this case the conductivity, is strictly non-negative. The argument is based on the deterministic nature of the dynamics in an external field. Each segment of a trajectory can be traversed in both directions, either with the field or against the field – often against the field is called the conjugate trajectory. Both paths exist and will give opposite contributions to a current, Δx_i and $\Delta x_{-i} = -\Delta x_i$, and have the same probability at zero field, but different probabilities when the external field is applied.

The average conductivity is the sum of pairs of contributions from time-reverse orbits. We ignore all orbits for which $\Delta x_i = 0$, as these make no contribution to the numerator of Equation (8.90). To determine the sign of the conductivity, it is sufficient to consider one pair of time-reverse periodic orbits. All of the orbits for which $\Delta x_i \neq 0$ are time reversible, so they can be either *forward* orbits, $\Delta x_i > 0$, or *backward* orbits, $\Delta x_i < 0$. The two Lyapunov exponents are $\lambda_{1i} > 0 > \lambda_{2i} = \lambda_{1,-i}$. The expanding rate for the forward orbit is given by $\exp(\lambda_{1i}\tau_i)$ and for the backward orbit by $\exp[\lambda_{1,-i}\tau_i] = \exp[-\lambda_{2i}\tau_i]$. Considering the contribution from forward and backward orbits to the numerator of Equation (8.126) we have:

$$\sum_i \Lambda_{1i}^{-1} \Delta x_i = \sum_{i>0} \left(\Lambda_{1i}^{-1} \Delta x_i + \Lambda_{1,-i}^{-1} \Delta x_{-i} \right)$$

$$= \sum_{i>0} (\exp[-\lambda_{1i}\tau_i]\Delta x_i - \exp[\lambda_{2i}\tau_i]\Delta x_i). \qquad (8.127)$$

The conductivity for the Lorentz gas is then the average current $\langle J_x(F_e)\rangle$ divided by the field F_e (in the limit as $F_e \to 0$). The current is essentially the average displacement divided by the time, so can be written as:

$$L_{xx} = \lim_{F_e \to 0} \frac{\langle J_x(F_e)\rangle}{F_e} = \lim_{F_e \to 0} \frac{1}{F_e} \frac{\sum_i \Lambda_{1i}^{-1}\Delta x_i}{\sum_i \Lambda_{1i}^{-1}\tau_i} = \lim_{F_e \to 0} \frac{F_e \sum_i \Lambda_{1i}^{-1}\Delta x_i}{F_e^2 \sum_i \Lambda_{1i}^{-1}\tau_i}. \qquad (8.128)$$

For the Lorentz gas, the two nonzero Lyapunov exponents are related to the external field and the displacement by $(\lambda_{1i} + \lambda_{2i})\tau_i = -F_e\Delta x_i$. This is the Lyapunov sum rule for the exponents and is exact (Lloyd *et al.* 1995). In Equation (8.128) the denominator is strictly positive, so the sign of L_{xx} is controled by the numerator. Combining Equations (8.127) and (8.128) with the Lyapunov sum rule gives the numerator as:

$$F_e \sum_i \Lambda_{1i}^{-1}\Delta x_i = F_e \sum_{i>0} \Delta x_i (\exp[-\lambda_{1i}\tau_i] - \exp[\lambda_{2i}\tau_i])$$

$$= F_e \sum_{i>0} \Delta x_i \exp[-\lambda_{1i}\tau_i](1 - \exp[(\lambda_{1i} + \lambda_{2i})\tau_i]) \qquad (8.129)$$

$$= \sum_{i>0} F_e \Delta x_i \exp[-\lambda_{1i}\tau_i](1 - \exp[-F_e\Delta x_i]).$$

As $\Lambda_i^{-1} = \exp[-\lambda_1 \tau]$ is strictly positive, and $\alpha(1 - \exp[-\alpha])$ is *never negative* ($\alpha = F_e\Delta x_i$), all contributions to the average conductivity are non-negative, and hence the conductivity is positive (that is, the current is either zero or in the same direction as the field). The observed irreversibility in this microscopically reversible system is due to the different stability weights assigned to forward and backward periodic orbits, but these stability weights are derived from the microscopic (reversible) dynamics. (Ruelle, 1999) has recently proposed formulas for the entropy production rate for Gaussian thermostatted systems and can sometimes prove that the entropy production is positive.

This result can be generalized (Rondoni and Morriss, 1997a; 1997b) to N-particle systems that have a periodic-orbit representation and a phase variable current with the same structure as that in Equation (8.128). A system of hard spheres subject to a color field (see Section 6.2) is a good example. The same two features are required: time-reversible paths and the Lyapunov sum rule.

Again the integral of the current along a periodic path is called $\Delta X_i = \int_0^{\tau_i} J(x_i(s)) \mathrm{d}s$ and the Lyapunov sum rule is:

$$\sum_{i=1}^{2d(N-1)} \lambda_i \tau_i = -F_e \Delta X_i. \tag{8.130}$$

The same argument applies again, with the only complication being that there are more than one expanding and contracting directions, so more than two Lyapunov exponents. The transport coefficient is a *conductivity* calculated by a periodic-orbit expansion of the form:

$$L = \lim_{F_e \to 0} \frac{\langle J \rangle}{F_e} = \lim_{F_e \to 0} \frac{F_e \sum_i \Lambda_{i,e}^{-1} \Delta X_i}{F_e^2 \sum_i \Lambda_{i,e}^{-1} \tau_i}. \tag{8.131}$$

We ignore periodic orbits for which $\Delta X_i = 0$ as these do not contribute. All the other period orbits come in pairs with $\Delta X_i = -\Delta X_{-i}$. The stability weight for the forward orbit with $\Delta X_i > 0$ is the product of expanding directions:

$$\Lambda_{i,e}^{-1} = \exp\left[-\tau_i \sum_{\{j:\lambda_{i,j}>0\}} \lambda_{i,j} \right], \tag{8.132}$$

where $\lambda_{i,j}$ is the jth Lyapunov exponent for orbit i. The stability weight for the backward orbit with $\Delta X_i < 0$ is:

$$\Lambda_{-i,e}^{-1} = \exp\left[-\tau_i \sum_{\{j:\lambda_{i,j}>0\}} \lambda_{-i,j} \right] = \exp\left[\tau_i \sum_{\{j:\lambda_{i,j}<0\}} \lambda_{i,j} \right]. \tag{8.133}$$

Combining Equations (8.132) and (8.133) gives:

$$\Lambda_{i,e} \Lambda_{-i,e}^{-1} = \exp\left[\tau_i \sum_j \lambda_{i,j} \right] = \exp\left[-F_e \Delta X_i \right]. \tag{8.134}$$

The contribution to Equation (8.131) from a pair of forward and backward orbits is then

$$F_e \sum_i \Lambda_{i,e}^{-1} \Delta X_i = F_e \sum_{i>0} \left(\Lambda_{i,e}^{-1} \Delta X_i + \Lambda_{-i,e}^{-1} \Delta X_{-i} \right)$$

$$= F_e \sum_{i>0} \Delta X_i \Lambda_{i,e}^{-1} \left(1 - \Lambda_{i,e} \Lambda_{-i,e}^{-1} \right) \tag{8.135}$$

$$= \sum_{i>0} F_e \Delta X_i \Lambda_{i,e}^{-1} \left(1 - e^{-F_e \Delta x_i} \right),$$

and all contributions are of the form $\alpha(1 - \exp[-\alpha])$, which are strictly non-negative. Substituting into Equation (8.131) it follows that the transport coefficient L is non-negative.

9

Nonequilibrium fluctuations

9.1 Introduction

Nonequilibrium steady states are fascinating systems to study. Although there are many parallels between these states and equilibrium states, a convincing theoretical description of steady states, particularly far from equilibrium, has yet to be found. Close to equilibrium, linear response theory and linear irreversible thermodynamics provide a relatively complete treatment, (Sections 2.1 to 2.3). However, in systems where local thermodynamic equilibrium has broken down, and thermodynamic properties are not the same local functions of thermodynamic state variables that they are at equilibrium, our understanding is very primitive indeed.

In Section 7.3 we gave a statistical-mechanical description of thermostatted, nonequilibrium steady states far from equilibrium – the *transient time-correlation function* (TTCF) and *Kawasaki* formalisms. The transient time-correlation function is the nonlinear analog of the Green–Kubo correlation functions. For linear transport processes the Green–Kubo relations play a role which is analogous to that of the partition function at equilibrium. Like the partition function, Green–Kubo relations are highly nontrivial to evaluate. They do, however, provide an exact starting point from which one can derive exact interrelations between thermodynamic quantities. The Green–Kubo relations also provide a basis for approximate theoretical treatments as well as being used directly in equilibrium molecular-dynamics simulations.

The TTCF and Kawasaki expressions may be used as nonlinear, nonequilibrium partition functions. For example, if a particular derivative commutes with the thermostatted, field-dependent propagator, then one can formally differentiate the TTCF and Kawasaki expressions for steady-state phase averages, yielding fluctuation expressions for the so-called *derived properties* The key point in such derivations is that the particular derivative should commute with the relevant propagators. If this is not so, one cannot derive tractable or useful results.

In order to constrain thermodynamic variables, two basic feedback mechanisms can be employed: the integral feedback mechanism employed, for example, in the Nosé–Hoover thermostat, (Section 5.2) and the differential mechanism employed in the Gaussian thermostat. A third mechanism, the proportional mechanism, has not found much use either in simulations or in theory, because it necessarily employs irreversible equations of motion.

In this chapter we will derive fluctuation expressions for the derivatives of steady-state phase averages. We will derive expressions for derivatives with respect to temperature, pressure, and the mean value of the dissipative flux. Applying these derivatives, in turn, to averages of the internal energy, the volume and the thermodynamic driving force yield expressions for the heat capacity and the compressibility. In order to ensure the commutivity of the respective derivatives and propagators, we will employ the Gaussian feedback mechanism exclusively. Corresponding derivations using Nosé–Hoover feedback are presently unknown. Rather than giving a general, but necessarily formal, derivation of the fluctuation formulae, we will instead concentrate on two specific systems: planar Couette flow and color conductivity. By concentrating on these specific systems we hope to make the discussion more concrete and simultaneously illustrate particular applications of the theory of nonequilibrium steady states discussed in Chapter 7.

A major new research area that has developed recently is the study of what has been called the fluctuation theorem. This result concerns the relative probability of fluctuations of different signs, that is, for example, the probability of a current $\overline{J}_\tau$ compared with a current $-\overline{J}_\tau$. The fluctuation theorem has application to small systems like nano-machines and leads to new methods of calculating free-energy differences. The methods are known as the Jarzynski equality and the Crooks relation. These results are particularly useful for the calculation of free-energy differences in single molecule unzipping or unfolding processes. These results are very new, so we will concentrate on the principle derivations and refer the interested reader to the literature for a more detailed account.

9.2 The specific heat

In this section we illustrate the use of the Kawasaki distribution function and the transient time-correlation function formalism by deriving formally exact expressions for the temperature derivative of nonequilibrium averages. Applying these expressions to the internal energy, we obtain two formulae (Evans and Morriss, 1987) for the isochoric heat capacity, one of which shows that the heat capacity can be calculated by analyzing fluctuations in the steady state. The second formula relates the steady-state heat capacity to the transient response observed when an external field perturbs an ensemble of equilibrium systems.

Transient time-correlation function approach

For a system undergoing planar Couette flow, the transient correlation function expression for the canonical ensemble average of a phase variable B is:

$$\langle B(t) \rangle = \langle B(0) \rangle - \beta \gamma V \int_0^t ds \langle B(s) P_{xy}(0) \rangle. \tag{9.1}$$

This expression relates the nonequilibrium value of a phase variable B at time t to the integral of a transient time-correlation function (the correlation between the shear stress in the equilibrium starting state $P_{yx}(0)$, and B at time s after the field is turned on). The temperature implied by the β is the temperature of the initial equilibrium ensemble. The steady state is *tied* to the initial ensemble by the constraint of constant peculiar kinetic energy. For generic nonequilibrium systems that exhibit mixing, Equation (9.1) can be rewritten as:

$$\langle B(t) \rangle = \langle B(0) \rangle - \beta \gamma V \int_0^t ds \langle \Delta B(s) J(0) \rangle, \tag{9.2}$$

where J is the nonequilibrium current and the variable $\Delta B(s)$ is defined as the difference between the phase variable at s and its *average value at s*, that is, $\Delta B(s) = B(s) - \langle B(s) \rangle$. If a system is turbulent, or has quasiperiodic oscillations, we may expect that it is not mixing.

The Gaussian thermostat fixes the kinetic energy of a system, but the Gaussian isokinetic Liouville operator is *independent* of the value of the temperature of the initial distribution. For each member of the ensemble, the Gaussian thermostat simply constrains the peculiar kinetic energy to be constant. As the Liouvillean, and the propagator in Equation (9.2), are independent of the *value* of the temperature we can calculate the temperature derivative very easily. The result is:

$$\frac{\partial}{\partial T} \langle B(t) \rangle = k_B \beta^2 \langle \Delta(B(0)) \Delta(H_0(0)) \rangle - k_B \beta (\langle B(t) \rangle - \langle B(0) \rangle)$$

$$- k_B \beta^3 F_e \int_0^t ds \langle \Delta(B(s) J(0)) \Delta(H_0(0)) \rangle. \tag{9.3}$$

The first term on the right-hand side is the equilibrium contribution. This is easily seen by setting $t = 0$. The second and third terms are nonequilibrium terms. This equation is not only valid in the steady-state limit $t \to \infty$, but also for all intermediate times t, which correspond to transient states along the path from the initial equilibrium state to the final nonequilibrium steady state.

If we choose to evaluate the temperature derivative of the internal energy H_0, we can calculate the heat capacity at constant volume and external field, C_{V,F_e}.

The result is (Evans and Morriss, 1987):

$$C_{V,F_e} = k_B\beta^2\langle(\Delta H_0(0))^2\rangle - k_B\beta(\langle H_0(t)\rangle - \langle H_0(0)\rangle)$$

$$- k_B\beta^3 F_e \int_0^t ds\langle\Delta(H_0(s)J(0))\Delta(H_0(0))\rangle. \tag{9.4}$$

Again the first term on the right-hand side is easily recognized as the equilibrium heat capacity. The second and third terms are nonlinear nonequilibrium terms. They signal the breakdown of local thermodynamic equilibrium. In the linear regime, for which linear response theory is valid, they are, of course, both zero. The third term takes the form of a transient time-correlation function. It measures the correlations of equilibrium energy fluctuations $\Delta H_0(0)$ with the transient fluctuations in the composite-time variable, $\Delta(H_0(s)J(0))$. The second term can, of course, be rewritten as the integral of a transient time-correlation function using Equation (9.1).

Kawasaki representation

Consider the Schrödinger form:

$$\langle B(t)\rangle = \int d\Gamma B(\Gamma)f(t). \tag{9.5}$$

The thermostatted Kawasaki form for the N-particle distribution function is:

$$f(t) = \exp\left[-\beta F_e \int_0^t ds J(-s)\right]f(0). \tag{9.6}$$

Since $f(t)$ is a distribution function it must be normalized. We guarantee this by dividing the right-hand side of Equation (9.6) by its phase integral. If we take the initial ensemble to be *canonical*, we find:

$$f(t) = \frac{\exp[-\beta(H_0 + F_e \int_0^t ds J(-s))]}{\int d\Gamma \exp[-\beta(H_0 + F_e \int_0^t ds J(-s))]}. \tag{9.7}$$

The exponents contain a divergence due to the fact that the time average of $J(-s)$ is nonzero. This secular divergence can be removed by multiplying the numerator and the denominator of Equation (9.7) by $\exp[+\beta F_e \int_0^t ds\langle J(-s)\rangle]$. This has the effect of changing the dissipative flux that normally appears in the Kawasaki exponent from $J(-s)$ to $\Delta J(-s)$, in both the numerator and denominator. The removal of the secular divergence has *no* effect on the results computed in this chapter and is included here for largely aesthetic reasons:

$$f(t) = \frac{\exp[-\beta(H_0 + F_e \int_0^t ds\Delta J(-s))]}{\int d\Gamma \exp[-\beta(H_0 + F_e \int_0^t ds\Delta J(-s))]}. \tag{9.8}$$

The average of an arbitrary phase variable $B(\Gamma)$ in the renormalized Kawasaki representation is:

$$\langle B(t) \rangle = \langle B(0) \rangle + \frac{\int d\Gamma \Delta B \exp\left[-\beta\left(H_0 + F_e \int_0^t ds \Delta J(-s)\right)\right]}{\int d\Gamma \exp\left[-\beta\left(H_0 + F_e \int_0^t ds \Delta J(-s)\right)\right]}. \tag{9.9}$$

To obtain the temperature derivative of Equation (9.9) we differentiate with respect to β. This gives:

$$\frac{\partial \langle B(t) \rangle}{\partial \beta} = -\int d\Gamma B\left(H_0 + F_e \int_0^t ds \Delta J(-s)\right) f(t)$$

$$+ \left(\int d\Gamma B f(t)\right)\left(\int d\Gamma \left(H_0 + F_e \int_0^t ds \Delta J(-s)\right) f(t)\right). \tag{9.10}$$

Using the Schrödinger–Heisenberg equivalence, we transfer the time dependence from the distribution function to the phase variable in each of the terms in Equation (9.10). This gives:

$$\frac{\partial \langle B(t) \rangle}{\partial \beta} = -\left\langle \Delta B(t) \Delta\left(H_0(t) + F_e \int_0^t ds \Delta J(t-s)\right)\right\rangle. \tag{9.11}$$

Substituting the internal energy for B in Equation (9.11) and making a trivial change of variable in the differentiation ($\beta \to T$) and integration ($t - s \to s$), we find that the heat capacity can be written as:

$$C_{V,F_e}(t) = k_B \beta^2 \langle (\Delta H_0(t))^2 \rangle + k_B \beta^2 F_e \int_0^t ds \langle \Delta H_0(t) \Delta J(s) \rangle. \tag{9.12}$$

The first term gives the steady-state energy fluctuations and the second term is a steady-state time correlation function. As $t \to \infty$, the only times s, which contribute to the integral, are times within a relaxation time of t, so that in this limit the time-correlation function has no memory of the time at which the field was turned on.

These theoretical results for the specific heat capacity $c_V = C_V/N$ of nonequilibrium steady states have been tested in nonequilibrium molecular-dynamics simulations of isothermal planar Couette flow (Evans, 1986; Evans and Morriss, 1987).

The steady-state heat capacity in Table 9.1 was calculated in three ways: from the transient correlation function expression, Equation (9.4), from the Kawasaki expression, Equation (9.12), and by direct numerical differentiation of the internal energy with respect to the initial temperature. Although we have been unable to prove the result theoretically, the numerical results suggest that the integral appearing on the right-hand side of Equation (9.4) is zero. All of our simulation results, within error bars, are consistent with this. As can be seen in Table 9.1 the transient correlation expression for the heat capacity predicts that it decreases as we go away

Table 9.1 *Specific heat capacity for the Lennard–Jones triple point. State point* $\gamma = 1$, $N = 108$, $r_c = 2.5$

Equilibrium $c_{V,\,\gamma=0} = 2.622 \pm 0.044$		$c_{V,\gamma=1}$
Transient correlation results: 200 K timesteps		
$(\langle E_\gamma \rangle_{ss} - \langle E_\gamma \rangle_{\gamma=0})/NT$	0.287 ± 0.0014	
$(\gamma/\rho T^3) \int_0^\infty ds \langle \Delta(H_0(s)P_{xy}(0))\Delta(H_0(0))\rangle$	-0.02 ± 0.05	2.395 ± 0.06
Kawasaki correlation results: 300 K timesteps		
$\langle \Delta(E)^2 \rangle_{ss}/NT^2$	3.307 ± 0.02	
$(\gamma/\rho T^2) \int_0^\infty ds \langle \Delta E(s)\Delta P_{xy}(0)\rangle_{ss}$	-1.050 ± 0.07	2.257 ± 0.09
Direct NEMD calculation: 100 K timesteps		2.35 ± 0.05

from equilibrium. The predicted heat capacity at a reduced shear rate $\gamma = 1$ is some 11 % smaller than the equilibrium value. This behavior of the heat capacity was first observed numerically by Evans (1983a).

The results obtained from the Kawasaki formula show that although the fluctuations in internal energy are greater than at equilibrium, the heat capacity decreases as the shear rate increases. The integral of the steady-state energy–stress fluctuations more than compensates for the increase in internal energy fluctuations. The Kawasaki prediction for the heat capacity is in statistical agreement with the transient correlation results. Both sets of results also agree with the heat capacity obtained by direct numerical differentiation of the internal energy. Table 9.2 shows a similar set of comparisons based on published data (Evans, 1983). Once again there is good agreement between results predicted on the basis of the transient correlation formalism and the direct NEMD method.

As a final comment of this section, we should stress that the heat capacity as we have defined it, refers only to the derivative of the internal energy with respect to

Table 9.2 *Comparison of the specific heat capacity of a soft sphere fluid under shear flow as a function of shear rate γ. State point: $N = 108$, $r_c = 1.5$, $T = 1.0877$, $\rho = 0.7$*

γ	E/NT	$c_{V,\gamma}$ (direct)	$c_{V,\gamma}$ (transient)
0.0	4.400	2.61	2.61
0.4	4.441	2.56	2.57
0.6	4.471	2.53	2.53
0.8	4.510	2.48	2.49
1.0 ± 0.01	4.550 ± 0.001	2.43	2.46 ± 0.002

Note: In these calculations, the transient time-correlation function integral, Equation (9.5), was assumed to be zero. Data from Evans (1983a; 1986; Evans and Morriss, 1987).

the temperature of the *initial* ensemble (or equivalently, with respect to the non-equilibrium *kinetic* temperature). Thus far, our derivations say nothing about the thermodynamic temperature of the steady state. We will return to this subject in Chapter 10.

9.3 The compressibility and isobaric specific heat

In this section we calculate formally exact fluctuation expressions for other derived properties, including the heat capacity at constant pressure and external field, C_{p,F_e}, and the compressibility, $\chi_{T,F_e} \equiv -(\partial \ln V/\partial p)_{T,F_e}$. The expressions are derived using the isothermal Kawasaki representation for the distribution function of an isothermal isobaric steady state.

The results indicate that the compressibility is related to nonequilibrium volume fluctuations in exactly the same way that it is at equilibrium. The isobaric heat capacity, C_{p,F_e}, on the other hand, is not simply related to the mean square of the enthalpy fluctuations, as it is at equilibrium. In a nonequilibrium steady state, these enthalpy fluctuations must be supplemented by the integral of the steady-state time cross-correlation function of the dissipative flux and the enthalpy.

We begin by considering the isothermal–isobaric equations of motion considered in Section 6.7. The obvious nonequilibrium generalization of these equations is:

$$\dot{\mathbf{q}}_i = \frac{\mathbf{p}_i}{m} + \dot{\varepsilon}\mathbf{q}_i + \mathbf{C}(\Gamma)F_e(t),$$

$$\dot{\mathbf{p}}_i = \mathbf{F}_i - \dot{\varepsilon}\mathbf{p}_i + \mathbf{D}(\Gamma)F_e(t) - \alpha(\Gamma, t)\mathbf{p}_i, \qquad (9.13)$$

$$\dot{V} = 3V\dot{\varepsilon}.$$

In these equations $\dot{\varepsilon}$ is the dilation rate required to precisely fix the value of the hydrostatic pressure, $3pV = \sum(p_i^2/m + \mathbf{q}_i \cdot \mathbf{F}_i)$. α is the usual Gaussian thermostat multiplier used to fix the peculiar kinetic energy, K. Simultaneous equations must be solved to yield explicit expressions for both multipliers. We do not give these expressions here, since they are straightforward generalizations of the field-free ($F_e = 0$) equations given in Section 6.7.

The external-field terms are assumed to be such as to satisfy the usual AIΓ condition. We define the dissipative flux, J, as the obvious generalization of the usual isochoric case:

$$\left(\frac{dI_0}{dt}\right)^{ad} \equiv -J(\Gamma)F_e. \qquad (9.14)$$

This definition is consistent with the fact that in the field-free adiabatic case, the enthalpy $I_0 \equiv H_0 + pV$ is a constant of the equations of motion given in

Equation (9.14). It is easy to see that the isothermal isobaric distribution, f_0, is preserved by the field-free thermostatted equations of motion:

$$f_0 = \frac{\exp[-\beta I_0]}{\int_0^\infty dV \int d\Gamma \exp[-\beta I_0]}. \tag{9.15}$$

It is a straightforward matter to derive the Kawasaki form of the N-particle distribution for the isothermal–isobaric steady state. The normalized version of the distribution function is:

$$f(t) = \frac{\exp\left[-\beta\left(I_0 + \int_0^t ds J(-s)F_e\right)\right]}{\int_0^\infty dV \int d\Gamma \exp\left[-\beta\left(I_0 + \int_0^t ds J(-s)F_e\right)\right]}. \tag{9.16}$$

The calculation of derived quantities is a simple matter of differentiation with respect to the variables of interest. As was the case for the isochoric heat capacity, the crucial point is that the field-dependent isothermal–isobaric propagator implicit in the notation $f(t)$ is independent of the pressure and the temperature of the entire ensemble. This means that the differential operators $\partial/\partial T$ and $\partial/\partial p_0$ commute with the propagator.

The pressure derivative is easily calculated as:

$$\left(\frac{\partial \langle B(t)\rangle}{\partial p_0}\right)_{T,F_e} = -\beta\langle B(t)V(t)\rangle + \beta\langle B(t)\rangle\langle V(t)\rangle. \tag{9.17}$$

If we choose B to be the phase variable corresponding to the volume then the expression for the isothermal, fixed-field compressibility takes on a form which is formally identical to its equilibrium counterpart:

$$\chi_{T,F_e} = \lim_{t\to\infty} -\frac{\beta}{V}\langle \Delta V(t)^2\rangle. \tag{9.18}$$

The limit appearing in Equation (9.18) implies that a steady-state average should be taken. This follows from the fact that the external field was *turned on* at $t = 0$.

The isobaric temperature derivative of the average of a phase variable can again be calculated from Equation (9.16):

$$\frac{\partial}{\partial \beta}\langle B(t)\rangle = \int_0^\infty dV \int d\Gamma f(t)B(0)\left(I_0 + \int_0^t ds J(-s)F_e\right)$$

$$- \langle I_0(t)\rangle \int_0^\infty dV \int d\Gamma f(t)\left[-\beta\left(I_0 + \int_0^t ds J(-s)F_e\right)\right]. \tag{9.19}$$

In deriving Equation (9.19) we have used the fact that $\int dV \int d\Gamma f(t)B(0) = \langle B(t)\rangle$. Equation (9.19) can clearly be used to derive expressions for the expansion coefficient. However, setting the test variable B to be the enthalpy and remembering that:

$$C_{p,F_e} = \left(\frac{\partial I_0}{\partial T}\right)_{p,F_e}, \tag{9.20}$$

leads to the isobaric heat capacity:

$$C_{p,F_e} = \lim_{t\to\infty} \frac{1}{k_B T^2} \left\{ \langle \Delta I_0(t)^2 \rangle + F_e \int_0^t ds \langle \Delta I_0(t) \Delta J(s) \rangle \right\}. \tag{9.21}$$

This expression is, of course very similar to the expression derived for the isochoric heat capacity in Section 9.2.

In contrast to the situation for the compressibility, the expressions for the heat capacities are not simple generalizations of the corresponding equilibrium fluctuation formulae. Both heat capacities also involve integrals of steady-state time-correlation functions involving cross correlations of the appropriate energy with the dissipative flux. Although the time integrals in Equations (9.12) and (9.21) extend back to $t = 0$, when the system was at equilibrium, for systems which exhibit mixing only the steady-state portion of the integral contributes. This is because $\langle \Delta B(t) \Delta J(0) \rangle \to \langle \Delta B(t) \rangle \langle \Delta J(0) \rangle = 0$ as $t \to \infty$. These correlation functions are therefore comparatively easy to calculate in computer simulations.

9.4 The fluctuation theorem

The mid-1990s saw the introduction of two important fluctuation theorems. The initial work of Evans *et al.* (1993a; 1993b) proposed a fluctuation theorem based on a dynamical weight function derived from the periodic orbit measure (in Section 8.6, Equation 8.89) and supported by molecular dynamics simulation results. Evans and Searles (1994) derived a fluctuation theorem for systems evolving from equilibrium toward a nonequilibrium steady state, and Gallavotti and Cohen (1995a; 1995b) developed a fluctuation theorem for steady-state systems. These two works were motivated by numerical evidence obtained previously by Evans, Cohen, and Morriss. An excellent introductory review is given in Bustamante *et al.* (2005) and a thorough technical review can be found in Evans and Searles (2002).

The first fluctuation theorem obtained by Evans *et al.* (1993a; 1993b) was based on the periodic orbit measures of Ruelle (1978) and Grebogi *et al.* (1988) discussed in Section 8.6. Of course, there was no reason to presume that measures derived for periodic orbits could be used for arbitrary segments of a generic trajectory. But, given that assumption, the fluctuation theorem could be derived and results for isoenergetic shear-flow simulations were presented which demonstrated the correctness of the result. In Figure 9.1a we see the probability of seeing a particular value of the integral of the shear stress $\overline{P}_{yx}$, for trajectory segments of length $t = 0.1$, in a simulation of 56 WCA disks at energy per particle of 1.56, a density of 0.8 and a shear rate of 0.5. Clearly, segments which have a negative value for $\overline{P}_{yx}$ are more probable in the distribution, which is consistent with a negative average shear stress. Figure 9.1b shows the same data as Figure 9.1a, but plotted to demonstrate the correctness of the fluctuation theorem.

The transient fluctuation theorem

For a system with internal energy H, we can split the rate of change of internal energy into two components: an adiabatic (or work) $\dot{W}(t)$ component, and a heat-transfer component $\dot{Q}(t)$. Thus the total change in the internal energy, by the first law of thermodynamics, is given by $\dot{H} = \dot{W} + \dot{Q}$. In the microscopic description of a system these interactions may be through thermal boundary conditions or through the action of an external field and thermostat.

To make the development more concrete, we will use the generic example of iso-energetic SLLOD equations of motion for shear flow, where a Gaussian thermostat applies the constant internal-energy constraint. This was the original example considered by Evans and Searles (1994) and is a little more challenging than other more straightforward currents. The equations of motion for a N-particle system subject to a steady external shear rate γ and a Gaussian thermostat (from Equation 5.20) can be written as:

$$\dot{\mathbf{q}}_i = \tfrac{1}{m}\mathbf{p}_i + \hat{\mathbf{x}}\gamma y_i,$$
$$\dot{\mathbf{p}}_i = \mathbf{F}_i - \hat{\mathbf{x}}\gamma p_{yi} - \alpha\mathbf{p}_i. \tag{9.22}$$

The rate of change in internal energy H is given by:

$$\dot{H} = -\gamma\sum_{i=1}^{N}\left(\tfrac{1}{m}p_{xi}p_{yi} + F_{xi}y_i\right) - \alpha\sum_{i=1}^{N}\tfrac{1}{m}\mathbf{p}_i^2 = -\gamma P_{yx}(\Gamma)V - 2K\alpha$$

$$= \dot{W} + \dot{Q}. \tag{9.23}$$

Here we have made the following identifications: $\dot{W} = -\gamma P_{yx}(\Gamma)V$ is the rate at which work is done on the system by the external field γ and $\dot{Q} = -2K\alpha$ is the rate at which heat is added to the system by the Gaussian thermostat. As the system is dissipative, on average, heat will be removed from the system so we will find that $\langle\dot{Q}\rangle < 0$. At this stage we have not specified the property that is thermostatted, but from the form of the equations of motion (9.22), the rate at which heat

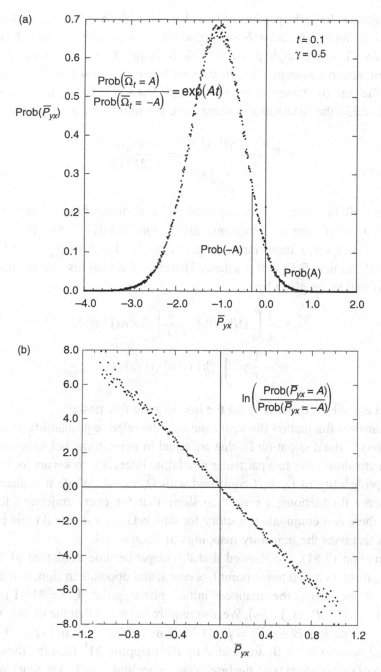

Figure 9.1 (a) The probability of fluctuations in the value of the shear stress for trajectory segments. Note that $\overline{\Omega}_t = -\gamma V \, \overline{P}_{yx}$. These are the original data from Evans *et al.* (1993a); (b) The same data as 9.1a, illustrating that the left-hand side of the fluctuation formula is linear in $\overline{P}_{yx}$

is added appears to be the product of two phase variables as $\dot{Q} = -2K(\Gamma)\alpha(\Gamma)$. This is not the case for a Gaussian or Nosé thermostat as the denominator of $\alpha(\Gamma)$ is always equal to $2K(\Gamma)$, so the single phase variable is simply the numerator of $\alpha(\Gamma)$.

For our generic example, the thermostat has been chosen to be isoenergetic, keeping the rate of change of internal energy fixed to zero at all times, so, from Equation (9.23), the instantaneous value of the thermostatting multiplier is given by:

$$\alpha(\Gamma) = \frac{-\gamma P_{yx}(\Gamma)V}{\sum\limits_{i=1}^{N} \frac{1}{m}\mathbf{p}_i^2} = \frac{-\gamma P_{yx}(\Gamma)V}{2K(\Gamma)}. \tag{9.24}$$

Here, we will be interested in the total reduced dissipation along a trajectory of length t, where the instantaneous dissipation is $\Omega(\Gamma) = 3N\alpha(\Gamma) = -\gamma\beta(\Gamma)$ $P_{yx}(\Gamma)V$. The inverse temperature is given by $\beta(\Gamma) = 3N/\sum p_i^2$, and for isoenergetic dynamics, $\beta(\Gamma)$ will fluctuate. Therefore the total dissipation along a trajectory from 0 to t is given by:

$$\overline{\Omega}_t = \frac{1}{t}\int_0^t \Omega(\Gamma(s))\mathrm{d}s = \frac{1}{t}\int_0^t 3N\alpha(\Gamma(s))\mathrm{d}s$$

$$= -\frac{1}{t}\gamma V \int_0^t \beta(\Gamma(s))P_{yx}(\Gamma(s))\mathrm{d}s. \tag{9.25}$$

The third equality is exact only for the isoenergetic thermostat.

The transient fluctuation theorem examines the relative probability of observing fluctuations in the dissipation $\overline{\Omega}_t$ that are equal in magnitude, but opposite in sign. These fluctuations refer to a particular fixed time interval t, so we are looking at the relative probability of $\overline{\Omega}_t = A$ compared with $\overline{\Omega}_t = -A$. While it is clear that we will observe fluctuations, we need to show that for every trajectory for which $\overline{\Omega}_t = A$, there is a conjugate trajectory for which $\overline{\Omega}_t = -A$. We do this by a construction that uses the trajectory mappings of Section 7.4.

In Equation (7.51), we showed that the *negative* time evolution of the shear stress P_{yx} from an initial phase point Γ is equal, and opposite in sign, to the *positive* time evolution from the mapped initial phase point $\Gamma^K = \mathbf{M}^K(\Gamma)$, that is $P_{yx}(-s, \Gamma, \gamma) = -P_{yx}(s, \Gamma^K, \gamma)$. We also require the result that the inverse temperature is invariant thus $\beta(-s, \Gamma) = \beta(s, \Gamma^K)$. This is illustrated in Figure 9.2. When two initial points (at $t = 0$) are related by the mapping $\mathbf{M}^K$ then the forward time evolution of one of them (say, the brown curve beginning at Γ^K) is equal and opposite to the backward time evolution of the other (Γ, the blue curve beginning at Γ), and for one of them $\overline{\Omega}_t = A$ and for the other $\overline{\Omega}_t = -A$. For a typical equilibrium distribution, both Γ and Γ^K have the same probability. To make the blue curve a conjugate trajectory all that is needed is a time translation so that it becomes the

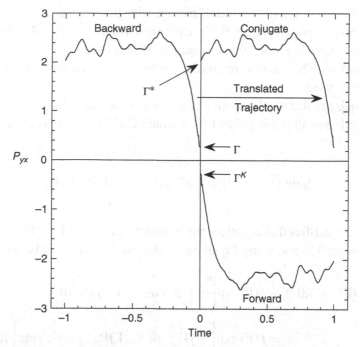

Figure 9.2 The brown curve is the evolution of the shear stress from the initial phase point $\Gamma^K = \mathbf{M}^K(\Gamma)$, the blue curve is the backward time evolution from the initial point Γ. The red curve is the conjugate trajectory obtained by a time translation. The initial phase point for the conjugate trajectory is Γ^*.

red curve in Figure 9.2. Then we have two conjugate trajectories beginning at initial points Γ^K and Γ^*. The point Γ^K under forward time evolution gives $\overline{\Omega}_t = A$, and the point Γ^* under forward time evolution gives $\overline{\Omega}_t = -A$.

In Section 7.2, we derived the Kawasaki distribution function for an arbitrary phase point Γ at time t, given the initial probability of the same phase point Γ in the equilibrium ensemble. Here we ask a different question, if the phase point $\Gamma(t)$ evolves from the point Γ under the system dynamics, what is the probability of $\Gamma(t)$ in the dynamically evolved system relative to the initial probability of Γ. In other words, how does the probability evolve for a comoving point in phase space? The Liouville equation in the comoving frame is Equation (7.17), $\frac{\mathrm{d}}{\mathrm{d}t} \ln f = -\Lambda$, and its solution gives:

$$f(\Gamma(t), t) = \exp\left[-\int_0^t \Lambda(\Gamma(s))\mathrm{d}s\right] f(\Gamma, 0)$$

$$= \exp\left[+\int_0^t 3N\alpha(\Gamma(s))\mathrm{d}s\right] f(\Gamma, 0). \tag{9.26}$$

The only change in the distribution function with time comes about through the change in phase-space volume in the exponential term. Notice that because the path is the comoving one, the dissipation is integrated along the path from $\Gamma \equiv \Gamma(0)$ to $\Gamma(t)$. This result is referred to as the Lagrangian form of the Kawasaki distribution function.

The probability that the dissipation takes on a particular value $\overline{\Omega}_t = -A$ is given by an integral over all phase points Γ^* for which $\overline{\Omega}_t(\Gamma^*) = -A$. Thus, using a delta function:

$$\text{Prob}(\overline{\Omega}_t = -A) = \int d\Gamma^* f(\Gamma^*)\delta(\overline{\Omega}_t(\Gamma^*) + A). \qquad (9.27)$$

Every Γ^* that satisfies this equation can be generated from a Γ by the construction given in Figure 9.2, and using Equation (9.26), the measure can be written as:

$$f(\Gamma^*, -t)d\Gamma^* = f(\Gamma)\exp\left[+\int_0^t 3N\alpha(-s, \Gamma, \gamma)ds\right]d\Gamma$$

$$= f(\Gamma)\exp\left[-\gamma V \int_0^t \beta(-s, \Gamma)P_{yx}(-s, \Gamma, \gamma)ds\right]d\Gamma$$

$$= f(\Gamma^K)\exp\left[\gamma V \int_0^t \beta(s, \Gamma^K)P_{yx}(s, \Gamma^K, \gamma)ds\right]d\Gamma^K$$

$$= f(\Gamma^K)\exp[-t\overline{\Omega}_t(\Gamma^K)]d\Gamma^K, \qquad (9.28)$$

where the second line follows, using Equation (9.25). The next equality uses Equation (7.52) and changes the initial phase point from Γ to Γ^K. Changing variables from Γ^* to Γ^K in Equation (9.27) (with unit Jacobian), and using the fact that the forward time trajectory from Γ^K has $\overline{\Omega}_t(\Gamma^K) = A$, gives:

$$\text{Prob}(\overline{\Omega}_t = -A) = \int d\Gamma^* f\left(\Gamma^*\right)\delta(\overline{\Omega}_t(\Gamma^*) + A)$$

$$= \int d\Gamma^K f(\Gamma^K)\exp[-t\overline{\Omega}_t(\Gamma^K)]\delta(\overline{\Omega}_t(\Gamma^K) - A)$$

$$= \exp(-At)\int d\Gamma^K f(\Gamma^K)\delta(\overline{\Omega}_t(\Gamma^K) - A)$$

$$= \exp(-At)\text{Prob}(\overline{\Omega}_t = A), \qquad (9.29)$$

and we have the probability that $\overline{\Omega}_t = A$ compared with the probability that $\overline{\Omega}_t = -A$. Rearranging Equation (9.29) we obtain the transient fluctuation theorem:

$$\frac{1}{t} \ln\left[\frac{\text{Prob}(\overline{\Omega}_t = A)}{\text{Prob}(\overline{\Omega}_t = -A)}\right] = A. \tag{9.30}$$

In this derivation the initial phase point has been arbitrary, except that the path leading from it, of length t, has $\Omega(\Gamma) = -\beta\gamma V\overline{P}_{yx}(\Gamma) = A$. For every such path there exists a conjugate path for which $\overline{\Omega}_t = -A$.

It is worth listing the assumptions used in deriving Equation (9.30):

(1) The initial distribution $f_0(\Gamma)$ is symmetric under the mapping, so for Couette flow $f_0(\Gamma) = f_0(\Gamma^K)$. An equilibrium distribution is sufficient, but not a necessary condition.
(2) The equations of motion must be reversible so that the conjugate trajectory exists. However, if the reverse trajectory always exists, then this enables the proof to be extended to stochastic dynamics (Kurchan, 1998; Lebowitz and Spohn, 1999; Maes, 1999; Searles and Evans, 1999) and some quantum systems where the concept of a path can be accommodated (De Roeck and Maes, 2004; Esposito and Mukamel, 2006).
(3) The initial ensemble and the subsequent dynamics are ergodically consistent:

$$f(M^T[\Gamma(t)], 0) \neq 0, \quad \forall \ \Gamma(0)$$

Ergodic consistency requires that the initial ensemble must actually contain time-reversed phases of all possible trajectory end points. Ergodic consistency would be violated, for example, if the initial ensemble was microcanonical but the subsequent dynamics were adiabatic and therefore did not preserve the energy of the system. Later, we will treat the Jarzynski and Crooks cases where the dynamics are not ergodically consistent and thereby obtain expressions for Helmholtz free-energy differences between systems.

It has been proposed (Searles and Evans, 2000) that the dissipation function $\Omega(\Gamma)$ be defined to be:

$$\int_0^t ds \Omega(\Gamma(s)) \equiv \ln\left(\frac{f(\Gamma(0), 0)}{f(\Gamma(t), 0)}\right) - \int_0^t \Lambda(\Gamma(s))ds = \overline{\Omega}_t t, \tag{9.31}$$

then Equation (9.30) applies to any valid combination of ensemble and dynamics, although the precise expression for $\overline{\Omega}_t$ does depend on the ensemble and dynamics. The original derivation of the transient fluctuation theorem was for homogeneously isoenergetic dynamics with an initial microcanonical ensemble. If the equations of motion are the homogeneous isoenergetic SLLOD equations of motion for planar Couette flow, the dissipative flux is P_{yx}, and the external field is the shear rate, γ, and we have:

$$\frac{\text{Prob}(\overline{\beta P_{yx_t}}\gamma = A)}{\text{Prob}(\overline{\beta P_{yx_t}}\gamma = -A)} = \exp[-AVt]. \tag{9.32}$$

Note that in this case the dissipation function Ω is precisely the thermodynamic entropy production $-P_{yx}(\Gamma)\gamma V$, divided by the absolute temperature, $k_B T(\Gamma)$, and also (because the system is at constant energy) it is equal to the entropy absorbed from the system by the thermostat, $\alpha(\Gamma)\sum p_i^2/mk_B T$. For a positive shear rate $\langle P_{yx}\rangle < 0$ as the viscosity must be positive (in accord with the Second Law of Thermodynamics). The fluctuation theorem is consistent with the Second Law, but it goes further in treating the fluctuations in a nonequilibrium steady state. The fluctuation theorem predicts that the probability of observing $\langle P_{yx}\rangle_\tau > 0$ along a trajectory segment of length τ is not zero, but much less probable than observing $\langle P_{yx}\rangle_\tau < 0$. Further, for a fixed value of the shear rate, it becomes exponentially more unlikely to observe positive values for $\langle P_{yx}\rangle_\tau$ as either the system size or the observation time is increased.

It is natural to expect that the transient fluctuation theorem, in some suitable long time limit, should yield a fluctuation theorem for steady states. This is the subject of a forthcoming paper by Searles, Rondoni, and Evans (Journal of Statistical Physics, submitted).

9.5 Gallavotti and Cohen fluctuation theorem

Irreversible heat transfers between the system and its environment, typically a thermal bath, characterize nonequilibrium systems. In steady-state systems, an external agent continuously does work on the system that produces heat, and that heat is transferred to the bath. The average amount of heat $\langle Q \rangle$ so produced implies an increase in the total average entropy of the system plus environment. Fluctuation theorems embody recent developments toward a unified treatment of arbitrarily large fluctuations in small systems. At equilibrium, in time-reversal-invariant systems, no net heat is transferred from the system to the bath. Therefore the probability of absorbing a given amount of heat Prob(Q) must be identical to the probability of releasing it Prob$(-Q)$, and the ratio Prob$(Q)/$Prob$(-Q) = 1$. Thus these transfers of heat to the reservoir are directly related to fluctuations within the system.

Motivated by the previous numerical support for the fluctuation theorem of Evans *et al.* (1993a), Gallavotti and Cohen (1995a) established an explicit mathematical expression that holds under general conditions for the ratio $P_t(\sigma)/P_t(-\sigma)$. The entropy production rate σ is the amount of heat Q produced within the system per unit time t, divided by the temperature of the heat bath T. Thus the entropy production is $\sigma = Q/Tt$. Associated with the entropy production is a time-dependent probability distribution $P_t(\sigma)$:

$$\lim_{t\to\infty}\frac{1}{t}\ln\left(\frac{P_t(\sigma)}{P_t(-\sigma)}\right) = \sigma. \tag{9.33}$$

Although this expression involves a limit of infinite time, a similar expression without the limit should be valid, to good approximation, as long as t is much greater than the decorrelation time, which is, roughly speaking, the recovery time of the steady state after it is slightly perturbed.

Equation (9.33) indicates that a steady-state system is more likely to deliver heat to the bath (σ is positive) than it is to absorb an equal quantity of heat from the bath (σ is negative). Nonequilibrium steady-state systems always dissipate heat on average. For macroscopic systems, the heat is an extensive quantity and therefore the ratio of probabilities $P(\sigma)/P(-\sigma)$ grows exponentially with the system size. That is to say, the probability of heat absorption by a macroscopic system is insignificant. For small systems such as molecular motors (Seifert, 2005b) that move along a molecular track, however, the probability of absorbing heat can be significant. On average, molecular motors produce heat, but it may be that they move by rectifying thermal fluctuations – a process that would imply the occasional capture of heat from the bath (Gaspard, 2006).

In 1995, Gallavotti and Cohen (1995a) derived a fluctuation relation using the framework of modern dynamical systems theory and a unifying principle originally proposed by Ruelle (1976; 1978) for turbulent motion. For a dynamical system in phase space Γ whose time evolution is governed by a map F, they assumed the following properties:

(1) *Chaoticity.* The chaotic hypothesis holds and we can treat the system (Γ, f^t) as a transitive Anosov system. The chaotic hypothesis that they proposed states the following (p. 935 of Gallavotti and Cohen, 1995b):

A reversible many-particle system in a stationary state can be regarded as a transitive Anosov system for the purpose of computing the macroscopic properties of the system.

Anosov systems (Gallavotti, 2000) are a paradigm for more general chaotic systems. A map $x \rightarrow f^t(x)$ is an Anosov map if at every point x of a bounded phase space Γ one can set up local coordinates with origin at x, continuously dependent on x and covariant under the action of f^t, and such that in this comoving coordinate system the point x appears as a hyperbolic fixed point of f^t. For a continuous time motion the local system of coordinates includes the phase-space velocity $\dot{x}$ and the motion in that direction cannot be expanding (because the velocity is bounded). Further, all neutral directions must be eliminated from Γ. In general, the motion of an Anosov system approaches one of infinitely many invariant closed sets $C_1, \ldots, C_q$ each of which contains a dense orbit, and each attractive set is smooth in phase space, with only one of them attractive.

The chaotic hypothesis extends the ergodic hypothesis to nonequilibrium systems. Therefore if a dynamical system is assumed to obey the chaotic hypothesis, then for all observables that are smooth functions on phase space, a probability distribution μ_{SRB} can be defined by:

$$\frac{1}{T} \int_0^T B(f^t(x)) \mathrm{d}t \xrightarrow[T \to \infty]{} \int_M B(y) \mu_{SRB}(\mathrm{d}y) \qquad (9.34)$$

(2) *Dissipation*. The phase-space volume undergoes contraction at a rate that, on average, equals to $D\langle\sigma(x)\rangle_+$, where $2D$ is the phase-space dimension and $\sigma(x)$ is a model-dependent "rate" per degree of freedom. This means that the divergence of the equations of motion and its time average will be non-negative:

$$D\langle\sigma(x)\rangle_+ = \int_M \sigma(y)\mu_{SRB}(\mathrm{d}y).\qquad(9.35)$$

This phase-space volume contraction or dissipation is a generalization of the usual entropy production rate of Section 2.2.

(3) *Dynamical Reversibility*. There is a map M in phase space such that if $f^t(x) = x(t)$ then $f^t(Mx) = Mx(t) = x(-t)$ and $M^2 = I$ is the identity. This requires that for every forward time path there exists a conjugate path, starting from a different initial phase point. We have already seen examples of such maps in Section 7.4, at Equation (7.39). The time reversal map $\mathbf{M}^T$ is the dynamical reversibility map for a Hamiltonian system and the Kawasaki map $\mathbf{M}^K$ in Table 7.3 is the appropriate map for shearing systems.

Gallavotti and Cohen then proved the following *"Fluctuation theorem."*

Let (Γ, f^t) satisfy the properties (1–3) (*chaoticity, dissipativity* and *reversibility*) then the probability $\mathrm{Prob}_\tau(p)$ that the total (extensive) entropy production $D\tau\sigma_\tau(x)$ over a time interval τ takes a value $D\tau\langle\sigma(x)\rangle_+p$ satisfies the large-deviation relation:

$$\frac{\mathrm{Prob}_\tau(p)}{\mathrm{Prob}_\tau(-p)} = \exp[D\tau\langle\sigma\rangle_+p],\qquad(9.36)$$

with an error in the argument of the exponential which can be estimated to be independent of p and τ. This means that a plot of the logarithm of the left-hand side as a function of p is linear (with increasing precision as $\tau \rightarrow \infty$). This theorem is known as the Gallavotti–Cohen fluctuation theorem and it is applicable to steady-state systems. It only refers to the phase-space compression rate (called an entropy production rate in Gallavotti and Cohen, 1995a). There is no apparent requirement that the system should be maintained at constant energy, constant kinetic energy, or even that it be maintained at constant volume. This is an illusion, as fluctuations in thermodynamic properties at equilibrium are dependent on the ensemble and therefore on the constraints on the system (we will return to this in the next section). For the isokinetic system at equilibrium, using the methods of Chapter 5, we can see that the instantaneous phase-space contraction is not zero. It fluctuates about an average of zero. Therefore, there exist regions in the phase space where the comoving flow is not Anosov (or not hyperbolic), as the number of expanding directions exceeds the number of contracting directions, and viceversa.

In a separate paper (Gallavotti, 1995) it was pointed out that the p in Equation (9.36) should be within an interval $(-p^*, p^*)$ where p^* is determined dynamically. While it is obvious that realistic model systems will not be Anosov in general, it may be very useful to suggest that deviations from the transitive

Anosov property cannot be observed at the macroscopic level. It has been shown (Evans *et al.*, 2005) that the value of $p*$ is proportional to the square of the field that perturbs the system from equilibrium so that, in the limit as the field goes to zero, the region of validity of this fluctuation theorem also goes to zero. Therefore the Gallavotti–Cohen fluctuation theorem cannot apply to any equilibrium system in which the energy is not a strict constant. Recently van Zon and Cohen (2004) have shown that the phase function for their generalized fluctuation theorem fails to satisfy the Gallavotti–Cohen theorem.

9.6 The Jarzynski equality

The maximum work theorem is a consequence of the Second Law of Thermodynamics. It states that the work performed by a system during a transformation from an initial state A to a final state B is less than the free-energy difference between the two states (Callen, 1985). If the transformation is done reversibly, then the *reversible work* is equal to the free-energy difference. Therefore, in general:

$$\langle \Delta W \rangle \geq \Delta F \tag{9.37}$$

where the average $\langle .. \rangle$ is over all irreversible paths beginning from an equilibrium state. The numerical calculation of free-energy differences was inspired in Zwanzig's formulation of thermodynamic perturbation theory (Zwanzig, 1954a; 1954b). Indeed, when it is applied to the solute–solvent contribution to the free energy change in a liquid (Raineri *et al.*, 2005), with Hamiltonian $H^D = H_0^D + \Psi^D$ where $D = A, B$, with A and B the two states, and Ψ^D is the solute–solvent potential energy of interaction, then the result is $\exp[-\beta \Delta F] = \langle \exp[-\beta \Delta \Psi] \rangle_0$ where the subscript indicates an equilibrium average. An alternative procedure, often called thermodynamic integration, is based on the coupling parameter formalism of Kirkwood (1968), where a parameter ξ, switching from zero to one, turns on the interaction leading to the change in free energy (Frenkel and Smit, 1996). The first method that we treat has a very similar flavor.

A novel treatment of dissipative processes in nonequilibrium systems was introduced by Jarzynski when he reported a nonequilibrium work relation (Jarzynski, 1997), now called the Jarzynski equality. This equality gives a practical way to determine free-energy differences using dynamical paths that are not quasistatic. The first result was obtained for a system weakly coupled to a thermal environment, but that has later been generalized to the strongly coupled case (Jarzynski, 2004).

Consider a system, in contact with a bath at temperature T, whose equilibrium state is determined by a control parameter λ. Initially, the control parameter is set at λ_A and the system is in equilibrium state A. An N-particle system in contact with a heat reservoir can be modeled using the canonical ensemble with the canonical phase-space distribution function given by $f(\Gamma) = \exp[-\beta H(\Gamma)]/Z(\beta)$ where $Z(\beta)$ is the partition function, and the Helmholtz free energy F is given by $\beta F = -\ln Z(\beta)$. We define the work function W to be the change in work between the two equilibrium states A and B given by:

$$\Delta W(t) = W_B - W_A = H_B(t) - H_A(0) + \int_0^\tau ds\, 2K\alpha(s) \tag{9.38}$$

where $K\alpha(s)$ is the kinetic energy multiplied by the phase-space-dependent Gaussian or Nosé multiplier. This is an integrated version of Equation (9.23), identifying $H_B(t) - H_A(0)$ as the change in internal energy and the integral of $-2K\alpha(s)$ as the heat added to the system. It is a very strong constraint on the dynamical path to insist that the system be in an equilibrium state at both ends of the process, A and B, and that these both be canonically distributed with the same temperature.

The free-energy difference between a system at temperature T with Hamiltonian H_B and another at the same temperature with Hamiltonian H_A (Lechner *et al.*, 2006; Scholl-Paschinger and Dellago, 2006) is given by:

$$\Delta F = F_B - F_A = -kT \ln \frac{Z_B}{Z_A} = -kT \ln \frac{\int d\Gamma \exp[-\beta H_B(\Gamma)]}{\int d\Gamma \exp[-\beta H_A(\Gamma)]}. \tag{9.39}$$

Introducing a parameter-dependent Hamiltonian $H(\Gamma, \lambda)$, defined so that $H(\Gamma, \lambda_A) = H_A(\Gamma)$ and $H(\Gamma, \lambda_B) = H_B(\Gamma)$, we can switch between states A and B continuously over some time period τ. If the control parameter changes so that λ moves along the same *path* (or *protocol* $\lambda(t)$), from initial state λ_A to final state λ_B, then, in general, the path in the combined phase space and parameter space $\Gamma(t), \lambda(t)$ will take the system out of equilibrium. From Equation (9.39), the change in Helmholtz free energy along the path is:

$$\exp[-\beta\Delta F] = \frac{\int d\Gamma(\tau)\exp[-\beta H_B(\Gamma(\tau))]}{Z_A}$$

$$= \int d\Gamma(\tau) \frac{\exp[-\beta H(\Gamma, \lambda_A)]}{Z_A} \exp\left[-\beta\left(\Delta W - \int_0^\tau ds\, 2K\alpha(s)\right)\right]$$

$$= \int d\Gamma(0) \left|\frac{\partial\Gamma(\tau)}{\partial\Gamma(0)}\right| \exp\left[+\beta\int_0^\tau ds\, 2K\alpha(s)\right] f_A(\Gamma, 0)\exp[-\beta\Delta W]. \tag{9.40}$$

The second equality comes from the expression for $H_B(\tau)$ given in Equation (9.38). The third equality uses Equation (9.26) to show that the Jacobian of the transformation from $\Gamma(\tau) \to \Gamma(0)$ is the inverse of the exponential that follows it. Here we use the fact that for a Gaussian or Nosé thermostat, $\alpha(s) = \alpha_{num}(s)/2K(s)$, where α_{num}

is the numerator in the expression for α. That is:

$$\left|\frac{\partial\Gamma(\tau)}{\partial\Gamma(0)}\right|\exp\left[+\beta\int_0^\tau ds 2K\alpha(s)\right]$$

$$= \exp\left[-\int_0^\tau ds 3N\alpha(s)\right]\exp\left[+\beta\int_0^\tau ds\alpha_{num}(s)\right]$$

$$= \exp\left[-\int_0^\tau ds\left(\frac{3N}{2K(s)}-\beta\right)\alpha_{num}(s)\right]. \tag{9.41}$$

For a Gaussian isokinetic thermostat $K(s) = K_0$ and $\frac{3N}{2K_0} = \beta$ for all time, so, in this case, this factor collapses to unity. The average reduces to the canonical distribution at A, which is canonical with inverse temperature β. The average is over all paths beginning in state A following the same parameter path (or protocol) to state B:

$$\exp[-\beta\Delta F] = \int d\Gamma(0)\exp[-\beta\Delta W]f_A(\Gamma(0),0)$$

$$= \langle\exp[-\beta\Delta W]\rangle_A. \tag{9.42}$$

In general, ΔW will depend on the initial phase point. Equation (9.42) is the Jarzynski equality.

Frequently, the Jarzynski equality is recast in the form $\langle\exp[-\beta\Delta W_{diss}]\rangle = 1$ where $\Delta W_{diss} = \Delta W - \Delta F$ is the dissipated work along a given trajectory. The exponential average appearing in the Jarzynski equality implies that $\langle\Delta W\rangle \geq \Delta F$ or equivalently, $\langle\Delta W_{diss}\rangle \geq 0$, which for macroscopic systems, is the statement of the Second Law of Thermodynamics in terms of free energy and work. An important consequence of the Jarzynski equality is that, although on average, $\Delta W_{diss} \geq 0$, the equality can only hold if there exist nonequilibrium trajectories with $\Delta W_{diss} \leq 0$. These trajectories, sometimes referred to as transient violations of the Second Law, represent work fluctuations that ensure the microscopic equations of motion are time-reversal invariant. The Jarzynski equality implies that one can determine the free-energy difference between initial and final equilibrium states, not just from a single reversible or quasistatic processes that connects those states, but also via an ensemble of nonequilibrium, irreversible processes that connects them. The ability to bypass reversible paths is of great practical importance.

This analysis assumes that both states A and B are equilibrium canonical states. It requires that all paths from state A to B are time reversible. The result shows that free-energy differences can be computed by averaging over nonequilibrium thermodynamic paths between the states. Cohen and Mauzerall (2004; 2005) have questioned the derivation and exactness of the Jarzynski equality. Their objections relate to two points. The first is the temperature used to weight the irreversible work in $\beta\Delta W$, and second relates to the relaxation of the irreversible state B^* (when the parameter change is complete) to the canonical state B. The derivation given above shows that for an isokinetic thermostat, the Jarzynski equality is exact and the temperature of the initial state determines the value of the kinetic energy along the entire path from A to B, although the system may be far from equilibrium. An initial canonical distribution is preserved by isokinetic dynamics, so there will, in general, be a heat transfer from the system while work is done. This would suggest that the relaxation $B^* \to B$ is not a significant part of the process. The requirement that state B is canonical has only been used to obtain the free-energy difference term in Equation (9.42) and has not been explicitly linked with the dynamics.

9.7 The Crooks relation

Crooks (1998; 1999; 2000) provided a significant generalization of the fluctuation theorem obtained earlier by Jarzynski. The Jarzynski equality relates the change in free energy ΔF between two equilibrium states to an appropriate work average calculated over irreversible paths. In the Jarzynski scenario, and also in Crooks' relation, the system is initially in thermal equilibrium, but is then driven out of equilibrium by the action of an external agent. Let $x_F(s)$ denote a time-dependent nonequilibrium *forward* process for which the variable s runs from $0 \to t$. The forward process initially acts on an equilibrium state A and it ends at a state B that is not at equilibrium. In the reverse process, the initial state B is allowed to reach equilibrium and the system evolves to a nonequilibrium state A. The nonequilibrium protocol for the reverse process $x_R(s)$ is time reversed with respect to the forward one, $x_R(s) = x_F(t - s)$ so that both processes last for the same time. For a dynamical process, rather than a stochastic process, it is natural to treat the case where the forward and backward processes both begin and end in equilibrium canonical states (Evans, 2003). Here we restrict our treatment to that case.

Let $\mathrm{Prob}_{A \to B}(\Delta W)$ and $\mathrm{Prob}_{B \to A}(\Delta W)$ stand for the work probability distributions along the forward and reversed processes respectively (Figure 9.3). We want to determine the ratio of the probabilities for a transition $A \to B$ where the change in the work function is equal to b, and a transition $B \to A$ where the change in work is equal to $-b$. At the two end-points of the process the system is an equilibrium state characterized by a canonical distribution, either $f_A(\Gamma) = \exp[-\beta H_A(\Gamma)]/Z_A(\beta)$ or $f_B(\Gamma) = \exp[-\beta H_B(\Gamma)]/Z_B(\beta)$. Then:

$$\frac{\mathrm{Prob}_{A \to B}(\Delta W = b)}{\mathrm{Prob}_{B \to A}(\Delta W = -b)} = \frac{\int d\Gamma_1 \delta(\Delta W(\Gamma_1) - b) \exp[-\beta H_A(\Gamma_1)]/Z_A(\beta)}{\int d\Gamma_2 \delta(\Delta W(\Gamma_2) + b)) \exp[-\beta H_B(\Gamma_2)]Z_B(\beta)}$$

$$= \exp[-\beta \Delta F] \frac{\int d\Gamma_1 \delta(\Delta W(\Gamma_1) - b) \exp[-\beta H_A(\Gamma_1)]}{\int d\Gamma_2 \delta[\Delta W(\Gamma_2) + b] \exp[-\beta H_B(\Gamma_2)]}.$$

$$(9.43)$$

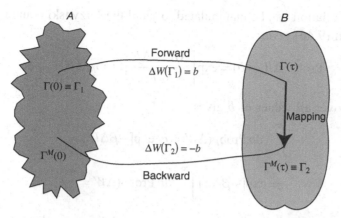

Figure 9.3 The forward and backward paths between systems A and B are constructed as before using a mapping

Here we have the probability of the forward path $A \to B$, which has $\Delta W(\Gamma_1) = b$ in the distribution at A, divided by the reverse path $B \to A$, which has $\Delta W(\Gamma_2) = -b$ in the distribution at B. Reversing the second path so that $\Delta W = +b$, and assigning it the probability it has in the initial distribution at A gives:

$$H_B(\Gamma_2) = H_A(\Gamma_1) + \Delta W(\Gamma_1) - \int_0^t ds 2K\alpha(s). \qquad (9.44)$$

Combining Equations (9.43) and (9.44) gives:

$$\frac{\text{Prob}_{A \to B}(\Delta W = b)}{\text{Prob}_{B \to A}(\Delta W = -b)}$$

$$= \exp[-\beta \Delta F] \frac{\int d\Gamma_1 \delta(\Delta W(\Gamma_1) - b) \exp[-\beta H_A(\Gamma_1)]}{\int d\Gamma_2 \delta(\Delta W(\Gamma_1) - b) \exp[-\beta(H_A(\Gamma_1) + \Delta W - \int_0^t ds 2K\alpha(s))]}$$

$$= \exp[-\beta \Delta F] \frac{\int d\Gamma_1 \delta(\Delta W(\Gamma_1) - b) \exp[-\beta H_A(\Gamma_1)]}{\exp[-\beta b] \int d\Gamma_2 \exp[+\beta \int_0^t ds 2K\alpha(s)] \delta(\Delta W(\Gamma_1) - b) \exp[-\beta H_A(\Gamma_1)]}$$

$$= \exp[-\beta \Delta F] \frac{\int d\Gamma_1 \delta(\Delta W(\Gamma_1) - b) \exp[-\beta H_A(\Gamma_1)]}{\exp[-\beta b] \int d\Gamma_1 \delta(\Delta W(\Gamma_1) - b) \exp[-\beta H_A(\Gamma_1)]}$$

$$= \exp[-\beta \Delta][\beta b]. \qquad (9.45)$$

Then the Crooks relation asserts:

$$\frac{\text{Prob}_F(W)}{\text{Prob}_R(-W)} = \exp\left[\frac{W - \Delta F}{k_B T}\right]. \qquad (9.46)$$

The Crooks relation can be manipulated to yield the Jarzynski equality. Rearranging Equation (9.46) gives:

$$\text{Prob}_F(\Delta W = b) = \exp\left[\frac{\Delta W - \Delta F}{k_B T}\right] \text{Prob}_R(\Delta W = -b). \qquad (9.47)$$

Integrating over all values of b gives:

$$\int_{-\infty}^{\infty} db \, \text{Prob}_F(\Delta W = b) \exp[-\beta \Delta W]$$

$$= \exp[-\beta \Delta F] \int_{-\infty}^{\infty} db \, \text{Prob}_R(\Delta W = -b), \qquad (9.48)$$

and as both probability distributions are normalized, this gives Jarzynski's equality, Equation (9.42):

$$\langle \exp[-\beta \Delta W] \rangle = \exp[-\beta \Delta F]. \qquad (9.49)$$

9.8 Experimental verification

The fluctuation theorems and work relations derived in the previous sections have consequences that can be measured experimentally. Some of the experiments have been preformed on a dynamical system in which the interpretation of the results is very difficult (Ciliberto and Laroche, 1998; Ciliberto *et al.*, 2004; Feitosa and Menon, 2004). Another test of the fluctuation theorems is observing the motion of a colloidal particle in an optical trap. In a series of papers (Wang *et al.*, 2002; Carberry *et al.*, 2004a; 2004b; Reid *et al.*, 2004; 2005; Wang *et al.*, 2005) the transient fluctuation theorem has been verified experimentally for a colloidal particle in an optical trap. Another experiment uses electrical circuits driven out of equilibrium by injecting a small current (van Zon *et al.*, 2004; Garnier and Ciliberto, 2005; Douarche *et al.*, 2006).

Using fluorescence spectroscopy, (Schuler *et al.*, 2005; Tietz *et al.*, 2006) measure the entropy production of a single two-level system that is an optically driven defect center in natural IIa-type diamond. Using a definition of the entropy of a single stochastic trajectory (Seifert, 2005a), they demonstrate that the total entropy production obeys some exact relations for finite length trajectories.

10

Thermodynamics of steady states

10.1 The thermodynamic temperature

Equilibrium thermodynamics provides a very useful connection between mechanical and thermal properties of fluids and solids. The predicted relationships between different quantities measured under different thermodynamic conditions are a fundamental consequence of thermodynamics. It is natural to attempt to develop a similar thermodynamic treatment of nonequilibrium systems, at least for steady states. At present, there are a number of different treatments: the extended irreversible thermodynamics (Jou *et al.*, 2001); the approach to microscopic relaxation processes (Öttinger, 2005); and the approach that we follow here. It is fair to say that, at present, there is no consensus on the correctness of any of these approaches, and indeed some debate about whether it is even possible to define the usual thermodynamic quantities for a nonequilibrium system. Clearly then, it is necessary to limit the types of nonequilibrium processes to which we apply thermodynamics. As an example of a system where a thermodynamic treatment may be successful, consider a steady-state Poiseuille flow system where we can define a local temperature and local shear rate at each point in the fluid. There will be gradients in both the shear rate and the temperature that determine the local streaming velocity profile and the conduction of heat to the boundary. If the fluid is near a phase boundary, then we want to use nonequilibrium thermodynamics to predict the stable phase at each point in the fluid.

In equilibrium thermodynamics, the familiar and intuitive concept of temperature is defined by the Zeroth Law of Thermodynamics. We consider systems whose volume V and pressure P can be determined. Then the mathematical postulate is that (Callen, 1960; Thompson, 1972):

Associated with each pair of systems A and B there is a function $f_{AB}(P_A, V_A; P_B, V_B)$ of the states (P_A, V_A) of A and (P_B, V_B) of B such that A and B are in equilibrium if and only if:

$$f_{AB}(P_A, V_A; P_B, V_B) = 0. \tag{10.1}$$

The first observation is that this implies that the states of two systems in equilibrium cannot be specified arbitrarily. The physical statement is a little different in character as it involves a third system. This goes to the question of how to measure temperature:

When any two bodies are each separately in thermal equilibrium with a third, they are also in equilibrium with each other.

Once again this can be stated in a more mathematical form: *A* state of equilibrium exists and is transitive. That is, if a system *A* is in equilibrium with a system *B*, and *B* is in equilibrium with a system *C*, then *A* is in equilibrium with *C*.

This makes good physical sense if we consider system *B* to be a thermometer. Then this means that if the temperature of *A*, as measured by thermometer *B*, is the same as the temperature of *C*, as measured by thermometer *B*, then *A* is at the same temperature as *C*, and therefore these two systems are at equilibrium with each other. If *A* is placed in contact with *C* there will be no net energy transfer between the two systems.

This means that from Equation (10.1), if:

$$f_{AB}(P_A, V_A; P_B, V_B) = 0 \quad \text{and} \quad f_{BC}(P_B, V_B; P_C, V_C) = 0 \qquad (10.2)$$

then it follows that:

$$f_{AC}(P_A, V_A; P_C, V_C) = 0. \qquad (10.3)$$

To derive the notion of temperature we consider *B* to be a test system (or thermometer). We assume that the two Equations (10.2) can be solved for P_B. From the implicit function theorem f must be continuous and have continuous partial derivatives. Therefore:

$$P_B = \Theta_A(P_A, V_A; V_B) = \Theta_C(P_C, V_C; V_B). \qquad (10.4)$$

Now, since *B* is a fixed system, we can ignore the V_B dependence of Θ_A and Θ_C. Therefore $\Theta_A(P_A, V_A)$ is then defined to be the *empirical temperature* of the system *A*, *as measured by the thermometer B*. A necessary and sufficient condition for two systems to be in equilibrium is that they have the same empirical temperature (as measured by a thermometer).

Given that the physical concept of the temperature can be defined in a sufficiently rigorous way from the Zeroth Law, the next question is how do we calculate the temperature in statistical mechanics? Thermodynamics introduces the concept of the entropy as the variable conjugate to the temperature. We will work from thermodynamics to obtain a microscopic representation for the temperature. Rugh (1997; 1998) has proposed a method of determining the temperature in a Hamiltonian dynamical system. It begins with the definition of the entropy S as the (canonically invariant) weighted area of the energy surface Ω (the level set of the Hamiltonian $H(\mathbf{q}, \mathbf{p})$), under the assumption that the dynamical system is ergodic in Ω. The usual thermodynamic temperature $T(E)$ is then:

$$\frac{1}{T(E)} = \frac{dS}{dE}.$$

(10.5)

Rugh defines the phase variable:

$$\Psi(\Gamma) = \frac{\partial}{\partial \Gamma} \cdot \left(\frac{\frac{\partial H}{\partial \Gamma}}{\frac{\partial H}{\partial \Gamma} \cdot \frac{\partial H}{\partial \Gamma}} \right) = \nabla_\Gamma \cdot \left(\frac{\nabla_\Gamma H}{\nabla_\Gamma H \cdot \nabla_\Gamma H} \right),$$

(10.6)

whose time average:

$$\lim_{t \to \infty} \frac{1}{t} \int_0^t d\tau \, \Psi(\Gamma(\tau)) = \frac{1}{T(E)},$$

(10.7)

equals the inverse of the thermodynamic temperature $T(E)$ of the system with total energy E. Physically, the phase-space derivative with respect to Γ is not dimensionally consistent, so we can write:

$$\frac{\partial}{\partial \Gamma} \equiv \left(a \frac{\partial}{\partial \mathbf{q}_1}, \ldots, a \frac{\partial}{\partial \mathbf{q}_N}, b \frac{\partial}{\partial \mathbf{p}_1}, \ldots, b \frac{\partial}{\partial \mathbf{p}_N} \right) \equiv (a\nabla_q, b\nabla_p) \equiv \nabla_\Gamma,$$

(10.8)

where the pre-factors a and b make all quantities dimensionless, and allow different weights to be applied to coordinates and momenta. We use the del operator ∇_Γ for the phase-space derivative, ∇_q for the coordinate space derivative and ∇_p for the momentum space derivative.

Consider a d-dimensional, N-particle system with Hamiltonian $H = \sum_{i=1}^{N} \frac{1}{2m_i} \mathbf{p}_i^2 + \Phi$, where $\Phi(\mathbf{q}_1, \ldots, \mathbf{q}_N)$ is the total potential energy. If the centre of mass, total momentum, and the kinetic energy are fixed, then there are $dN - d$ independent coordinates and $dN - d - 1$ independent momenta. Including only the independent coordinates and momenta in the phase-space derivative, the phase variable Ψ is:

$$\Psi_E = \frac{a^2 \nabla_q^2 \Phi + b^2 (dN - d - 1)}{a^2 (\nabla_q \Phi \cdot \nabla_q \Phi) + b^2 \sum_{i=1}^{(N-1)'} \mathbf{p}_i^2}$$

$$- 2 \frac{\left(a^4 \nabla_q \Phi \cdot \nabla_q^2 \Phi \cdot \nabla_q \Phi + b^4 \sum_{i=1}^{(N-1)'} \mathbf{p}_i^2 \right)}{\left(a^2 (\nabla_q \Phi \cdot \nabla_q \Phi) + b^2 \sum_{i=1}^{(N-1)'} \mathbf{p}_i^2 \right)^2}, \qquad (10.9)$$

where the subscript E indicates that the energy is conserved and $(N - 1)'$ means that the sum only includes the $dN - d - 1$ independent terms. In the ideal gas limit $\Phi \to 0$, the first term in Equation (10.9) reduces to:

$$\Psi_E = \frac{(dN - d - 1)}{\sum_{i=1}^{(N-1)'} \mathbf{p}_i^2}, \qquad (10.10)$$

which is the usual definition of the kinetic temperature. In the limit as $b \to 0$, Equation (10.9) gives the configurational temperature of (Butler *et al.*, 1998).

If we call T_1 the temperature obtained from the first term in Equation (10.9) and T the temperature obtained from both terms, then, from Figure 10.1, we observe that at $w = 1$, $T_1 \to T_K = 2$ from below as $N \to \infty$, while $T \to T_K$ from above as $N \to \infty$. As a function of w, T is closer than T_1 to the correct value at large values of w, while for small w, $T \to 2N/(N - 1)$ rather than to $T_K = 2$, due to the counting degrees of freedom. We conclude that the dynamic definition of temperature Equation (10.9), as well as T_1, yield the correct values for equilibrium systems in the thermodynamic limit $N \to \infty$.

Inspired by the approach of Rugh, a particularly useful generalization obtained by Jepps *et al.* (2000) showed that an arbitrary vector field $\mathbf{V}(\Gamma)$ with finite, nonzero divergence, $0 < |\langle \nabla H \cdot \mathbf{V}(\Gamma) \rangle| < \infty$ and $\langle \nabla H \cdot \mathbf{V}(\Gamma) \rangle$ growing more slowly than $\exp[N]$ as $N \to \infty$, then the thermodynamic temperature is given by:

$$kT = \frac{\langle \nabla_\Gamma H \cdot \mathbf{V}(\Gamma) \rangle}{\langle \nabla_\Gamma \cdot \mathbf{V}(\Gamma) \rangle}. \qquad (10.11)$$

Here the angled brackets refer to ensemble averages over either the microcanonical, canonical, or various molecular dynamics ensembles.

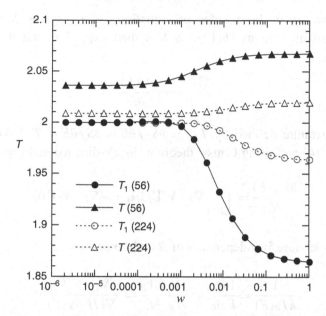

Figure 10.1 The temperatures calculated for a two-dimensional soft sphere system at $\rho = 0.9$, from Ψ_E as a function $w = a^2$ ($2^{-19} \leq a \leq 1$ and $b = 1$), for different numbers of particles N. $T_1(N)$ is the temperature obtained from averaging only the first term in (10.9), whereas $T(N)$ is the average of the full expression Ψ_E.

For the microcanonical ensemble, the measure on the energy surface is uniform, and we can write the Boltzmann entropy (ignoring semiclassical corrections) as:

$$\exp[S(E)/k] = \int_{A(E)} d\mu_E. \tag{10.12}$$

The average of an arbitrary phase variable $B(\Gamma)$ in the microcanonical ensemble is:

$$\langle B(\Gamma) \rangle_E = \frac{\int_{A(E)} B(\Gamma) d\mu_E}{\int_{A(E)} d\mu_E} = \exp[-S(E)/k]/k \int_{A(E)} B(\Gamma) d\mu_E. \tag{10.13}$$

If we choose the phase variable to be $B(\Gamma) = \nabla_\Gamma H \cdot V(\Gamma)$, where $V(\Gamma)$ is the vector field, and define a new function:

$$W_B(E) = \exp[S_B(E)/k] = \int_{A(E)} B(\Gamma) d\mu_E, \tag{10.14}$$

in analogy to Equation (10.12), then from Equation (10.13), we have·

$$\langle B(\Gamma) \rangle_E = \frac{\exp[S_B(E)/k]}{\exp[S(E)/k]} = \exp[(S_B(E) - S(E))/k]. \tag{10.15}$$

Taking logarithms of both sides of Equation (10.13) gives $S_B(E) - S(E) = k \ln\langle B(\Gamma)\rangle_E$, so as long as $\langle B(\Gamma)\rangle_E$ is less than $\exp[N]$ in the thermodynamic limit, then:

$$\frac{\partial S_B(E)}{\partial E} = \frac{\partial S(E)}{\partial E}, \tag{10.16}$$

and the temperature defined by $T_B^{-1} = \partial S_B/\partial E = \partial S/\partial E = T^{-1}$. Differentiating Equation (10.14) and using Gauss' theorem (in dN-dimensions) gives:

$$\frac{\partial W_B(E)}{\partial E} = \int_{A(E)} \nabla_\Gamma \cdot \mathbf{V}(\Gamma)d\mu_E = \langle \nabla_\Gamma \cdot \mathbf{V}(\Gamma)\rangle_E. \tag{10.17}$$

Then the temperature is independent of B and given by:

$$\frac{1}{kT_B(E)} = \frac{1}{k}\frac{\partial S_B}{\partial E} = \frac{1}{W_B}\frac{\partial W_B}{\partial E} = \frac{\langle \nabla_\Gamma \cdot \mathbf{V}(\Gamma)\rangle_E}{\langle \nabla_\Gamma H \cdot \mathbf{V}(\Gamma)\rangle_E}, \tag{10.18}$$

or Equation (10.11). Similar arguments can be used for the canonical and molecular-dynamics ensembles. The details are given in Jepps *et al.* (2000). This result and the result of Rugh are significant extensions of the definition of the temperature. In general, they both involve contributions from coordinate and momentum degrees of freedom, and have given rise to measures of the *configurational* temperature that do not involve the momentum (Butler *et al.*, 1998).

Apart from the Zeroth Law definition of the temperature, all other methods of obtaining the temperature require the definition of the entropy. As entropy and temperature are thermodynamic conjugate variables this should not surprise us.

Nonequilibrium systems

The principal difficulty in extending the equilibrium definition of temperature to systems out of equilibrium is that there is no simple statement of the Zeroth Law (Ritort, 2005). The only nonequilibrium states for which a temperature could be defined would be steady states, where all the system observables take well-defined average values that do not change with time. But to achieve a nonequilibrium steady state we require some field or boundary condition to perturb the system away from equilibrium, and a reservoir in contact with the system (or thermostat) that removes the dissipated heat. Thus simply achieving a nonequilibrium steady state implies the system has a steady energy flow in, and a steady, equal, energy flow out to the reservoir. All equilibrium constructions require that the only energy flow be between the system and thermometer.

In the absence of a defining principle like the Zeroth Law, it has been conventional to use *operational* definitions of the temperature. These are typically based on the extension of the equipartition theorem to nonequilibrium systems, or the application of Equation (3.149). The extension of the equipartition result essentially assumes that:

$$(dN - d - 1)\frac{kT}{2} = \sum_{i=1}^{N} \frac{1}{2} m_i v_i^2, \tag{10.19}$$

connects the temperature and the peculiar translational kinetic energy of the system. The peculiar velocity is given by Equation (3.123), as the particle velocity $\dot{\mathbf{r}}_i$ minus the local fluid streaming velocity $\mathbf{u}(\mathbf{r}_i, t)$. Clearly this will give a steady well-defined result in a nonequilibrium steady state, but this may not be equal to the true thermodynamics temperature. As both Equation (10.19) and the thermodynamic temperature have steady values, then using an *operational* definition such as this leads to a well-defined state point, even if the thermodynamics temperature is not known.

The absence of a definition of the nonequilibrium entropy rules out the two methods we used for equilibrium systems. The only exception to this is the case of isokinetic color diffusion, which has a Hamiltonian representation, Equation (5.37). Morriss and Rondoni (1999) have used this approach to calculate the thermodynamic temperature of a nonequilibrium steady state, using Equations (10.6) and (10.7). A strong correlation between the kinetic temperature orthogonal to the color current and the ratio of the averages of two given phase variables is observed.

For a system in a color field ε the equations of motion for the canonical coordinates $\Gamma \equiv (\mathbf{q}, \boldsymbol{\pi})$ can be obtained from the Hamiltonian, where β is a parameter introduced for greater generality, and:

$$\mathbf{\Phi} = \phi^{\text{int}} + \phi^{\text{ext}} = \phi^{\text{int}} - \varepsilon \sum_{i=1}^{N} c_i x_i, \tag{10.20}$$

is the total potential energy (including internal and external interactions). The total force on particle i is given by $\mathbf{F}_i^\varepsilon = -\partial \mathbf{\Phi}/\partial \mathbf{q} = \mathbf{F}^{\text{int}} + \varepsilon c_i \hat{\mathbf{x}}$. The canonical ($\pi_i$) and physical ($\mathbf{p}_i$) momenta are connected by $\pi_i \exp[\phi/2K] = \mathbf{p}_i$ and the two times by $\exp[\beta\phi/2K] = dt/d\lambda$, so choosing $\beta = 0$, we obtain the usual equations of motion for isokinetic color diffusion, Equation (6.18):

$$\dot{\mathbf{q}}_i = \mathbf{p}_i,$$
$$\dot{\mathbf{p}}_i = \mathbf{F}_i^\varepsilon - \alpha \mathbf{p}_i = \mathbf{F}_i + \hat{\mathbf{x}} \varepsilon c_i - \alpha \mathbf{p}_i. \tag{10.21}$$

When $\phi^{\text{ext}} = 0$, this model reduces to an equilibrium isokinetic system. For $\beta = 0$, the Hamiltonian for this system is Equation (5.37):

$$H_{\beta=0}(\mathbf{q}, \boldsymbol{\pi}) = \frac{1}{2} \exp[\Phi(\mathbf{q})] \sum_{i=1}^{N} \pi_i^2 - \frac{1}{2} \exp[-\Phi(\mathbf{q})]. \qquad (10.22)$$

We now differentiate $H_{\beta=0}$ to construct the function Ψ_K. Differentiating, using initial conditions and parameters which lead to the GIK equations, and then substituting into Equation (10.6), we get:

$$\Psi_K = e^{\Phi/2K} \frac{a^2 \exp[-\Phi/K]\nabla_q^2 \Phi + b^2 dN}{a^2 \exp[-\Phi/K](\nabla_q \Phi \cdot \nabla_q \Phi) + 2b^2 K}$$

$$\frac{-2e^{\Phi/2K} a^4 \exp[-2\Phi/K](\nabla_q \Phi \cdot (\nabla_q \nabla_q \Phi) \cdot \nabla_q \Phi) + 2a^2 b^2 \exp[-\Phi/K](\nabla_q \Phi \cdot \nabla_q \Phi) + 2b^4 K}{\left(a^2 \exp[-\Phi/K](\nabla_q \Phi \cdot \nabla_q \Phi) + 2b^2 K\right)^2},$$

$$(10.23)$$

where the subscript K indicates constant kinetic energy. If we consider the ideal gas limit $\Phi \to 0$, and substitute the first term of Equation (10.23) in Equation (10.7), we get

$$\frac{1}{T_1} = \langle \exp[\Phi/2K] \rangle \frac{1}{T_K}, \qquad (10.24)$$

where $T_K = 2K/dN$ is the fixed value of the kinetic temperature. Now, the arbitrarily chosen zero of the potential scale cannot influence the value of the temperature, hence we replace Φ by $\varphi = \Phi - \Phi_0$, with $\Phi_0 = 2K \ln\langle \exp[\Phi/2K] \rangle$, therefore $\langle \exp[\Phi/2K] \rangle = 1$ and $1/T_1 = 1/T_K$. We can now absorb the factor $\exp[-\Phi_0/2K]$ into a redefined value of a, where $a' = a\exp[-\Phi_0/2K]$. Then Ψ_K becomes

$$\Psi_K = \exp[\varphi/2K] \frac{a'^2 \exp[-\varphi/K]\nabla_q^2 \varphi + b^2 dN}{a'^2 \exp[-\varphi/K](\nabla_q \varphi \cdot \nabla_q \varphi) + 2b^2 K}$$

$$- 2e^{\varphi/2K} \frac{a'^4 \exp[-2\varphi/K](\nabla_q \varphi \cdot (\nabla_q \nabla_q \varphi) \cdot \nabla_q \varphi) + 2a'^2 b^2 \exp[-\varphi/K](\nabla_q \varphi \cdot \nabla_q \varphi) + 2b^4 K}{\left(a'^2 \exp[-\varphi/K](\nabla_q \varphi \cdot \nabla_q \varphi) + 2b^2 K\right)^2}.$$

$$(10.25)$$

Therefore, in the numerical simulations, it suffices to choose values of a and b, and to calculate the average of the right-hand side of Equation (10.23) and of $\exp[\phi/2K]$, in order to obtain the results corresponding to a'.

The results from a GIK system of two-dimensional soft disks at equilibrium are given in Figure 10.2. The weights are obtained by setting $b = 1$ and varying a. The results are very similar to those obtained from the constant energy simulations

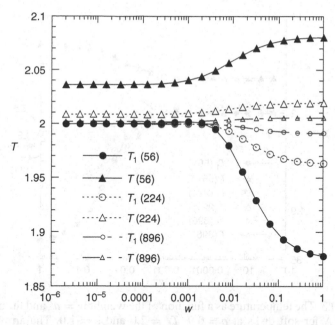

Figure 10.2 The temperature from Equations (10.7) and (10.25) as a function of $w = a^2$, and of the number of particles N, for soft disks at $\rho = 0.9$ and $T_1(N)$. $T_1(N)$ is from only the first term in Equation (10.25), whereas $T(N)$ is the average of the full expression

described in Figure 10.1, showing the same convergence, both with N and with w to the value of $T_K = 2.0$.

Equations (10.25) and (10.7) can now be used to define a nonequilibrium temperature. In Figure 10.3, for $N = 56$ at $a^2 = 1$, the temperature T is dominated by the potential contributions. The result 2.07 (assuming errors of at least 1 %) is rather close to the fixed kinetic value $T_K = 2.0$, and to the orthogonal kinetic temperature, defined by $\sum p_{i,y}^2 = dNT_y$, which is $T_y = 1.994$. For $N = 224$ at $a^2 = 1$, the temperature $T = 1.98$, which is in even better agreement with the corresponding $T_y = 1.966$. For $N = 896$, $T = 1.93$, which is lower than $T_y = 1.954$. We conclude that the temperature defined by Equation (10.25) agrees with the orthogonal part of the kinetic temperature, i.e. the part which is unaffected by the streaming motion. As such, it would appear to be a reasonable value for the temperature away from equilibrium. During these simulations, the value of T_y appeared to be strongly correlated to the quantity $X = \langle \nabla_q^2 \Phi \rangle / \langle \nabla_q \Phi \cdot \nabla_q \Phi \rangle$ in a way that is quite robust to changes in the density, field, and system size. Although this expression is similar to the first term in Equation (10.25), it is not the same. However, we now see that this is just a different choice of the vector field used to define the temperature.

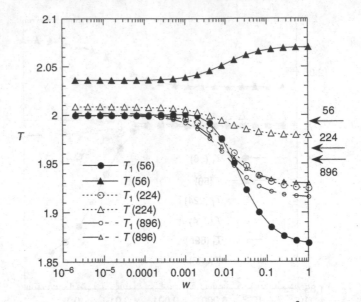

Figure 10.3 The temperature as a function of the weight $w = a^2$ and the number of particles N for soft disks at $\rho = 0.9$, $T_K = 2.0$ and $\varepsilon = 1.0$. The arrows on the right-hand side mark the average values of the kinetic temperature orthogonal to the flow T_y, for the system sizes indicated

10.2 Green's expansion for the entropy

The entropy of an equilibrium system is proportional to the logarithm of the accessible phase volume. Since the accessible phase for a steady state is a fractal of lower dimension than the equilibrium system, the calculation of the entropy for steady states is a formidable unsolved problem. The fine-grained entropy as computed from the full phase space has a number of difficulties. If the distribution function is fractal, there is no limit to the smallness of the phase-space structures and therefore no limit to the sensitivity of the full distribution function to uncontrolled external perturbations. In an experiment, averaging over an ensemble of possible external fluctuations would of course *wash out* the fine structure below a critical length scale. The precise cut-off value would be determined by the amplitude and spectrum of the external fluctuations. This *washing out* of fine structure provides an ansatz for the computation of the entropy for nonequilibrium steady states.

Evans (1989) described a method for computing the coarse-grained entropy of nonequilibrium steady states by decomposing the Gibbs entropy, into terms arising from the partial distribution functions involving correlations of 1, 2, 3, ... , N particles. If the expansion is carried out to order N, the fine-grained Gibbs entropy will be obtained. At equilibrium the expansion gives more than 90 % of the entropy from the singlet and pair contributions, for densities below 75 % of the freezing density, and appears to converge rapidly. Away from equilibrium, the expansion will consist of a series of finite terms until the dimension of the partial distribution function exceeds the dimension of the accessible phase space. Once this occurs all succeeding terms will be infinite. The method yields finite terms below this dimension, because all the lower dimensional integrals are carried out in the accessible phase space.

Green (1954) used Kirkwood's factorization of the N-particle distribution function to write an expansion for the entropy as infinite hierarchy of z-functions, as:

$$
\begin{aligned}
\ln f_1^{(i)} &\equiv z_1^{(i)}, \\
\ln f_2^{(ij)} &\equiv z_2^{(ij)} + z_1^{(i)} + z_1^{(j)}, \\
\ln f_3^{(ijk)} &\equiv z_3^{(ijk)} + z_2^{(ij)} + z_2^{(jk)} + z_2^{(ki)} + z_1^{(i)} + z_1^{(j)} + z_1^{(k)},
\end{aligned}
\tag{10.26}
$$

where the f-functions are the partial 1, 2, 3, . . . ,-body distribution functions. Then the Gibbs fine-grained entropy (Equation 8.1) can be written as an infinite series:

$$
S = -k_B \left\{ \frac{1}{1!} \int d\Gamma_1\, f_1 z_1 + \frac{1}{2!} \int d\Gamma_1 \int d\Gamma_2\, f_2 z_2 + \right\}.
\tag{10.27}
$$

From Equation (10.26), the entropy per particle is given by:

$$
\frac{S}{N} = -\frac{k_B}{\rho} \int d\mathbf{p}_1\, f_1(\mathbf{p}_1) \ln f_1(\mathbf{p}_1)
$$

$$
+ \frac{k_B}{2N} \int d\Gamma_1 \int d\Gamma_2\, f_2(\Gamma_1, \Gamma_2) \ln\left(\frac{f_2^{(12)}}{f_1^{(1)} f_1^{(2)}} \right) + \cdots
\tag{10.28}
$$

Assuming that the fluid is homogeneous enables the spatial integration to be performed in the first term. At equilibrium, the two-body distribution function factors into a product of kinetic and configurational parts, so for two-dimensional fluids, Equation (10.28) reduces to:

$$
\frac{S}{N} = 1 - k_B \ln\left(\frac{\rho}{2\pi m k_B T} \right) - \frac{k_B \rho}{2} \int d\mathbf{r}_{12} g(r_{12}) \ln g(r_{12}) + \cdots,
\tag{10.29}
$$

where $g(r_{12})$ is the equilibrium radial distribution function. Mountain and Raveche (1971) and Wallace (1987) have used experimental radial distribution function data to test Equation (10.29). They found that the Green expansion for the entropy, terminated at the pair level, gives a surprisingly accurate estimate of the entropy from the dilute gas to the freezing density. To our knowledge, Evans' work was the first use of the Green expansion in computer simulations. In the canonical ensemble, Green's entropy expansion is nonlocal so Evans' calculations of the entropy integrated the relevant distribution functions over the *entire* simulation volume. A reformulation of Equation (10.29) by Baranyai and Evans (1989), succeeds in developing a local expression for the entropy of the canonical ensemble, which is ensemble independent.

Evans (1989) tested Equation (10.29), truncated at the pair level, using a system of 32 soft disks (see Table 10.1). The kinetic temperature corrected for $O(1/N)$ factors, is denoted by T_K. The thermodynamic temperature T_{th} was calculated from Equation (10.29) using a finite difference approximation. The analytical expression for the kinetic contribution to the entropy was not used, but rather this contribution was calculated from simulation data by histograming the observed particle velocities and numerically integrating the single particle contribution. The numerical estimate for the kinetic contribution to the entropy was then compared to the theoretical expression (basically the Boltzmann H-function) and agreement was observed within the estimated statistical uncertainties.

Using the entropies calculated at $\rho = 0.6$ and $\rho = 0.7$, a finite difference approximation to $\partial s/\partial \rho^{-1}$ gives the pressure from the relation $p = T(\partial S/\partial V)|_U$, with the virial expression calculated directly from the simulation. The virial pressure at $e = 2.134$, $\rho = 0.65$ is 3.85, whereas the pressure calculated exclusively by numerical differentiation of the entropy is 3.72 ± 0.15. The

Table 10.1 *Equilibrium moderate density data (the uncertainties in the entropies are ± 0.005)*

ρ	e	s	T_K	T_{th}
0.6	1.921	3.200	1.552	1.614
	2.134	3.341		
	2.347	3.464		
0.625	1.921	3.034	1.499	1.500
	2.134	3.176		
	2.347	3.318		
0.65	1.921	2.889	1.445	1.454
	2.134	3.044		
	2.347	3.182		
0.675	1.921	2.754	1.306	1.374
	2.134	2.919		
	2.347	3.064		
0.7	1.921	2.889	1.326	1.291
	2.134	3.044		
	2.347	3.182		

largest source of error in these calculations is likely to be in the finite difference approximation to the various partial derivatives.

Away from equilibrium, the main difficulty in using even the first two terms in Equation (10.28) is the dimensionality of the required histograms. The nonequilibrium pair distribution function does *not* factorize into a product of kinetic and configurational parts, thus the full function of six variables is required. Evans used a density of $\rho \sim 0.1$ where the configurational contributions to the entropy should be unimportant. He evaluated the entropy of the same system of 32 soft disks, but now subject to isoenergetic planar Couette flow, using the SLLOD equations of motion. The thermostatting multiplier α, takes the form:

$$\alpha = -\frac{P_{xy}\gamma V}{\sum p_i^2/m}. \tag{10.30}$$

To check that the configurational parts of the entropy can be ignored, some checks on the equilibrium thermodynamic properties of the system were performed. Table 10.2 shows the thermodynamic temperature computed using a finite

Table 10.2 *Low density data (the uncertainties in the entropy are ± 0.005)*

ρ	γ	e	s	T_k	T_{th}
0.075	0.0	2.134	6.213		
		1.921	5.812		
0.1	0.0	2.134	5.917(27)	2.175	2.12(6)
		2.346	6.013		
0.125	0.0	2.134	5.686		
		1.921	5.744		
0.075	0.5	2.134	5.852	2.190	2.088
		2.347	5.948		
		1.921	5.539		
0.1	0.5	2.134	5.653	2.171	2.048
		2.346	5.747		
		1.921	5.369		
0.125	0.5	2.134	5.478	2.153	2.088
		2.347	5.573		
		1.921	5.380		
0.075	1.0	2.134	5.499	2.188	1.902
		2.347	5.604		
		1.921	5.275		
0.1	1.0	2.134	5.392	2.169	1.963
		2.346	5.492		
		1.921	5.157		
0.125	1.0	2.134	5.267	2.149	2.019
		2.347	5.368		

difference approximation to the derivative, $\partial e/\partial s$. At equilibrium, at a density of $\rho = 0.1$, the thermodynamic temperature is 2.12 ± 0.06 compared with the kinetic temperature of 2.17. At $e = 2.134$ the thermodynamic pressure is 0.22, in reasonable agreement with the virial pressure of 0.24. The disagreement between the thermodynamic and the kinetic expressions for both the temperature and the pressure arise from two causes; the absence of the configurational contributions, and the finite difference approximations for the partial derivatives.

Figure 10.4 shows the pair distribution function for a 32-particle system under shear, that is free of the singularities apparent in the two-particle system. The smoothness is due to averaging over all the other $N-2$ particles *washing out* the fine structure. Table 10.2 gives the kinetic contribution to the entropy as a function of energy, density and shear rate. For a given energy and density, the entropy is observed to be a monotonically *decreasing* function of the shear rate. As expected, the equilibrium state has the maximum entropy. It is clear that the entropy can

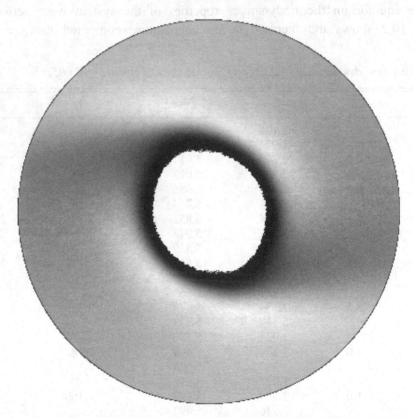

Figure 10.4 The pair distribution function for a 32-particle soft-disk fluid at a high shear rate of 2.0. The density is 0.1 and energy per particle is 1.921. The distribution is seemingly smooth. In spite of the high anisotropy of this distribution, the configurational contribution to the entropy is only about 0.4 %

be written as a function, $S = S(N, V, U, \gamma)$. Defining $T_{th} = \partial U / \partial S|_{V,\gamma}$, $p_{th} = T \partial S / \partial V|_{U,\gamma}$ as and $\zeta_{th} = T \partial S / \partial \gamma|_{U,V}$ we can write:

$$dU = T_{th} dS - p_{th} dV + \zeta_{th} d\gamma. \tag{10.31}$$

Some years ago Evans and Hanley (1980) proposed Equation (10.31) as a generalized Gibbs relation, however, at that time there was no way of directly computing the entropy or any of the free energies. This forced Evans and Hanley to postulate that the thermodynamic temperature was equal to the equipartition or kinetic temperature. Evans and Hanley observed that, away from equilibrium, although the pressure tensor is anisotropic, the thermodynamic pressure must be independent of the manner in which a virtual volume change is performed. The thermodynamic pressure must therefore be a scalar. They assumed that the thermodynamic pressure would be equal to the *simplest scalar invariant of the pressure tensor* that was also consistent with equilibrium thermodynamics. In two-dimensional systems they assumed that $p = \frac{1}{2}(P_{xx} + P_{yy})$. We can calculate the coarse-grained Gibbs entropy directly and check this postulate. We assume that the internal energy is the sum of the peculiar kinetic energy and the potential energy and that the thermodynamic entropy is equal to the coarse-grained Gibbs entropy. Table 10.1 shows a comparison of kinetic and thermodynamic temperatures for the 32-particle soft-disk system.

As has been known for some time (Evans, 1983a), $\partial T_k / \partial \gamma|_{V,U} < 0$ leading to a decrease in the kinetic temperature with increasing shear rate. For this low-density system, the effect is far smaller than has been seen for moderately dense systems. At a density of 0.1 the kinetic temperature drops by 0.3 % as the shear rate is increased to unity. The precision of the kinetic temperature is about 0.01 %. The thermodynamic temperature decreases as the shear rate is increased, but in a far more dramatic fashion. It decreases by 10 % over the same range of shear rates. The results clearly show that away from equilibrium the thermodynamic temperature is smaller than the kinetic or equipartition temperature. As the shear rate increases, the discrepancy grows larger.

Using the simulation data at $e = 2.134$, the thermodynamic pressure can be estimated as a function of shear rate. Table 10.3 shows the finite difference approximation for the thermodynamic pressure p_{th} the hydrostatic pressure $p_{th} = \frac{1}{2}(P_{xx} + P_{yy})$ and the largest and smallest eigenvalues of the pressure tensor p_1, p_2 respectively. As expected, the hydrostatic pressure increases with shear rate. This effect, known as shear dilatancy, is very slight at these low densities. The thermodynamic pressure shows a much larger effect, but it *decreases* as the shear rate is increased. In an effort to give a mechanical interpretation to the thermodynamic pressure, we calculated the two eigenvalues of the pressure tensor. Away from

Table 10.3 *Nonequilibrium pressure:* $e = 2.134$, $\rho = 0.1$

γ	p_{th}	p_{tr}	p_1	p_2
0.0	0.215(7)	0.244	0.244	0.244
0.5	0.145	0.245	0.361	0.130
1.0	0.085	0.247	0.397	0.096

equilibrium, the diagonal elements of the pressure tensor differ from one another and from their equilibrium values giving rise to normal stress effects. The eigenvalues are influenced by all the elements of the pressure tensor including the shear stress. One of the eigenvalues increases with shear rate, while the other decreases and within statistical uncertainties the latter is equal to the thermodynamic pressure.

Evans (1989) conjectured that the thermodynamic pressure is equal to the minimum eigenvalue of the pressure tensor, that is $p_{th} = p_2$. This relation is exact at equilibrium and is in accord with our numerical results. It is also clear that if the entropy is related to the minimum reversible work required to accomplish a virtual volume change in a nonequilibrium steady-state system, then $p_2 dV$ is the minimum work that is possible. Carrying out a virtual volume change by moving walls inclined at arbitrary angles with respect to the shear plane then the minimum virtual pV work (minimized over all possible inclinations of the walls) will be $p_2 dV$.

Figure 10.5 shows the kinetic contribution to the entropy as a function of shear rate for the 32-particle system at an energy $e = 2.134$ and a density $\rho = 0.1$. The entropy seems to be a linear function of shear rate for the range of shear rates covered by the simulations. Combining these results with those from Table 10.1 allows us to compute ζ_{th} as a function of shear rate. For $\gamma = 0.0$, 0.5, 1.0 we find that $\zeta_{th}/N = 1.22$, 1.08, 0.91 respectively. Most of the decrease in ζ is due to the decrease in the thermodynamic temperature with increasing shear rate. We have assumed that asymptotically s is linear in shear rate, as the shear rate tends to zero. It is always possible that, at shear rates which are too small for us to simulate, this linear dependence gives way to a quadratic variation.

Although these calculations are restricted to the low-density gas regime, the results suggest that a sensible definition for the nonequilibrium entropy can be given. A definition, based on Equation (10.28) avoids the divergences inherent in the fine-grained entropy due to the contraction of the nonequilibrium phase space. At low densities this entropy reduces to the Boltzmann entropy implicit in the Boltzmann H-function. This entropy is, for states of a specified energy and density, a maximum at equilibrium. For a recent update on the Green expansion of the entropy see Evans and Rondoni (2002).

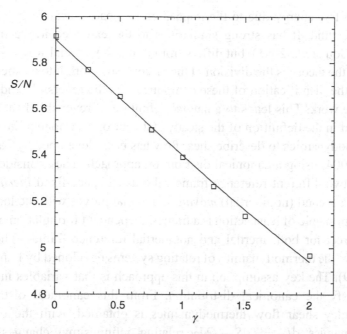

Figure 10.5 The kinetic contribution to the entropy as a function of shear rate. The system density is 0.1 and the energy per particle is 2.134. The entropy is approximately a linear function of shear rate. The derivative of the entropy with respect to shear rate gives ζT. ζ is positive, but decreases with shear rate, due to the decrease in the thermodynamic temperature with increasing shear rate

Defining a temperature on the basis of this entropy indicates that far from equilibrium there is no reason to expect that the kinetic temperature is equal to the thermodynamic temperature. Similarly there seems to be no reason to expect that the average of the diagonal elements of the pressure tensor will be equal to the thermodynamic pressure. The concept of minimum reversible virtual work, together with our numerical results suggests that the thermodynamic pressure is instead equal to the minimum eigenvalue of the pressure tensor. It remains to be seen whether the entropy so defined is a local maximum in nonequilibrium steady states. If this can be satisfactorily demonstrated, then we will have for the first time a fundamental basis for a generalized thermodynamics of steady states far from equilibrium.

10.3 Prospects

A detailed theory of the thermodynamics of linear viscoelastic fluids in steady shear has been given by Daivis and Matin (2003) and Daivis (2007). This theory is exact and gives simple and easily computed expressions for the change in the internal energy and free energy due to the imposition of steady shear for a general linear viscoelastic fluid. It has strong similarities to the extended irreversible thermodynamics (Jou *et al.*, 2001), but differs from it in a few crucial ways. An important element of the theory is the division of the viscoelastic work into elastic and viscous parts, and the identification of these components as the reversible and irreversible parts of the work. This leads to a natural definition of reversible heat transfer that can be used in the definition of the steady-state entropy. A nonequilibrium steady-state thermodynamics to describe shear flow has been developed by Taniguchi and Morriss (2004) using a canonical distribution approach. They consider a shearing system in two different reference frames, the usual space-fixed *inertial* reference frame and a second (noninertial) *moving* frame that moves with the local fluid velocity. The principle of least action is a frame-independent formulation of mechanics that is correct for both inertial and noninertial reference frames. The idea is an extension of the thermodynamics of rotating systems developed by Landau and Lifshitz (1959). The key assumption in this approach is that variables in the moving frame satisfy the canonical distribution. From this canonical distribution, the Evans–Hanley shear flow thermodynamics is obtained, with the First Law of Thermodynamics $dU = TdS - Qd\gamma$ relating infinitesimal changes in internal energy dU, entropy dS and shear rate $d\gamma$ with kinetic temperature T. These two approaches to the thermodynamics of nonequilibrium steady states are very different, The use of rheology and the theory of linear viscoelastic fluids is quite different to the statistical-mechanical route. It seems likely that if a thermodynamic treatment is possible, then it should apply to steady states. The difficulty is connected to the treatment of heat, work is much more straightforward. In equilibrium, the concept of temperature is straightforward and the use of Carnot cycles leads directly to the definition of the entropy as the state function conjugate to the temperature. Away from equilibrium, even for steady states, the definition of both the temperature and the entropy are problematic. The starting point is usually the equilibrium form, with either an extension to nonequilibrium, or the assumption that the equilibrium result is approximately applicable. Neither of these approaches is entirely satisfactory. The other difficulty is the absence of some suitable experimental test bed. Despite these difficulties, progress continues (Sasa and Tasaki, 2006).

References

Abraham, R. and Marsden, J. E. (1978). *Foundations of Mechanics*. Reading: Benjamin/
    Cummins.
Alder, B. J. and Wainwright, T. E. (1956). *The International Symposium on Statistical
    Mechanical Theory of Transport Processes*. Brussels: Interscience.
Allen, M. P. and Tildesley, D. J. (1987). *Computer Simulation of Liquids*. Oxford:
    Clarendon Press.
Andersen, H. C. (1980). *Journal of Chemical Physics*, **72**, 2384.
Artuso, R., Aurell, E., and Cvitanovic, P. (1990a). *Nonlinearity*, **3**, 325.
Artuso, R., Aurell, E., and Cvitanovic, P. (1990b). *Nonlinearity*, **3**, 361.
Artuso, R. and Cristadoro, G. (2004). *Journal of Physics A*, **37**, 85.
Baladi, V. (2000). *Positive Transfer Operators and Decay of Correlations*. Singapore:
    World Scientific.
Balatoni, J. and Renyi, A. (1956). *Publications of the Mathematics Institute of the
    Hungarian Academy of Science*, **1**, 9.
Baranyai, A. and Cummings, P. T. (1999). *Journal of Chemical Physics*, **110**, 42.
Baranyai, A. and Evans, D. J. (1989). *Physical Review*, **A40**, 3817.
Barker, J. A. and Henderson, D. (1976). *Reviews of Modern Physics*, **48**, 587.
Benettin, G., Galgani, L., Giorgilli, A., and Strelcyn, J.-M. (1980a). *Meccanica*, **15**, 9.
Benettin, G., Galgani, L., Giorgilli, A., and Strelcyn, J.-M. (1980b). *Meccanica*, **15**, 21.
Benettin, G., Galgani, L., Giorgilli, A., and Strelcyn, J.M. (1978). *Comptes Rendus
    Hebdomadaires Des Seances De L Academie Des Sciences Serie A*, **286**, 431.
Benettin, G., Galgani, L., and Strelcyn, J. M. (1976). *Physical Review A*, **14**, 2338.
Berne, B. J. (1977). *Statistical Mechanics Part B: Time Dependent Processes*,
    ed. B. J. Berne. New York: Plenum Press, Chapter 5.
Bird, R. B., Armstrong, R. C., and Hassager, O. (1977). *Dynamics of Polymeric Liquids*.
    New York: Wiley.
Boltzmann, L. (1964). *Lectures on Gas Theory*. New York: Cambridge University Press.
Bowen, R. (1970). *American Journal Mathematics*, **92**, 725.
Bowen, R. (1978). *On Axiom A Diffeomorphism*. CBMS Regional Conference Series
    in Mathematics. Providence: American Mathematical Society.
Brophy, J. J. (1966). *Basic Electronics for Scientists*. New York: McGraw-Hill.
Brown, D. and Clarke, J. H. R. (1986). *Physical Review A*, **34**, 2093.
Brown, R. (1828a). *Philosophical Magazine*, **4**, 161.
Brown, R. (1828b). *Annalen der Physik und Chemie*, **14**, 294.
Bustamante, C., Liphardt, J., and Ritort, F. (2005). *Physics Today*, **58**, 43.

Butler, B. D., Ayton, O., Jepps, O. G., and Evans, D. J. (1998). *Journal of Chemical Physics*, **109**, 6519.

Callen, H. B. (1960). *Thermodynamics*. New York: Wiley.

Callen, H. B. (1985). *Thermodynamics and an Introduction to Thermostatistics*. New York: Wiley.

Carberry, D. M., Reid, J. C., Wang, G. M. *et al.* (2004a). *Physical Review Letters*, **92**, 140601.

Carberry, D. M., Williams, S. R., Wang, G. M., Sevick, E. M. and Evans, D. J. (2004b). *Journal of Chemical Physics*, **121**, 8179.

Ciccotti, G. and Jacucci, G. (1975). *Physical Review Letters*, **35**, 789.

Ciccotti, G., Jacucci, G., and McDonald, I. R. (1976). *Physical Review A*, **13**, 426.

Ciccotti, G., Jacucci, G., and McDonald, I. R. (1979). *Journal of Statistical Physics*, **21**, 1.

Ciliberto, S., Garnier, N., Hernandez, S. *et al.* (2004). *Physica A*, **340**, 240.

Ciliberto, S. and Laroche, C. (1998). *Journal De Physique Iv*, **8**, 215.

Cohen, E. G. D. (1983). *Physica*, **118A**, 17.

Cohen, E. G. D. and Mauzerall, D. (2004). *Journal of Statistical Mechanics*, P07006.

Cohen, E. G. D. and Mauzerall, D. (2005). *Molecular Physics*, **103**, 2923.

Crooks, G. E. (1998). *Journal of Statistical Physics*, **90**, 1481.

Crooks, G. E. (1999). *Physical Review E*, **60**, 2721.

Crooks, G. E. (2000). *Physical Review E*, **61**, 2361.

Cvitanovic, P., Artuso, R., Mainieri, R., Tanner, G., and Vattay, G. (2005). *Chaos: Classical and Quantum*. Copenhagen: Neils Bohr Institute.

Cvitanovic, P., Gaspard, P., and Schreiber, T. (1992). *Chaos*, **2**, 85.

Daivis, P. J. (2007). *Journal of Non-Newtonian Fluid Mechanics*, in press.

Daivis, P. J. and Matin, M. L. (2003). *Journal of Chemical Physics*, **118**, 11111.

Daivis, P. J., Travis, K. P., and Todd, B. D. (1996). *Journal of Chemical Physics*, **104**, 9651.

de Groot, S. R. and Mazur, P. (1962). *Non-Equilibrium Thermodynamics*. Amsterdam: North-Holland.

De Roeck, W. and Maes, C. (2004). *Physical Review E*, **69**.

de Schepper, I. M., Haffmans, A. F. E. M., and van Beijeren, H. (1986). *Physical Review Letters*, **57**, 1715.

Dettmann, C. P. and Morriss, G. P. (1996a). *Physical Review E*, **54**, 2495.

Dettmann, C. P. and Morriss, G. P. (1996b). *Physical Review E*, **53**, R5545.

Dettmann, C. P., Morriss, G. P., and Rondoni, L. (1997). *Chaos Solitons & Fractals*, **8**, 783.

Devaney, R. L. (1986). *An Introduction to Chaotic Dynamical Systems*. Menlo Park: Benjamin/Cummings.

Doetsch, G. (1961). *Guide to the Applications of Laplace Transforms*. New York: Van Nostrand.

Dorfman, J. R. (1999). *An Introduction to Chaos in Nonequilibrium Statistical Mechanics*. Cambridge: Cambridge University Press.

Dorfman, J. R. and Cohen, E. G. D. (1965). *Physics Letters*, **16**, 124.

Dorfman, J. R. and Cohen, E. G. D. (1972). *Physical Review A*, **6**, 776.

Douarche, F., Joubaud, S., Garnier, N. B., Petrosyan, A., and Ciliberto, S. (2006). *Physical Review Letters*, **97**, 140603.

Dufty, J. W. and Lindenfeld, M. J. (1979). *Journal of Statistical Physics*, **20**, 259.

Dyson, F. J. (1949). *Physical Review*, **75**, 486.

Eckmann, J.-P. and Procaccia, I. (1986). *Physical Review A*, **34**, 659.

Eckmann, J. P. and Ruelle, D. (1985). *Reviews of Modern Physics*, **57**, 617.

Edberg, R., Evans, D. J., and Morriss, G. P. (1986). *Journal of Chemical Physics*, **84**, 6933.

Edberg, R., Morriss, G. P., and Evans, D. J. (1987). *Journal of Chemical Physics*, **86**, 4555.

Einstein, A. (1905). *Annalen der Physik*, **17**, 549.

Ernst, M. H., Cichocki, B., Dorfman, J. R., Sharma, J., and van Beijeren, H. (1978). *Journal of Statistical Physics*, **18**, 237.

Erpenbeck, J. J. (1984). *Physical Review Letters*, **52**, 1333.

Erpenbeck, J. J. (1989). *Physical Review A*, **39**, 4718.

Erpenbeck, J. J. and Wood, W. W. (1977). *Statistical Mechanics B. Modern Theoretical Chemistry*. Vol. 6, ed. B. J. Berne. New York: Plenum, p. 1.

Erpenbeck, J. J. and Wood, W.W. (1981). *Journal of Statistical Physics*, **V24**, 455.

Esposito, M. and Mukamel, S. (2006). *Physical Review E*, **73**, 046129.

Evans, D. (1993). *Molecular Physics*, **80**, 221.

Evans, D. J. (1980). *Physical Review A*, **23**, 1988.

Evans, D. J. (1981). *Physical Review A*, **23**, 2622.

Evans, D. J. (1982a). *Physics Letter*, **91A**, 457.

Evans, D. J. (1982b). *Molecular Physics*, **47**, 1165.

Evans, D. J. (1983a). *Journal of Chemical Physics*, **78**, 3297.

Evans, D. J. (1983b). *Physica*, **118A**, 51.

Evans, D. J. (1984). *Physics Letter*, **101A**, 100.

Evans, D. J. (1985). *Physical Review A*, **32**, 2923.

Evans, D. J. (1986). Fourth Australian National Congress on Rheology (Australian Society of Rheology).

Evans, D. J. (1989). *Journal of Statistical Physics*, **57**, 745.

Evans, D. J. (2003). *Molecular Physics*, **101**, 1551.

Evans, D. J., Cohen, E. G. D., and Morriss, G. P. (1990). *Physical Review A*, **42**, 5990.

Evans, D. J., Cohen, E. G. D., and Morriss, G. P. (1993a). *Physical Review Letters*, **71**, 2401.

Evans, D. J., Cohen, E. G. D., and Morriss, G. P. (1993b). *Physical Review Letters*, **71**, 3616.

Evans, D. J., Cohen, E. G. D., Searles, D. J., and Bonetto, F. (2000). *Journal of Statistical Physics*, **101**, 17.

Evans, D. J. and Hanley, H. J. M. (1980). *Physics Letters*, **80A**, 175.

Evans, D. J. and Hanley, H. J. M. (1989). *Molecular Physics*, **68**, 97.

Evans, D. J., Hoover, W. G., Failor, B. H., Moran, B., and Ladd, A. J. C. (1983). *Physical Review A*, **28**, 1016.

Evans, D. J. and Macgowan, D. (1987). *Physical Review A*, **36**, 948.

Evans, D. J. and Morriss, G. P. (1983a). *Chemical Physics*, **77**, 63.

Evans, D. J. and Morriss, G. P. (1983b). *Physical Review Letters*, **51**, 1776.

Evans, D. J. and Morriss, G. P. (1983c). *Physics Letters A*, **98**, 433.

Evans, D. J. and Morriss, G. P. (1984a). *Chemical Physics*, **87**, 451.

Evans, D. J. and Morriss, G. P. (1984b). *Computer Physics Reports*, **1**, 297.

Evans, D.J. and Morriss, G. P. (1984c). *Physical Review A*, **30**, 1528.

Evans, D. J. and Morriss, G. P. (1985). *Physical Review A*, **31**, 3817.

Evans, D. J. and Morriss, G. P. (1986). *Physical Review Letters*, **56**, 2172.

Evans, D. J. and Morriss, G. P. (1987). *Molecular Physics*, **61**, 1151.

Evans, D. J. and Morriss, G. P. (1988a). *Molecular Physics*, **64**, 521.

Evans, D. J. and Morriss, G. P. (1988b). *Physical Review A*, **38**, 4142.

Evans, D. J. and Morriss, G. P. (1990a). *Molecular Physics*, **70**, 347.

Evans, D. J. and Morriss, G. P. (1990b). *Statistical Mechanics of Nonequilibrium Liquids*. London: Academic.

Evans, D. J. and Rondoni, L. (2002). *Journal of Statistical Physics*, **109**, 895.

Evans, D. J. and Searles, D. J. (1994). *Physical Review E*, **50**, 1645.

Evans, D. J. and Searles, D. J. (2002). *Advances in Physics*, **51**, 1529.

Evans, D. J., Searles, D. J., and Rondoni, L. (2005). *Physical Review E*, **71**, 056120.

Farmer, J. D. (1982). *Physica D*, **4**, 366.

Farmer, J. D., Ott, E., and Yorke, J. A. (1983). *Physica D*, **7**, 153.

Feigenbaum, M. J. (1978). *Journal of Statistical Physics*, **19**, 25.

Feitosa, K. and Menon, N. (2004). *Physical Review Letters*, **92**, 164301.

Fermi, E., Pasta, J. G., and Ulam, S. M. (1955). Los Alamos Report No. LA-1940.

Ferziger, J. H. and Kaper, H. G. (1972). *Mathematical Theory of Transport Processes in Gases*. Amsterdam: North-Holland.

Feynman, R. P. (1951). *Physical Review*, **84**, 108.

Frascoli, F., Searles, D. J., and Todd, B. D. (2006). *Physical Review E*, **73**, 046206.

Frenkel, D. and Smit, B. (1996). *Understanding Molecular Simulation. From Algorithms to Applications*. San Diego: Academic Press.

Frenkel, J. (1955). *Kinetic Theory of Fluids*. New York: Dover.

Gallavotti, G. (1995). *Mathematical Physics Electronic Journal*, **1**, 1.

Gallavotti, G. (2000). *Journal of Mathematical Physics*, **41**, 4061.

Gallavotti, G. and Cohen, E. G. D. (1995a). *Physical Review Letters*, **74**, 2694.

Gallavotti, G. and Cohen, E. G. D. (1995b). *Journal of Statistical Physics*, **80**, 931.

Garnier, N. and Ciliberto, S. (2005). *Physical Review E*, **71**, 060101.

Gaspard, P. (1998). *Chaos, Scattering and Statistical Mechanics*. Cambridge: Cambridge University Press.

Gaspard, P. (2006). *Physica A*, **369**, 201.

Gauss, K. F. (1829). *Journal für die Reine und Angew and te Mathematik (Crelle's Journal)*, **IV**, 232.

Gear, W. C. (1971). *Numerical Initial Value Problems in Ordinary Differential Equations*. London: Prentice Hall.

Gibbs, J. W. (1902). *Elementary Principles in Statistical Mechanics*. New Haven: CT, Yale University Press.

Gillan, M. J. and Dixon, M. (1983). *Journal of Physics C*, **16**, 869.

Goldstein, H. (1980). *Classical Mechanics*. Reading: Addison-Wesley.

Gosling, E. M., McDonald, I. R., and Singer, K. (1973). *Molecular Physics*, **26**, 1475.

Grassberger, P. (1983). *Physics Letters A*, **97**, 227.

Grassberger, P. and Procaccia, I. (1983a). *Physical Review A*, **28**, 2591.

Grassberger, P. and Procaccia, I. (1983b). *Physica D*, **9**, 189.

Grebogi, C., Ott, E., and Yorke, J. A. (1988). *Physical Review A*, **37**, 1711.

Green, M. S. (1954). *Journal of Chemical Physics*, **22**, 398.

Haken, H. (1983). *Physics Letters A*, **94**, 71.

Halsey, T. C., Jensen, M. H., Kadanoff, L. P., Procaccia, I., and Shraiman, B. I. (1986). *Physical Review A*, **33**, 1141.

Hansen, D. and Evans, D. (1994). *Molecular Physics*, **81**, 767.

Hansen, J.-P. and Verlet, L. (1969). *Physical Review*, **184**, 151.

Hardy, G. H., Littlewood, J. E., and Pólya, G. (1934). *Inequalities*. Cambridge: Cambridge University Press.

Hentschel, H. G. E. and Procaccia, I. (1983). *Physica D*, **8**, 435.

Hess, S. (1987). *Journal of Non-Newtonian Fluid Mechanics*, **23**, 305.

Hess, S., Hanley, H. J. M., and Herdegen, N. (1984). *Physics Letters A*, **105**, 238.

Holian, B. L. and Evans, D. J. (1983). *Journal of Chemical Physics*, **78**, 5147.

Holian, B. L. and Evans, D. J. (1985). *Journal of Chemical Physics*, **83**, 3560.

Hood, L. M., Evans, D. J., and Morriss, G. P. (1987). *Molecular Physics*, **62**, 419.

Hoover, W. G. (1985). *Physical Review A*, **31**, 1695.

Hoover, W. G. (1986). *Molecular Dynamics*. Springer.

Hoover, W. G. and Ashurst, W. T. (1975). *Theoretical Chemistry: Advances and Perspectives*, Vol. 1, ed. H. Eyring and D. Henderson. New York: Academic. p. 1.

Hoover, W. G., Evans, D. J., Hickman, R. B. *et al.* (1980). *Physical Review A*, **22**, 1690.

Hoover, W. G., Ladd, A. J. C., and Moran, B. (1982). *Physical Review Letters*, **48**, 1818.

Hoover, W. G. and Posch, H. A. (1985). *Physics Letters A*, **113**, 82.

Hoover, W. G. and Posch, H. A. (1987). *Physics Letters A*, **123**, 227.

Huang, K. (1963). *Statistical Mechanics*. New York: Wiley.

Hunt, T. A. and Todd, B. D. (2003). *Molecular Physics*, **101**, 3445.

Hurewicz, W. and Wallman, H. (1948). *Dimension Theory*. Princeton: Princeton University Press.

Irving, J. H. and Kirkwood, J. G. (1950). *Journal of Chemical Physics*, **18**, 817.

Jarzynski, C. (1997). *Physical Review Letters*, **78**, 2690.

Jarzynski, C. (2004). *Journal of Statistical Mechanics*, P09005.

Jensen, M. H., Kadanoff, L. P., and Procaccia, I. (1987). *Physical Review A*, **36**, 1409.

Jepps, O. G., Ayton, G., and Evans, D. J. (2000). *Physical Review E*, **62**, 4757.

José, J. V. and Saletan, E. J. (1998). *Classical Dynamics: A Contemporary Approach*. Cambridge: Cambridge University Press.

Jou, D., Casas-Vazquez, J., and Lebon, G. (2001). *Extended Irreversible Thermodynamics*. Berlin: Springer.

Jourdain, P. (1908). *Arch. Math. Physik*, **14**, 289.

Kadanoff, L. P. (1983). *Physics Today*, 46.

Kaplan, J. and Yorke, J. A. (1979). *Functional Differential Equations and Approximation of Fixed Points*, Vol. 730, ed. H.-O. Peitgen and H.-O. Walter. Berlin: Springer-Verlag, 204.

Kawasaki, K. and Gunton, J. D. (1973). *Physical Review A*, **8**, 2048.

Kirkpatrick, T. R. (1984). *Physical Review Letters*, **53**, 1735.

Kirkwood, J. G. (1968). *Theory of Liquids*. ed. B. J. Alder. New York: Gordon & Breach.

Klages, R. and Dorfman, J. R. (1995). *Physical Review Letters*, **74**, 387.

Kraynik, A. M. and Reinelt, D. A. (1992). *International Journal of Multiphase Flow*, **18**, 1045.

Kreuzer, H. J. (1981). *Nonequilibrium Thermodynamics and its Statistical Foundations*. Oxford: Oxford University Press.

Kubo, R. (1957). *Journal of Physical Society Japan*, **12**, 570.

Kubo, R. (1982). *International Journal of Quantum Chemistry*, **16**, 25.

Kurchan, J. (1998). *Journal of Physics A*, **31**, 3719.

Ladd, A. J. C. and Hoover, W. G. (1985). *Journal of Statistical Physics*, **38**, 973.

Landau, L. D. and Lifshitz, E. M. (1959). *Statistical Mechanics*. Oxford: Pergamon.

Langevin, P. (1908). *Comptes Rendus Hebdomadaires des Seances de L' Academie Des Sciences*, **146**, 530.

Lebowitz, J. L. and Spohn, H. (1999). *Journal of Statistical Physics*, **95**, 333.

Lechner, W., Oberhofer, H., Dellago, C., and Geissler, P. L. (2006). *Journal of Chemical Physics*, **124**, 044113.

Lees, A. W. and Edwards, S. F. (1972). *Journal of Physics C*, **5**, 1921.

Levesque, D. and Verlet, L. (1970). *Physical Review A*, **2**, 2514.

Lloyd, J., Niemeyer, M., Rondoni, L., and Morriss, G. P. (1995). *Chaos*, **5**, 536.

Lorenz, E. N. (1963). *Journal of Atmospheric Science*, **20**, 130.

Macgowan, D. and Evans, D. J. (1986a). *Physics Letters A*, **117**, 414.

Macgowan, D. and Evans, D. J. (1986b). *Physical Review A*, **34**, 2133.

Maes, C. (1999). *Journal of Statistical Physics*, **95**, 367.

Magnus, W. (1954). *Commun. Pure Appl. Math.*, **7**, 649.

Mandelbrot, B. (1983). *The Fractal Geometry of Nature*. San Francisco: Freeman.

Maxwell, J. C. (1873). *Proceedings of the Royal Society*, **148**, 48.

Misner, C. W., Thorne, K. S., and Wheeler, J. A. (1970). *Gravitation*. San Francisco: Freeman.

Monaghan, D. R. J. and Morriss, G. P. (1997). *Physical Review E*, **56**, 476.

Mori, H. (1965a). *Progress in Theoretical Physics*, **33**, 423.

Mori, H. (1965b). *Progress in Theoretical Physics*, **34**, 399.

Mori, H. (1980). *Progress in Theoretical Physics*, **63**, 1044.

Morriss, G. P. (1985). *Physics Letters A*, **113**, 269.

Morriss, G. P. (1987). *Physics Letters A*, **122**, 236.

Morriss, G. P. (1988). *Physical Review A*, **37**, 2118.

Morriss, G. P. (1989a). *Physics Letters A*, **134**, 307.

Morriss, G. P. (1989b). *Physical Review A*, **39**, 4811.

Morriss, G. P. (2002). *Physical Review E*, **65**, 017201.

Morriss, G. P. and Dettmann, C.P. (1998). *Chaos*, **8**, 321.

Morriss, G. P. and Evans, D. J. (1985). *Molecular Physics*, **54**, 629.

Morriss, G. P. and Evans, D. J. (1987). *Physical Review A*, **35**, 792.

Morriss, G. P. and Evans, D. J. (1989). *Physical Review A*, **39**, 6335.

Morriss, G. P. and Evans, D. J. (1991). *Computer Physics Communications*, **62**, 267.

Morriss, G. P., Evans, D. J., Cohen, E. G. D., and van Beijeren, H. (1989). *Physical Review Letters*, **62**, 1579.

Morriss, G. P. and Rondoni, L. (1999). *Physical Review E*, **59**, R5.

Mountain, R. D. and Raveché, H. J. (1971). *Journal of Chemical Physics*, **29**, 2250.

Noll, W. (1955). *Journal of Rational Mechanics and Analysis*, **4**, 627.

Nosé, S. (1984a). *Molecular Physics*, **52**, 255.

Nosé, S. (1984b). *Journal of Chemical Physics*, **81**, 511.

Onsager, L. (1931). *Physical Review*, **37**, 405.

Oseledec, V. I. (1968). *Transactions of the Moscow Mathematical Society*, **19**, 179.

Ott, E. (2002). *Chaos in Dynamical Systems*. Cambridge: Cambridge University Press.

Öttinger, H. C. (2005). *Beyond Equilibrium Thermodynamics*. New York: Wiley.

Panja, D. and van Zon, R. (2002). *Physical Review E*, **65**, 060102.

Parrinello, M. and Rahman, A. (1980a). *Journal of Chemical Physics*, **76**, 2662.

Parrinello, M. and Rahman, A. (1980b). *Physical Review Letters*, **45**, 1196.

Parrinello, M. and Rahman, A. (1981). *Journal of Applied Physics*, **52**, 7182.

Parry, W. (1986). *Communications in Mathematical Physics*, **106**, 267.

Parry, W. E. (1973). *The Many-Body Problem*. Oxford: Clarendon Press.

Pars, L. A. (1968). *A Treatise on Analytical Dynamics*. London: Heineman.

Pechukas, P. and Light, J. C. (1966). *Journal of Chemical Physics*, **44**, 3897.

Poisson (1829). *Journal de l'Ecole, Polytechnique tome*, **xiii**, xx.

Pomeau, Y. and Resibois, P. (1975). *Physics Reports*, **19**, 63.

Procaccia, I. (1985). *Physica Scripta*, **T9**, 40.

Raimes, S. (1972). *Many-Electron Theory*. Amsterdam: North-Holland.

Raineri, F. O., Stell, G., and Ben-Amotz, D. (2005). *Molecular Physics*, **103**, 3209.

Rainwater, J. C. and Hanley, H. J. M. (1985). *International Journal of Thermophysics*, **6**, 595.

Rainwater, J. C., Hanley, H. J. M., and Paskiewicz, T. (1985). *Journal of Chemical Physics*, **83**, 339.

Reid, J. C., Carberry, D. M., Wang, G. M. *et al.* (2004). *Physical Review E*, **70**, 016111.

Reid, J. C., Sevick, E. M., and Evans, D. J. (2005). *Europhysics Letters*, **72**, 726.

Ritort, F. (2005). *Journal of Physical Chemistry B*, **109**, 6787.

Rondoni, L. and Morriss, G. P. (1997a). *Journal of Statistical Physics*, **86**, 991.

Rondoni, L. and Morriss, G. P. (1997b). *Physics Reports*, **290**, 173.

Rowlinson, J. S. and Widom, B. (1982). *Molecular Theory of Capillarity*. Oxford: Clarendon Press.

Ruelle, D. (1976). *American Journal of Mathematics*, **98**, 619.

Ruelle, D. (1978). *Thermodynamic Formalism*. Reading: Addison-Wesley.

Ruelle, D. (1999). *Journal of Statistical Physics*, **95**, 393.

Rugh, H. H. (1997). *Physical Review Letters*, **78**, 772.

Rugh, H. H. (1998). *Journal of Physics A*, **31**, 7761.

Saltzmann, B. (1961). *Journal of Atmospheric Science*, **19**, 329.

Sarkovskii, A. N. (1964). *Ukranianski Mathematicheskii Zhumal*, **16**, 61.

Sasa, S. I. and Tasaki, H. (2006). *Journal of Statistical Physics*, **125**, 125.

Schol-Paschinger, E. and Dellago, C. (2006). *Journal of Chemical Physics*, **125**, 054105.

Schuler, S., Speck, T., Tietz, C., Wrachtrup, J., and Seifert, U. (2005). *Physical Review Letters*, **94**, 180602.

Schuster, H. G. (1988). *Deterministic Chaos: An Introduction*. Weinheim: VCH.

Searles, D. J. and Evans, D. J. (1999). *Physical Review E*, **60**, 159.

Searles, D. J. and Evans, D. J. (2000). *Journal of Chemical Physics*, **113**, 3503.

Seifert, U. (2005a). *Physical Review Letters*, **95**, 040602.

Seifert, U. (2005b). *Europhysics Letters*, **70**, 36.

Shimada, I. and Nagashima, T. (1979). *Progress in Theoretical Physics*, **61**, 1605.

Sparrow, C. (1982). *The Lorentz Equations: Bifurcations, Chaos, and Strange Attractors*. New York: Springer-Verlag.

Sprott, J. C. (2003). *Chaos and Time Series Analysis*. Oxford: Oxford University Press.

Szydlowski, M. and Biesiada, M. (1991). *Physical Review D*, **44**, 2369.

Taniguchi, T. and Morriss, G. P. (2002). *Physical Review E*, **66**, 066203.

Taniguchi, T. and Morriss, G. P. (2004). *Physical Review E*, **70**, 056124.

Tenenbaum, A., Ciccotti, G., and Gallico, R. (1982). *Physical Review A*, **25**, 2778.

Thompson, C. J. (1972). *Mathematical Statistical Mechanics*. New York: Macmillan.

Tietz, C., Schuler, S., Speck, T., Seifert, U., and Wrachtrup, J. (2006). *Physical Review Letters*, **97**, 050602.

Todd, B. D. and Daivis, P. J. (1998). *Physical Review Letters*, **81**, 1118.

Todd, B. D. and Daivis, P. J. (1999). *Computer Physics Communications*, **117**, 191.

Todd, B. D. and Daivis, P. J. (2000). *Journal of Chemical Physics*, **112**, 40.

Todd, B. D. and Daivis, P. J. (2007). *Molecular Simulations*, **33**, 189.

Todd, B. D. and Evans, D. J. (1995). *Journal of Chemical Physics*, **103**, 9804.

Todd, B. D., Evans, D. J., and Daivis, P. J. (1995). *Physical Review E*, **52**, 1627.

Tolman, R. C. (1979). *The Principles of Statistical Mechanics*. New York: Dover.

van Beijeren, H. (1984). *Physics Letter*, **105A**, 191.

van Kampen, N. G. (1971). *Physica Norvegica*, **5**, 279.

van Zon, R., Ciliberto, S., and Cohen, E. G. D. (2004). *Physical Review Letters*, **92**, 130601.

van Zon, R. and Cohen, E. G. D. (2004). *Physical Review E*, **69**, 056121.

Vance, W. N. (1992). *Physical Review Letters*, **69**, 1356.

Visscher, W. M. (1974). *Physical Review A*, **10**, 2461.

Wallace, D. C. (1987). *Journal of Chemical Physics*, **87**, 2282.

Wang, G. M., Reid, J. C., Carberry, D. M. *et al.* (2005). *Physical Review E*, **71**, 046142.

Wang, G. M., Sevick, E. M., Mittag, E., Searles, D. J., and Evans, D. J. (2002). *Physical Review Letters*, **89**, 050601.

Whittaker, E. T. (1961). *A Treatise on Analytical Dynamics of Particles and Rigid Bodies*. Cambridge: Cambridge University Press.

Yamada, T. and Kawasaki, K. (1967). *Progress in Theoretical Physics*, **38**, 1031.

Yamada, T. and Kawasaki, K. (1975a). *Progress in Theoretical Physics*, **53**, 111.

Yamada, T. and Kawasaki, K. (1975b). *Progress in Theoretical Physics*, **53**, 437.

Zwanzig, R. (1954a). *Journal of Chemical Physics*, **22**, 2099.

Zwanzig, R. (1954b). *Journal of Chemical Physics*, **22**, 1420.

Zwanzig, R. (1961). *Lectures in Theoretical Physics, Vol. III*. Boulder: Wiley, p. 135.

Zwanzig, R. (1965). *Annual Review of Physical Chemistry*, **16**, 67.

Zwanzig, R. (1982). Remark made at conference on Nonlinear Fluid Behaviour, June, 1982, Boulder, CO.

Index

Printed in the United States
By Bookmasters